AF716758

RNA Editing

Current Research and Future Trends

Edited by

Stefan Maas

Division of Genetics and Developmental Biology
National Institute of General Medical Sciences
National Institutes of Health
Bethesda, MD
USA

Caister Academic Press

Copyright © 2013

Caister Academic Press
Norfolk, UK

www.caister.com

British Library Cataloguing-in-Publication Data
A catalogue record for this book is available from the British Library

ISBN: 978-1-908230-23-2

Description or mention of instrumentation, software, or other products in this book does not imply endorsement by the author or publisher. The author and publisher do not assume responsibility for the validity of any products or procedures mentioned or described in this book or for the consequences of their use.

All rights reserved. No part of this publication may be reproduced, stored in a retrieval system, or transmitted, in any form or by any means, electronic, mechanical, photocopying, recording or otherwise, without the prior permission of the publisher. No claim to original U.S. Government works.

Cover design adapted from Figure 1.1.

Printed and bound in Great Britain

Contents

Contributors

Jorge Cruz-Reyes
Department of Biochemistry and Biophysics
Texas A&M University
College Station, TX
USA

cruzrey@tamu.edu

Federica Galeano
Oncohaematology Department
Bambino Gesù Children's Hospital
IRCCS
Rome
Italy

federica.galeano@opbg.net

Angela Gallo
Oncohaematology Department
Bambino Gesù Children's Hospital
IRCCS
Rome
Italy

angela.gallo@opbg.net

Miguel Holmgren
Molecular Neurophysiology Section
Porter Neuroscience Research Center
National Institute of Neurological Disorders and Stroke
National Institutes of Health
Bethesda, MD
USA

holmgren@ninds.nih.gov

Heather A. Hundley
Medical Sciences Program
Indiana University
Bloomington, IN
USA

hahundle@indiana.edu

Jane E. Jackman
Center for RNA Biology
Molecular, Cellular and Developmental Biology Program
and
Department of Biochemistry
The Ohio State University
Columbus, OH
USA

jackman.14@osu.edu

Jonatha M. Gott
Center for RNA Molecular Biology
Case Western Reserve University
Cleveland, OH
USA

jmg13@case.edu

Jin Billy Li
Department of Genetics
Stanford University School of Medicine
Stanford, CA
USA

jin.billy.li@stanford.edu

Franco Locatelli
Oncohaematology Department
Bambino Gesù Children's Hospital
IRCCS
Rome
and
Università di Pavia
Pavia
Italy

franco.locatelli@opbg.net

Stefan Maas
Division of Genetics and Developmental Biology
National Institute of General Medical Sciences
National Institutes of Health
Bethesda, MD
USA

stefan.maas@nih.gov

Blaine H.M. Mooers
Department of Biochemistry and Molecular Biology
University of Oklahoma Health Sciences Center
Biomedical Research Center
Oklahoma City, OK
USA
blaine-mooers@ouhsc.edu

Kazuko Nishikura
Gene Expression and Regulation Program
The Wistar Institute
Philadelphia, PA
USA

kazuko@wistar.org

Bhalchandra S. Rao
Center for RNA Biology
Molecular, Cellular and Developmental Biology Program
and
Department of Biochemistry
The Ohio State University
Columbus, OH
USA

rao.105@buckeyemail.osu.edu

Laurie K. Read
Department of Microbiology and Immunology
University at Buffalo
State University of New York
Buffalo, NY
USA

lread@buffalo.edu

Joshua J.C. Rosenthal
Institute of Neurobiology
andDepartment of Biochemistry
University of Puerto Rico
San Juan
Puerto Rico

rosenthal.joshua@gmail.com

Harold C. Smith
University of Rochester School of Medicine and Dentistry
Department of Biochemistry and Biophysics
Center for RNA Biology and Wilmot Cancer Center
Rochester, NY
USA
harold.smith@rochester.edu

Meng How Tan
Department of Genetics
Stanford University School of Medicine
Stanford, CA
USA

menghow.tan@gmail.com

Sara Tomaselli
Oncohaematology Department
Bambino Gesù Children's Hospital
IRCCS
Rome
Italy

sara.tomaselli@opbg.net

Bjorn-Erik Wulff
Department of Biochemistry
Stanford University School of Medicine
Stanford, CA
USA

wulff@stanford.edu

Preface

The discovery of RNA editing more than 25 years ago uncovered a new layer of genetic information that provides organisms the flexibility to explore sequence space that is not readily accessible through DNA mutations. As we have learned about individual cases of recoding RNA editing and how gene functions are governed or modulated by the editing status of their gene products, important insights on gene regulation, RNA biology, and evolutionary mechanisms have followed.

Recently, the revolution of DNA sequencing technologies has further invigorated the RNA editing field facilitating and accelerating the analysis of established RNA editing targets and their functions and also enabling the tackling of new questions and hypotheses.

All of these aspects are illustrated and captured by each of the ten chapters presented in this book. Even as they span a diverse set of organisms (squid, slime moulds, trypanosomes, insects, rodents, human) and editing mechanisms (nucleotide insertion/deletion, substitution, modification) and targets (mRNAs, rRNAs, miRNAs, siRNAs, viral RNAs), early advances in one field have informed the work in others. This emphasizes that to understand the broader implications of the RNA editing phenomenon for normal physiology and disease mechanisms as well as evolution, it is essential to appreciate the different modalities and implementations of this process in biological systems. Some organisms rely on RNA editing to produce functional versions of many essential proteins while others seem to utilize it mainly to fine-tune protein properties and organismal behaviours.

It is also clear that RNA editing is often highly integrated with other gene regulatory processes that impact the stability, translatability, structure, localization, specificity or coding potential of RNA molecules (see especially Chapters 4, 6 and 9). Finally, the diversity of cellular pathways impacted by RNA editing makes it a rich field that allows and often requires new collaborations and interdisciplinary approaches to unravel the physiological meaning of individual RNA editing events. Without doubt, we have just begun to unravel the intricacies of post-transcriptional gene regulation with RNA editing playing various supporting-to-leading roles.

Stefan Maas

Regulation of Ion Channel and Transporter Function Through RNA Editing

1

Miguel Holmgren and Joshua J.C. Rosenthal

Abstract

A large proportion of the recoding events mediated by RNA editing are in mRNAs that encode ion channels and transporters. The effects of these events on protein function have been characterized in only a few cases. In even fewer instances are the mechanistic underpinnings of these effects understood. This chapter focuses on how RNA editing affects protein function and higher order physiology. In mammals, particular attention is given to the GluA2, an ionotropic glutamate receptor subunit, and $K_v1.1$, a voltage-dependent K^+ channel, because they are particularly well understood. In addition, work on cephalopod K^+ channels and Na^+/K^+-ATPases has also provided important clues on the rules used by RNA editing to regulate excitability. Finally, we discuss some of the emerging targets for editing and how this process may be used to regulate nervous function in response to a variable environment.

Introduction

The recent improvements in DNA sequencing technologies have led to an explosion in the discovery of new RNA editing sites arising from adenosine deamination. Recent reports suggest that there are thousands of RNA editing sites in the human brain transcriptome, although only a small fraction of these are in open reading frames and recode amino acids (Li *et al.*, 2009, 2011; Peng *et al.*, 2012). Editing in invertebrates appears to be even more extensive. A recent analysis of transcriptomes from *Drosophila melanogaster* at different developmental stages uncovered over 600 recoding events (Graveley *et al.*, 2011), and studies on a handful of individual mRNAs from cephalopods have uncovered close to a hundred such sites (Patton *et al.*, 1997; Rosenthal and Bezanilla, 2002b; Palavicini *et al.*, 2009; Colina *et al.*, 2010; Garrett and Rosenthal, 2012). A disproportionately large number of these edits occur within mRNAs that encode proteins directly involved in excitability. Although it is now clear that RNA editing is modifying the primary sequences of many target proteins, surprisingly few of the editing sites have been characterized on a functional level. Of those that have been studied, the majority lie within mRNAs encoding ion channels and transporters.

This chapter aims to give the reader an up-to-date accounting of the best studied examples of functional changes caused by editing. First we would like to stress that a full functional characterization of an editing site is a quite difficult undertaking because, to be complete, it must be carried on many different organizational levels. For example, an edit in an ion channel could affect a specific biophysical property, which would then affect the action potential, which could affect a neural circuit, which could then affect a specific behaviour. For the most part, studies to date have focused on the beginning of this sequence, and to greatly different depths. In most cases, studies have been descriptive, defining a physiological property that is altered by RNA editing. In some cases the descriptions have been in exceptional detail, but they are descriptions nonetheless. In two cases the mechanistic underpinnings of specific changes are well understood. Only a few studies have explored the relevance of an editing event on higher order function.

This chapter focuses most heavily on the functional consequences of editing ion channel transcripts. Ion channels are exceptionally diverse, and, for non-aficionados, the jargon associated with their physiology can be arcane. However, all ion channel function shares certain common features, and we will try to focus on these in our discussions. For example, channels open and close, or gate, in response to an extrinsic factor, such as changes in the transmembrane voltage or the concentration of a specific ligand. Voltage-dependent K^+ channels are a good example of the former, and ionotropic glutamate receptors of the latter. The kinetics associated with opening and closing, in addition to a channel's sensitivity to an extrinsic factor, are physiologically relevant characteristics. Once open, all channels permit ions to pass through them in response to an electrochemical gradient. The degree of selectivity for one ion species over another varies greatly between channels and is an important determinant for nerve physiology. Finally, open channels often spontaneously close, even when the extrinsic factor that commands them to stay open is still present. For voltage-dependent channels this process is known as inactivation, and for ligand-gated channels it is known as desensitization. Here too the kinetics of the entry and exit from these processes can be important. Depending on the target, almost all of these physiological characteristics can be modified by RNA editing.

Regulation of excitatory neurotransmission by recoding glutamate receptors

In the mammalian brain most fast excitatory neurotransmission is generated by ionotropic glutamate receptors located in the postsynaptic membrane. When glutamate is released into the synaptic cleft, these receptors open a cation selective pore, allowing extracellular Na^+, and sometimes Ca^{2+}, to enter the cell, causing a transient depolarization. Repeated stimulation by their agonist often makes the receptor less sensitive to further stimulation, a process known as desensitization. Because they play such a critical role in generating and shaping synaptic potentials, glutamate receptors are an exceptionally diverse gene family. Early studies identified three basic groups of ionotropic glutamate receptors based on pharmacological differences: α-amino-3-hydroxy-5-methyl-4-isoxazolepropionic acid receptor (AMPA) receptors, kainate receptors, and *N*-methyl-D-aspartate (NMDA) receptors. Molecular sequencing has supported this classification. For the purpose of this chapter we will focus only on AMPA and kainate receptors because their mRNAs are edited, and those encoding NMDA receptors are not as far as we know. In the mammalian brain, AMPA receptors generate the bulk of fast, excitatory neurotransmission. Kainate receptors are thought to play a more subtle, modulatory, role.

Glutamate receptors are tetramers and the individual subunits share a common design (Laube *et al.*, 1998; Rosenmund *et al.*, 1998; Nakagawa *et al.*, 2005; Sobolevsky *et al.*, 2009). The monomer consists of four membrane spans (M1–4). M2, which lines the ion conduction pathway, is atypical in that it does not fully span the membrane, both entering from – and exiting to – the cytoplasm. In mammals, both AMPA and kainate receptors are diverse. There are four AMPA receptors subunits named GluA1–4 using the most up-to-date nomenclature, although they have also been termed GluR-A to D or GluR1–4 in past studies, and their genes are referred to as Gria1–4. These subunits form both homo – and heterotetramers, with different combinations in different brain regions (Keinanen *et al.*, 1990; Sommer *et al.*, 1990). Alternative exon splicing further increases AMPA receptor diversity, generating two forms, known as 'flip' and 'flop', which have altered sequences in the large extracellular domain immediately before TM4 (Sommer *et al.*, 1990), and have different kinetic properties (Monyer *et al.*, 1991; Lambolez *et al.*, 1996). There are five kainate receptors (GluK1–5), and GluK1 and GluK3 form both homo – and heteromultimers. Thus, for glutamate receptors, there is substantial molecular diversity even before RNA editing enters the equation.

There are numerous editing sites scattered throughout the mRNAs encoding both AMPA and kainate receptors, however, a single site stands above the rest in terms of the thoroughness with which it has been studied, and its importance to the host's physiology. Editing at codon 586 of GluA2, better known as the Q/R site, recodes a conserved glutamine within M2 to Arginine. This position lies at a critical juncture in the ion conduction pathway (Sommer *et al.*, 1991). In the adult mammalian brain, this site is edited with near-perfect efficiency in GluA2, whereas in GluA1, 3 and 4 the codon remains a Glutamine. An Arginine at this site renders the receptors impermeable to Ca^{2+} (Fig. 1.1) and decreases their conductance by an order of magnitude (Hollmann *et al.*, 1991; Burnashev *et al.*, 1992). Further mutagenesis of codon 586 demonstrated that the charge of the side chain was the main determinant in abolishing divalent permeability, thus the effect is likely electrostatic, however changes in side chain volume also influenced the relative permeability of Ca^{2+} and Mg^{2+}. Another interesting feature of the Q/R site in heteromultimers is that arginine's effect is dominant. Thus, because essentially 100% of GluA2 subunits are edited, we can conclude that this site is a major determinant for keeping Ca^{2+} from crossing the postsynaptic membrane into the cell during excitatory postsynaptic potentials.

The importance of editing the Q/R site of GluA2 was demonstrated in several elegant studies by Peter Seeburg and colleagues using transgenic mice. First, ADAR2 knockout mice suffered seizures and died soon after birth and Q/R site editing was very low in these animals (Higuchi *et al.*, 2000). Interestingly, in wild-type background, abolishing only editing at the Q/R site by disrupting the ECS structure was also lethal, even when just applied to a single allele (Brusa *et al.*, 1995). This phenotype could be rescued by hardwiring an arginine in codon 586 in the gene, thus obviating the need for editing the Q/R site (Higuchi *et al.*, 2000). The lethality associated with the inability to edit the Q/R site of GluA2 was not a developmental problem. A conditional knock-in of an uneditable GluA2 throughout the adult forebrain also led to severe epileptic seizures and lethality (Krestel *et al.*, 2004). Additionally, expressing different levels of the uneditable GluA2 led to a variety of neurological symptoms, from extreme lethargy to hyperactivity (Feldmeyer *et al.*, 1999). Taken together, these results underscore the importance of editing this site completely.

This idea is reinforced by more recent data relating Q/R editing of GluA2 with disease. In post-mortem studies, under-editing of the Q/R site has been associated with sporadic amyotrophic lateral sclerosis (ALS) (Kawahara *et al.*, 2004). Further studies indicated that the under editing could be attributed to a down-regulation of ADAR2 in motoneurons (Hideyama *et al.*, 2011). Significantly, the reduction in Q/R site editing was modest. Studies on mice showed that ALS-like symptoms, including motoneuron death, could be reproduced by conditional knockouts of ADAR2 in wild-type animals, but not in

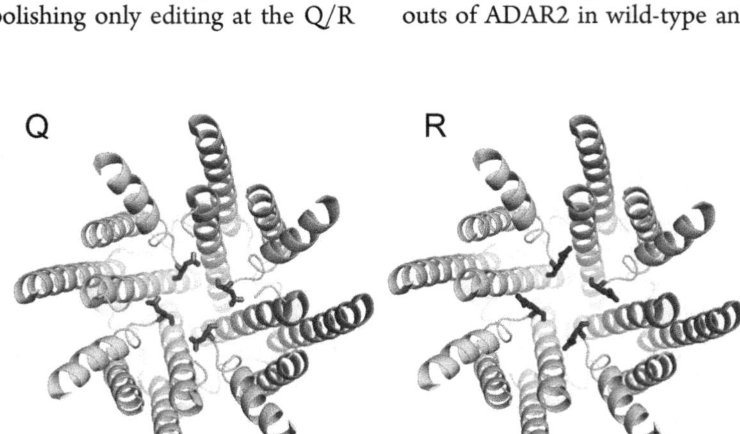

Figure 1.1 Two views of the ion conduction pathway of GluA2 with either a glutamine (grey) or a arginine (blue) at codon 586. Structures are from pdb 3KG2 (Sobolevsky *et al.*, 2009). The view shows only the transmembrane spans and is from the inside looking out. As can be seen, the Q/R site sits directly in the ion permeation pathway. The Q/R substitution was generated with Pymol software. The position of the R side chain was not calculated via energy minimization, but merely reflects the most common rotamer for R. A colour version of this figure is located in the plate section at the back of the book.

mice carrying pre-edited versions of GluA2 at the Q/R site (Hideyama *et al.*, 2010). These studies on humans and mice, in conjunction with a recent study that found few deficits in ADAR2-less mice hardwired with an arginine at the Q/R site of GluA2 (Horsch *et al.*, 2011), led to the conclusion that this site is far and away the most important for mammalian physiology. Remarkably, other glutamate receptor mRNAs are also edited at the Q/R site, apparently with less significant consequences.

As with GluA2, the kainate receptors GluK1 and GluK2 are also edited at the Q/R site (Herb *et al.*, 1996; Maas *et al.*, 1996), although far less efficiently (Belcher and Howe, 1997; Paschen *et al.*, 1997; Bernard *et al.*, 1999). In GluK1 and GluK2, the Q/R site causes very similar functional changes, namely a drastic reduction in Ca^{2+} permeability and a significant reduction in single channel conductance (Egebjerg and Heinemann, 1993; Kohler *et al.*, 1993; Swanson *et al.*, 1996). In addition, editing the Q/R site renders channels insensitive to block by polyamines, but makes them sensitive to block by specific fatty acids such as arachidonic acid (Wilding *et al.*, 2005, 2008, 2010). Thus, Q/R site editing can be used to regulate kainate receptors in a manner that is dependent on other cellular components. GluK2 mRNAs are also edited at two additional sites: codon 567, where an isoleucine is converted to a valine, and codon 571, where a tyrosine is converted to a cysteine (Kohler *et al.*, 1993). Little is known about the physiological effects of these edits, but it was suggested that they fine-tune Ca^{2+} permeability.

AMPA receptor mRNAs in rats, mice and humans also undergo editing at codon 764, recoding an exonically encoded arginine to a glycine (Lomeli *et al.*, 1994). This edit, at the end of the M3–M4 extracellular loop, occurs in GluA2–4, but not in GluA1. It has a marked and specific effect on receptor physiology; with a glycine at position 764, receptors recover from desensitization twice as quickly. This effect was apparent when receptors were activated by physiologically relevant applications of glutamate in terms of concentration and duration, suggesting that this edit could be important in shaping the magnitudes of excitatory postsynaptic potentials during repetitive firing. The overall physiological importance of editing glutamate receptor mRNAs cannot be understated. However, only in the case of the Q/R site in GluA2 are the consequences of not editing well understood. How the Q/R site in the kainate receptors GluK1–2, and the R/G site in GluA2–4 relate to higher-order physiological processes should prove to be fertile ground for further investigations.

Voltage-activated potassium channels

In neurons, voltage-activated potassium (K_V) channels play important roles in excitability. For example, once an action potential has been initiated and the membrane potential has depolarized, K_V channels are called into action to repolarize the membrane potential back towards the resting potential; therefore, their activity is a critical determinant of the action potential's shape (Aldrich *et al.*, 1979). More importantly, K_V channels play key roles in setting the firing properties of neurons (Connor and Stevens, 1971). Therefore, changes in K_V channel function can have significant consequences on a neuron's excitability status. Not surprisingly, K_V channels from both invertebrates and vertebrates have been extensively targeted by the process of RNA editing (Patton *et al.*, 1997; Rosenthal and Bezanilla, 2002b; Hoopengardner *et al.*, 2003; Ryan *et al.*, 2008; Ingleby *et al.*, 2009).

Mammalian KV1.1 channel inactivation is regulated by editing

In mammals, RNA editing targets a highly conserved isoleucine located in the lining of the K_V1.1 channel's permeation pathway and recodes it to valine (I400V). From functional (Liu *et al.*, 1997) and structural work (Long *et al.*, 2005), it is known that the side chain of this position faces the lumen of the ion pathway in a region called the intracellular cavity (Doyle *et al.*, 1998) (Fig. 1.2A). The conversion of I → V at this site is broadly understood. Editing is mediated by ADAR2, which recognizes a 114-nucleotide imperfect inverted repeat hairpin derived entirely from an exonic region (Bhalla *et al.*, 2004). Levels of editing vary between different regions of the nervous system of humans, with high incidence in the spinal cord, medulla and thalamus and comparatively low

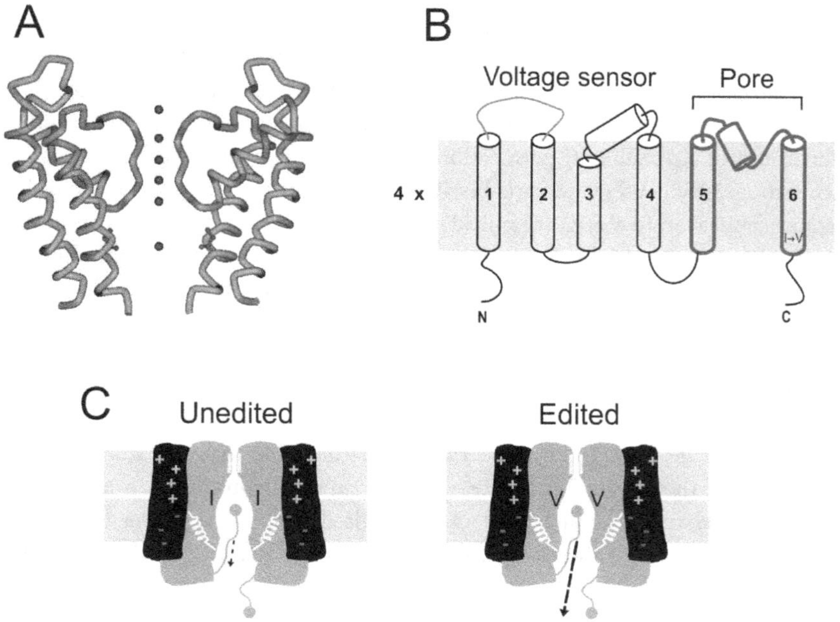

Figure 1.2 RNA editing of mammalian K_V channels. (A) Crystal structure of the last two transmembrane segments from rat K_V1.2 channels (accession ID 2A79). Green spheres represent K^+ along the ion permeation pathway. In red is shown the isoleucine edited by ADAR2. (B) Membrane topology of the K_V channel. A functional channel is a tetramer. In each subunit, the first four TM segments form the voltage-sensing domain, while the last two delimit the permeation pathway. The I→V edit is located at the intracellular end of the TM6. (C) Cartoon corresponding to the functional consequences of I→V substitution at the intracellular cavity of K_V channels. RNA editing targets exclusively the unbinding kinetics of the inactivation gate, increasing the off-rate by ~20-fold. A colour version of this figure is located in the plate section at the back of the book.

incidence in cortex, cerebellum or hippocampus (Hoopengardner *et al.*, 2003). Interestingly, a similar pattern was observed in rodents, suggesting that regulation of RNA editing at this site is evolutionarily conserved (Hoopengardner *et al.*, 2003). Another relevant biological feature of this site is that it has also evolved in mRNAs encoding K_V2 channels in squid and fruit fly (Patton *et al.*, 1997; Bhalla *et al.*, 2004).

Evolutionary convergence and the rather conspicuous location of this editing site were strong indicators that the I → V conversion would significantly alter functional properties. Generally, K_V channels assemble as homotetramers of α subunits (MacKinnon *et al.*, 1993), each subunit containing a voltage-sensing domain (formed by the first four transmembrane segments) and a pore domain (formed by the last two transmembrane segments and a re-entrant loop between them which includes the selectivity filter; Fig. 1.2B). In response to a depolarization, conformational changes within the voltage sensor drive the opening of an activation gate located at the intracellular end of the permeation pathway, about 7–10 Å from the editing site. Mammalian K_V1.1 channels co-assemble with β-subunits to give rise to a macromolecular complex of eight subunits per channel (Gulbis *et al.*, 2000). β-Subunits can confer fast inactivation to mammalian K_V1.1 channels (Rettig *et al.*, 1994), a process in which permeation is shut by an intracellular inactivation gate that binds the ion permeation pathway once the activation gate has opened. Which gating mechanism is altered by the I → V conversion?

Expressing the human K_V1.1 α subunit alone renders voltage-activated K^+ currents that do not inactivate, which allows the study of the activation gate mechanism in isolation. Edited and unedited versions of the hK_V1.1 α-subunits showed little difference, opening and closing with similar kinetics and voltage dependence (Bhalla *et al.*, 2004). Even though the editing site and the activation gate are just a few angstroms apart, these results are not surprising. For a long time it has been known that quaternary ammonium (QA) ions (~8–10 Å in diameter) can block (but not permeate) at the intracellular end of K_V channels' permeation pathway (Armstrong, 1966, 1969). Remarkably, if blocked channels are forced to close very quickly, the activation gate can shut the permeation pathway and trap the blocker inside (Armstrong, 1971). Activation gating in the absence and presence of the blocker are quite similar, suggesting that the architecture of the intracellular cavity remains relatively unchanged in open and closed channels (Holmgren *et al.*, 1997). Similarly, the intracellular cavity in the crystal structure of KcsA (a bacterial K^+ channel) with bound QA derivatives is almost identical to the cavity of an unbound channel (Zhou *et al.*, 2001).

Co-expressing edited or unedited forms of $hK_V1.1$ α subunits with the $hK_V1.1$ β-subunit conferred, in both cases, fast inactivation to the ionic currents. The simplest kinetic scheme to describe inactivation is:

$$C \rightleftharpoons O \underset{k_{off}}{\overset{k_{on}}{\rightleftharpoons}} I$$

where C, O and I denote closed, open and inactivated states, respectively. At voltages where the relative probability of opening is maximal (>0 mV), inactivation is determined by the on – and off-rate constants. The I → V substitution within the intracellular cavity of $hK_V1.1$ channels specifically targeted the unbinding kinetics (k_{off}), speeding it up by about 20-fold (Bhalla *et al.*, 2004) (Fig. 1.2C). The fact that the inactivation particle for $hK_V1.1$ resides on a β-subunit presented significant complications for further analysing the effects of the I400V edit. For example, the α–β stoichiometry is difficult to control in heterologous expression systems and the $hK_V1.1$ β-Subunit is known to be regulated by phosphorylation. However, the functional consequence of RNA editing could be transferred to *Shaker* $K_V1.1$ channels, whose inactivation particle is part of the α subunit. In fact, mimicking the editing event in Shaker produced an identical effect on the unbinding kinetics without changing other properties (Bhalla *et al.*, 2004). This observation turned out to be critical, allowing us to pursue the precise mechanism by which inactivation is altered by RNA editing.

Since isoleucine and valine are both aliphatic amino acids, and the side chain of the edited position is facing the lumen of the pore, it was hypothesized that a reduction in hydrophobicity accompanying a valine substitution was responsible for speeding up the unbinding kinetics. To test this hypothesis, we devised a strategy to change the chemistry at the edited position during an experiment. The *Shaker* position corresponding to I400 was mutated to cysteine in order to serve as a target for chemical modification with methanethiosulphonate (MTS) reagents. As expected, the more hydrophobic a moiety attached to the edited codon, the slower the unbinding kinetics became (Gonzalez *et al.*, 2011). In other words, once the inactivation particle is bound, hydrophobic interactions determine its off-rate. Such interactions demand close proximity between binding partners. Which specific amino acids of K_V 1.1 interacts with the edited position at the intracellular cavity of the protein?

Fast inactivation in *Shaker* $K_V1.1$ channels is a well-understood process. In their pioneering studies, Aldrich and coworkers (Hoshi *et al.*, 1990; Zagotta *et al.*, 1990) discovered that a mild intracellular treatment with trypsin removed the channels' ability to inactivate. By studying the inactivation properties of a series of deletion mutations at the N-terminus of the protein, they concluded that the inactivation gate is formed by the first ~20–25 amino acids of the protein. They also showed that inactivation is restored in the non-inactivating mutants by exposing the intracellular side of the protein to a synthetic peptide formed by the first 20 amino acids. It was also known that the inactivation gate (N-terminus) is an open channel blocker (Demo and Yellen, 1991) that can compete with intracellular, but not extracellular, QA ions, like tetraethylammonium (Choi *et al.*, 1991). Further, crystallography data show that QA ions bind at the intracellular cavity (Zhou *et al.*, 2001). All these results combined strongly suggest that to inactivate the current, the N-terminus enters the intracellular cavity once the channel has opened.

Is the very tip of the N-terminus entering deeply into the cavity of the channel? To approach this question, we extended our chemical approach by developing channel constructs in which an additional cysteine was substituted at positions near the N-terminus. The expectation was that if the two cysteines were close together in an inactivated channel, a disulfide bond would form, creating a stably blocked channel that required a reducing agent to recover. We made six double cysteine channels with one cysteine at the edited position and the second between positions 2 and 7 at the N-terminus. Only with the construct 2C–470C did we observed an irreversible current reduction when the channels were exposed to an oxidizing environment, indicating that position 2 is the binding partner for the edited codon (Gonzalez *et al.*, 2011). In summary, once valine substitutes for isoleucine, the intracellular cavity of $K_V1.1$ channels loses an important hydrophobic component to its association with the very tip of the N-terminus, an interaction that determines the *off* kinetics of the inactivation gate.

For a neuron, what are the implications of editing a $K_V1.1$ channels' intracellular cavity? Although as yet untested, we predict it to have a profound impact on excitability. Because this editing site accelerates the inactivation gate's *off* rate, a neuron possessing a large proportion of edited channels is poised to have a larger pool of available channels, particularly during periods or repetitive firing. Consequently, RNA editing could play an important role in regulating action potential shape during repetitive firing (Aldrich *et al.*, 1979), as well as helping to set the actual firing frequency of a neuron (Connor and Stevens, 1971). In fact, fast inactivating K^+ currents have been shown to play key roles in determining the firing properties in axons and dendrites (Debanne *et al.*, 1997; Hoffman *et al.*, 1997). Additionally, tissue-specific regulation of the RNA editing levels (Hoopengardner *et al.*, 2003; Decher *et al.*, 2010) can have important pharmacological implications when a drug interacts near an editing site (Decher *et al.*, 2010). It turns out that the apparent affinity of several open channel blockers which can be used as therapeutics is significantly reduced by RNA editing (Decher *et al.*, 2010). Similarly, highly unsaturated fatty acids can also bind to the intracellular cavity of $K_V1.1$ channels and block their currents in a process similar to inactivation. Their apparent affinities are also influenced by RNA editing (Decher *et al.*, 2010).

Cephalopod K^+ channels are recoded at unprecedented levels

In mammals, recoding events caused by RNA editing are exceptionally rare. The same cannot be said for cephalopods, where more recoding events have been discovered in mRNAs encoding K^+ channels alone than in the entire human brain transcriptome. The first K^+ channel mRNA to be analysed in squid encoded a K_V2 subfamily member expressed in the optic lobe, a large ganglion within the central nervous system (Patton *et al.*, 1997). In this mRNA only a short 360 nt region was examined for RNA editing and 12 editing sites were uncovered. These sites changed codons in the voltage-sensing fourth transmembrane span, and also in the helices encoding the pore. Two sites were selected for electrophysiological characterization: Y576C in the pore and I597V in S6. Y576C slowed both the rate of channel closing and slow inactivation. I597V had the opposite effect, increasing both. A subsequent study examined the entire mRNA encoding the squid delayed rectifier of the giant axon, a K_V1 family member (Rosenthal *et al.*, 1996; Rosenthal and Bezanilla, 2002b). Here 14 recoding events were identified, mostly in the channel's tetramerization domain, and S1 and S3 transmembrane spans. As with squid K_V2, many sites affected closing kinetics, and some also shifted the voltage-dependence of activation. One site in the tetramerization domain (R87G) reduced the α-subunits' ability to oligomerize into tetramers by drastically reducing their affinity for each other. This effect could be expected to influence overall K^+ conductance in the giant axon, a physiological property that is actively regulated between species of squid which inhabit different thermal environments (Rosenthal and Bezanilla, 2002a).

The bewildering array of RNA editing events in squid K^+ channels makes us speculate on why these organisms use this process so extensively used to regulate excitability. RNA editing provides an organism with physiological options. In theory, they can choose to edit a specific position when conditions are favourable. RNA editing of the octopus orthologue of squid K_V1

gives some substance to this speculation (Garrett and Rosenthal, 2012). Different octopus species inhabit a tremendous range of temperature environments, from the poles to the equator. As with squid, mRNAs encoding octopus K_V1 are extensively edited. For seven species studied, the gene sequences for the same K_V1 orthologue were virtually identical, however the mRNAs were edited at 19 sites, some shared, some species specific. Although many of these sites affected K_V channel function, one site, which recoded an isoleucine to a valine in the fifth transmembrane span (I321V), doubled the speed of channel closing and the extent to which it was edited correlated closely with the host species' thermal environment. I321V was edited almost to completion in Arctic and Antarctic species, yet scarcely edited in tropical species, and temperate species edited it to intermediate extents. Single-channel analysis revealed that the edit destabilizes the open ion conducting state, poising the channel on the edge of closing. These results were significant because they were the first to link RNA editing with an environmental factor. Temperature was chosen because it is easy to manipulate and has a direct, predictable effect on channel function. Future studies will help determine whether organisms can rapidly change the extent of RNA editing at specific, functionally relevant sites in acclimation to environmental pressures, or instead whether these sites evolve over generations by evolutionary adaptation. The relevant environmental pressure, and the molecular mechanisms underlying adaptation or acclimation, should prove interesting for further studies.

Drosophila KV channels are also edited extensively

mRNAs for K_V channels, including *shaker*, the most intensively studied channel in history, are edited in *Drosophila* (Ryan *et al.*, 2008; Ingleby *et al.*, 2009). These sites have a variety of functional effects, including changes to activation, deactivation and inactivation kinetics, and some small shifts to the channels' voltage sensitivity. In *shaker*, there are four editing sites, giving rise to 16 different combinations of edited isoforms (Hoopengardner *et al.*, 2003). Interestingly, the frequency of occurrence of these isoforms varied between different parts of the adult's anatomy (Ingleby *et al.*, 2009). It was also found that editing at different sites was linked, and the linkage varied between different regions. These editing patterns made a difference on a functional level. An interesting feature of editing in *Shaker* is that the physiological properties of different editing isoforms could not be predicted by the effects of the individual editing sites. This phenomenon, known as functional epistasis, means that the effect of editing a specific site is context dependent, greatly expanding the possibilities for regulation. On top of this, the *shaker* locus encodes numerous functionally different splice variants (Papazian *et al.*, 1987; Tempel *et al.*, 1987; Schwarz *et al.*, 1988; Timpe *et al.*, 1988), and the functional effects of editing in most of these has yet to be explored.

Voltage-activated calcium channels

Voltage-activated calcium (Ca_V) channels are broadly expressed in excitable cells, where they play diverse functions. In some muscle cells, like the pacemaker myocytes in the mammalian sinoatrial node, Ca_V channels are essential to the repetitive electrical activity of these cells (Mangoni *et al.*, 2006). However, because intracellular Ca^{2+} is a second messenger, Ca_V channels are also used as an entryway for Ca^{2+} to regulate diverse cellular processes, from gene expression to motion (Hille, 2001). The primary sequence of the Ca_V channel pore-forming α-subunit contains four repeated domains (Tanabe *et al.*, 1987; Catterall, 2000), each containing a voltage-sensing domain and a pore domain. Presumably, these repeats form a functional unit that should be similar to the assembly of the four individual α subunits of K_V channels (Long *et al.*, 2005). Using genomic approaches in *Drosophila*, several editing sites have been identified in Ca_V channels (Hoopengardner *et al.*, 2003; Graveley *et al.*, 2011). At present we have no clues on the biological or physiological significance of these sites.

Recently, three RNA editing sites were discovered in mRNAs encoding rodent $Ca_V1.3$ channels, clustered in four contiguous amino acids located in the cytoplasmic C-terminus (i.e. after the fourth transmembrane repeat) (Huang *et al.*, 2012).

These amino acids (**IQ**D**Y**; in bold are represented the amino acids recoded by RNA editing) are part of a calmodulin (CaM)-binding domain. The $Ca_V1.3$–CaM–Ca^{2+} complex influences the gating properties of the channel, particularly the Ca^{2+}-dependent inactivation kinetics (Shen *et al.*, 2006; Yang *et al.*, 2006; Huang *et al.*, 2012).

I → M, Q → R and Y → C are edited to different extents, but none by more than ~50%. Editing of this region is restricted to the central nervous system, with high levels of activity in the frontal cortex, hippocampus, medulla oblongata and cerebellum. Using a proteomic approach, the authors were able to identify peptides containing amino acid substitutions, demonstrating that all three possible conversions exist at the protein level. Further support for multiple variants was provided at the mRNA level, where seven out of the eight possible combinations of editing sites were identified. The most abundant variants were MQDY and MQDC. In ADAR2$^{-/-}$/GluR-B$^{R/R}$ knockout mice (Higuchi *et al.*, 2000) all three editing sites were absent. What are the functional consequences of these editing events?

Upon membrane depolarization, $Ca_V1.3$ channels open a gate that allows Ca^{2+} to flow into the cell, producing an inward current. Even if the depolarization is maintained, the Ca^{2+} current will inactivate. Kinetically, CDI is quite different than fast inactivation in K_V channels. On the one hand, the rate of inactivation depends on the levels of intracellular Ca^{2+} (Brehm and Eckert, 1978; Tillotson, 1979), a property which led this process to be named Ca^{2+}-dependent inactivation (CDI). On the other hand, Ca_V channels enter inactivation from the closed state (Tadross *et al.*, 2010). It has been previously shown that mutating the first two editing sites (IQ) to alanine almost completely abolishes CDI (Yang *et al.*, 2006). Therefore, CDI was examined in all single RNA editing conversions (Huang *et al.*, 2012). Only I → M and Q → R produced a substantial decrease in the levels of CDI, and the effect was additive when they both were edited (IQ → MR).

By transfecting primary hippocampal neurons with wild-type and edited channels, it was shown that all constructs reached the cell surface equally well (Huang *et al.*, 2012), providing confidence that any potential consequence of editing on an animal's behaviour would likely originate from the changes in channel function and not their density at the cell membrane. To explore the consequences of editing in neurons, the authors turned to a comparative study using wild-type and ADAR2$^{-/-}$/GluR-B$^{R/R}$ knockout mice. As a first attempt, they selected neurons from the suprachiasmatic nucleus (SCN), which are known to contain $Ca_V1.3$ channels involved in the electrical rhythmic activity that governs circadian rhythms (Pennartz *et al.*, 2002). SCN neurons do indeed edit the three sites, but to relatively low levels (< 20%). Remarkably, Ca^{2+} currents recorded from neurons taken from knockout animals showed noticeably more inactivation, consistent with the expected lack of edited channel variants. SCN neurons spontaneously fire action potentials (Inouye and Kawamura, 1979; Reppert and Weaver, 2001). L-type Ca^{2+} channels (Ca_V1 family) play an essential role in setting the frequency of the spontaneous firing (Pennartz *et al.*, 2002). Because neurons from knockout mice would have, at steady-state, fewer conducting Ca^{2+} channels available, they would be expected to fire at lower frequencies than neurons from wild-type animals, and this was found to be the case (Huang *et al.*, 2012).

Ion pumps and transporters

In contrast to ion channels, which allow ions to permeate at rates near diffusion, pumps and transporters do so at rates many orders of magnitude slower. This difference arises from the fundamental necessity of pumps and transporters to occlude, or trap, their substrates at some point during the transport process. For example, the Na^+/K^+-ATPase uses the energy derived from the hydrolysis of ATP to transport three Na^+ out of the cell and import two K^+, at rates of ~100/s. Both, Na^+ and K^+ are transported *against* their electrochemical gradients, so pumps cannot afford the cost of having a continuous ion permeation pathway at any time. Fig. 1.3A shows a schematic view of the transport mechanism of the Na^+/K^+-ATPase. The gates that control access of the ions to the inside or outside of the cell alternate opening. The extent to which the function of pumps and transporters are tuned by

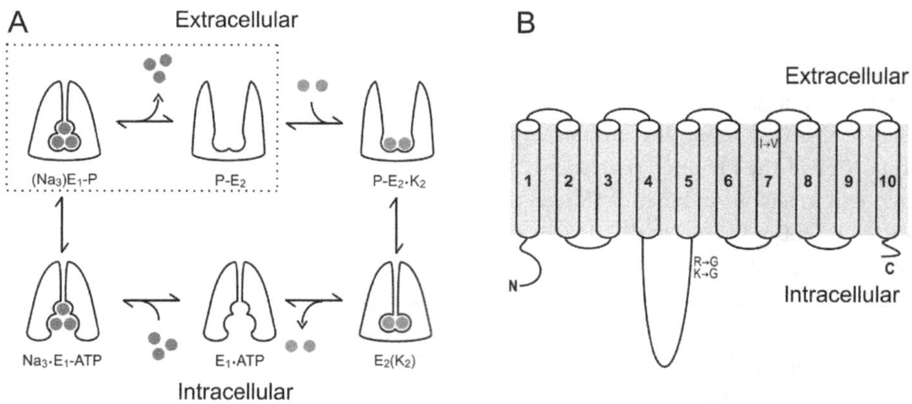

Figure 1.3 RNA editing of squid Na^+/K^+-ATPase. (A) Albers–Post transport cycle. Green and orange spheres represent Na^+ and K^+, respectively. The alternate gating mechanism with occlusion states for Na^+ (top-left corner) and K^+ (bottom-right corner) are essentials for transporting these ions against their electrochemical gradients. Dotted lines enclose the states associated with the extracellular release of Na^+, which were isolated by removing K^+ from all solutions and the absence of intracellular ADP. (B) Membrane topology of the α Na^+/K^+-ATPase subunit. A functional Na^+/K^+-ATPase is formed by the co assembly of α – and β-subunits. The α-subunit, however, contains all requirements for transport: ATP binding domain, phosphorylation site and permeation pathway. Two amino acid substitutions by RNA editing were identified in the phosphorylation domain located within the large intracellular loop between TM segments 4 and 5. An additional site was identified at the extracellular end of TM segment 7. A colour version of this figure is located in the plate section at the back of the book.

RNA editing appears to be substantially smaller than in ion channels. At present, editing sites have been identified in the Na^+/K^+-ATPase and the Na^+/Ca^{2+}–K^+ exchanger in *Drosophila* (Stapleton *et al.*, 2006; Graveley *et al.*, 2011) and the Na^+/K^+-ATPase of squid (Colina *et al.*, 2010). Only for the squid Na^+/K^+-ATPase, however, do we have information on what these editing sites do. Interestingly, as discussed in detail below, RNA editing has targeted the kinetics of Na^+ occlusion to speed up the pump's transport velocity (Colina *et al.*, 2010).

Squid pump mRNAs are edited in four sites, targeting three different amino acids. Positions R663 and K666 are in the phosphorylation domain, located within the large intracellular loop between transmembrane (TM) segments 4 and 5 (Fig. 1.3B). R663 had a single editing site, converting it to glycine. Both adenosines of the K666 codon (AAG) were found to be edited, giving rise to the possibility of three conversions: K→R, G or E. K→E, however, was never observed in any region of the squid nervous system. The fourth editing event is an I→V conversion at amino acid 877, which lies at the external end of the seventh TM segment (Fig. 1.3B). R663G is consistently edited at percentages >65% throughout the squid's nervous system. The other sites, however, showed high spatial regulation, suggesting that they might be edited differently depending on the demands of different types of neurons. What are the functional consequences of these editing events on the pump's function?

The function of the Na^+/K^+-ATPase is to maintain the Na^+ and K^+ gradients between the inside and outside of the cell. How effectively the pump can do its job depends, ultimately, on its turnover rate (i.e. the number of transport cycles the pump can undergo per unit of time). Because the pump moves an unequal number of ions per cycle, it generates an electrical current (I_p) that can be measured and used as a faithful metre of its turnover rate. We showed that the maximal turnover rates of the exonic and all edited versions of the pump were similar, suggesting that RNA editing does not target the rate limiting step when pumps are running at full speed. The pump's transport velocity, however, is voltage dependent, reaching a maximum at voltages >~0 mV and decreasing monotonically at negative potentials. Therefore, from a cell's perspective, the relevant issue is the pump's speed at the resting membrane potential (~ −80 mV), where it spends most of its time. We studied the voltage dependence of I_p in all constructs and observed that I877V pumps were able to pump faster at negative potentials, i.e. the I_p versus voltage curve was shifted towards more negative potentials. What is the mechanism of this change?

Before Na^+ ions are released to the external bulk solution, they must travel through a narrow access channel (Gadsby *et al.*, 1993; Hilgemann, 1994; Holmgren *et al.*, 2000), along which there is a voltage drop. Therefore, negative potentials will drive external Na^+ back to their binding sites, slowing down the turnover rate, as observed experimentally (Gadsby *et al.*, 1985; Gadsby and Nakao, 1989; Nakao and Gadsby, 1989; Sagar and Rakowski, 1994). In order to acquire precise mechanistic information about the transitions involving the binding/release of external Na^+ it is necessary to isolate them (Fig. 1.3A, enclosed by dashed lines). We achieved these conditions by removing K^+ from both the intracellular and extracellular solutions, while maintaining internal Na^+ and ATP, but not ADP. We also used the cut-open oocyte voltage-clamp technique, which allowed us access to both sides of the membrane (Taglialatela *et al.*, 1992). Fig. 1.4A shows examples of pump-mediated currents obtained in response to voltage steps from 0 mV to −198 mV (green), −78 mV (orange) and 42 mV (brown). As expected, the pump-mediated currents have only transient components, which represent the redistribution of the Na^+-bound states upon the fast voltage change. Clearly, these transient currents have multiple components. A careful dissection shows that the transient currents contain a fast ($\tau < 100\,\mu s$), medium ($\tau \sim 0.2$–0.5 ms) and a slow ($\tau \sim 1$–10 ms) component, which led us to propose that they represent the sequential and distinct interactions of each Na^+ with the pump (Holmgren *et al.*, 2000; Gadsby *et al.*, 2012), as shown in Fig. 1.4B. By estimating the amount of charge moved by the slow component (Q_s) at many voltages, we can infer the voltage dependence of the steady-state occupancy of the deeply occluded state (Fig. 1.4C). For genomic (filled cyan circles) and edited pumps (filled red circles), the distributions followed a Boltzmann function with the same steepness. However, RNA editing shifted the voltage dependence towards more negative potentials (Fig. 1.4C). This means that at resting membrane potentials, an edited pump will less likely be in the deep occluded state, therefore favouring forward pumping. Does RNA editing directly influence the kinetics of the slow component? To address this question we studied the voltage dependence of the relaxation rates of the slow component (Fig. 1.4D). The solid lines through the data represent the best fit to:

$$k_v = k_f + \frac{k_b}{1 + \frac{K_{0.5}(0)\exp(\lambda VF/RT)}{[Na]_0}}$$

which describes the occlusion (k_b) and deocclusion (k_f) rates as voltage independent, and the voltage dependence arising from the binding/unbinding reaction with an apparent affinity ($K_{0.5}(0)$) value at 0 mV. *F*, *R* and *T* are constants with their usual meaning and values. The parameter λ denotes the fractional 'electrical depth' of the access channel. In this model, the asymptote at positive potentials corresponds to k_f. As observed in Fig. 1.4D, RNA editing does not influence k_f; in both edited and unedited pumps it is ~ 100/s. The asymptote at negative potentials is the sum of k_f and k_b. Even though we cannot reach the voltages needed to observe the asymptotic value, we can confidently estimate it because λ was independently determined to be 0.67 for edited and unedited pumps. From the fits, k_b showed a small change from ~ 2000/s to ~ 1600/s for unedited and edited pumps, respectively. These results suggest that the kinetics of the occlusion/deocclusion transition remain largely unaltered by RNA editing. The most significant change predicted by the model was a substantial reduction of the apparent affinity in edited pumps. $K_{0.5}(0)$ was ~ 7.5 M for genomic pumps, as compared with ~ 13 M for the I877V pumps. What could be the underlying reason for this effect?

From Na^+/K^+-ATPase models based on the Ca^{2+}-ATPase crystal structure (Ogawa and Toyoshima, 2002) and Na^+/K^+-ATPase crystal structures (Morth *et al.*, 2007; Shinoda *et al.*, 2009) we know that the cation binding sites are

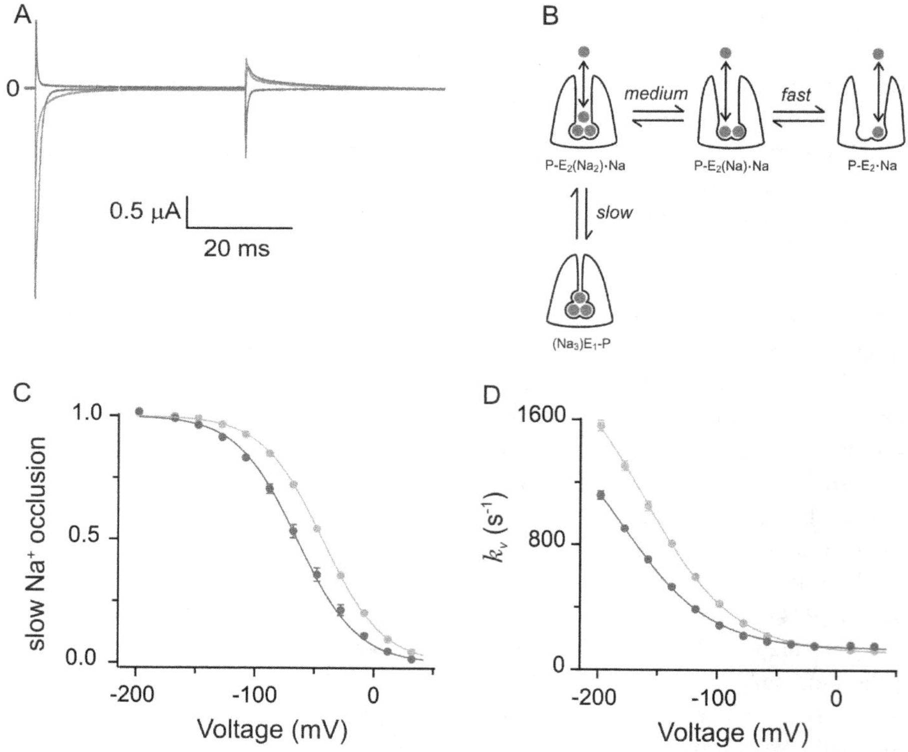

Figure 1.4 RNA editing of squid Na^+/K^+-ATPase: functional consequences. (A) Transient currents mediated by the Na^+/K^+-ATPase trapped in states linked to release of external Na^+. Green, orange and cyan current traces represent pump currents in response to fast voltage steps from 0 mV to −198, −78 and +42 mV, respectively. It is noticeable the multi component nature of these pump-mediated transient currents. (B) Cartoon representing the kinetic scheme that best describes the transient currents shown in (A). In this model, each of the Na^+ binding/release and occlusion/deocclusion transitions are kinetically distinct. The slow component symbolizes the deep occlusion transition, which is the rate limiting step of the transient pump currents. (C) Voltage distribution of the deep occlusion transition. Cyan and red circles represent deep occlusion of unedited (I877) and edited (V877) pumps. Solid lines represent Boltzmann fits. RNA editing shifted the distribution towards more negative potentials which translate in faster pump velocities at potentials near the resting state of a neuron. (D) Voltage dependence of the relaxation rate of the deep occlusion transition. Same colour scheme as C. Solid lines correspond to 'access channel model' fits. Edited pumps reduced their apparent affinity for Na^+ binding to ~13 M from a value of ~7 M in unedited pumps. A colour version of this figure is located in the plate section at the back of the book.

far away from position I877, so it is unlikely that a valine substitution would directly alter the way Na^+ binds. Another possibility would be that electrostatics influence the apparent affinity ($K_{0.5}(0)$) for external Na^+. In fact, this mechanism has been used by nature to help Na^+/K^+-ATPases from marine osmoconformers to adapt to the high concentrations of Na^+ found in sea water (Colina *et al.*, 2007). But I and V are both aliphatic amino acids, so it is also unlikely that they affect electrostatics. A more likely explanation is that the relative contribution of each component at different voltages is different between genomic and edited pumps. By quantifying the relative amounts of the slow, medium and fast charge components at many voltages, we discovered a clear shift in their distributions caused by the I877V edit: the fast component showed a greater fractional representation at the expense of the slow component.

Our studies clearly demonstrate that RNA editing tunes the Na^+/K^+-ATPase's transport velocity. Interestingly, the I877V editing site is spatially regulated through the squid's nervous system, suggesting that RNA editing regulates the specific Na^+ homeostasis demands of different types of neurons. As we observed with both human and octopus K^+ channels, the removal of a single methyl group by recoding I→V can produce a large effect, presumably because nature has selected key positions to edit. Many interesting questions remain concerning editing of the squid Na^+/K^+-ATPase. Mechanistically, how does this small change in primary structure at position 877 affect the larger-scale movements associated with Na^+ release? What are the functional consequences of editing codons R663 and K666? Finally, how do these edits relate to the physiological demands of different neurons?

Emerging targets for RNA editing

Besides the examples listed thus far, the functional effects of a handful of editing sites are just beginning to be explored. For example, an isoleucine can be recoded to a methionine in the third transmembrane span of the human GABA-gated Cl^- channel α3 subunit (Ohlson *et al.*, 2007). A methionine at this site causes the multimeric channel to activate more slowly and deactivate more rapidly. It also largely abolished outward rectification of the current–voltage relationship, suggesting that the site also affects ion selectivity (Rula *et al.*, 2008). The mechanisms underlying these effects are largely unknown. In *Drosophila*, a GABA-gated Cl^- channel is also edited, increasing the channel's sensitivity to its agonist (Jones *et al.*, 2009). A recent study on the *Drosophila* developmental transcriptome has predicted more than 600 recoding events due to adenosine deamination, and more than 100 of these occur in ion channels, transporters or the machinery for synaptic vesicle release (Graveley *et al.*, 2011). Recent efforts on the human transcriptome have uncovered many new sites that have yet to be characterized. Clearly, much work remains before we get a clear understanding of how RNA editing is used to regulate animal physiology in general, and excitability in particular.

References

Aldrich, R.W., Jr., Getting, P.A., and Thompson, S.H. (1979). Mechanism of frequency-dependent broadening of molluscan neurone soma spikes. J. Physiol. *291*, 531–544.

Armstrong, C.M. (1966). Time course of TEA^+-induced anomalous rectification in squid giant axons. J. Gen. Physiol. *50*, 491–503.

Armstrong, C.M. (1969). Inactivation of the potassium conductance and related phenomena caused by quaternary ammonium ion injection in squid axons. J. Gen. Physiol. *54*, 553–575.

Armstrong, C.M. (1971). Interaction of tetraethylammonium ion derivatives with the potassium channels of giant axons. J. Gen. Physiol. *58*, 413–437.

Belcher, S.M., and Howe, J.R. (1997). Characterization of RNA editing of the glutamate-receptor subunits GluR5 and GluR6 in granule cells during cerebellar development. Brain Res. Mol. Brain Res. *52*, 130–138.

Bernard, A., Ferhat, L., Dessi, F., Charton, G., Represa, A., Ben-Ari, Y., and Khrestchatisky, M. (1999). Q/R editing of the rat GluR5 and GluR6 kainate receptors *in vivo* and *in vitro*: evidence for independent developmental, pathological and cellular regulation. Eur. J. Neurosci. *11*, 604–616.

Bhalla, T., Rosenthal, J.J., Holmgren, M., and Reenan, R. (2004). Control of human potassium channel inactivation by editing of a small mRNA hairpin. Nat. Struct. Mol. Biol. *11*, 950–956.

Brehm, P., and Eckert, R. (1978). Calcium entry leads to inactivation of calcium channel in Paramecium. Science *202*, 1203–1206.

Brusa, R., Zimmermann, F., Koh, D.S., Feldmeyer, D., Gass, P., Seeburg, P.H., and Sprengel, R. (1995). Early-onset epilepsy and postnatal lethality associated with an editing-deficient GluR-B allele in mice. Science *270*, 1677–1680.

Burnashev, N., Monyer, H., Seeburg, P.H., and Sakmann, B. (1992). Divalent ion permeability of AMPA receptor channels is dominated by the edited form of a single subunit. Neuron *8*, 189–198.

Catterall, W.A. (2000). Structure and regulation of voltage-gated Ca^{2+} channels. Annu. Rev. Cell. Dev. Biol. *16*, 521–555.

Choi, K.L., Aldrich, R.W., and Yellen, G. (1991). Tetraethylammonium blockade distinguishes two inactivation mechanisms in voltage-activated K^+ channels. Proc. Natl. Acad. Sci. U.S.A. *88*, 5092–5095.

Colina, C., Rosenthal, J.J., Degiorgis, J.A., Srikumar, D., Iruku, N., and Holmgren, M. (2007). Structural basis of Na^+/K^+-ATPase adaptation to marine environments. Nat. Struct. Mol. Biol. *14*, 427–431.

Colina, C., Palavicini, J.P., Srikumar, D., Holmgren, M., and Rosenthal, J.J. (2010). Regulation of Na^+/K^+ ATPase transport velocity by RNA editing. PLoS Biol. *8*, e1000540.

Connor, J.A., and Stevens, C.F. (1971). Prediction of repetitive firing behaviour from voltage clamp data on an isolated neurone soma. J. Physiol. *213*, 31–53.

Debanne, D., Guerineau, N.C., Gahwiler, B.H., and Thompson, S.M. (1997). Action-potential propagation gated by an axonal I(A)-like K^+ conductance in hippocampus. Nature *389*, 286–289.

Decher, N., Streit, A.K., Rapedius, M., Netter, M.F., Marzian, S., Ehling, P., Schlichthorl, G., Craan, T., Renigunta, V., Kohler, A., *et al.* (2010). RNA editing modulates the binding of drugs and highly unsaturated fatty acids to the open pore of Kv potassium channels. EMBO J. *29*, 2101–2113.

Demo, S.D., and Yellen, G. (1991). The inactivation gate of the Shaker K^+ channel behaves like an open-channel blocker. Neuron 7, 743–753.

Doyle, D.A., Morais Cabral, J., Pfuetzner, R.A., Kuo, A., Gulbis, J.M., Cohen, S.L., Chait, B.T., and MacKinnon, R. (1998). The structure of the potassium channel: molecular basis of K^+ conduction and selectivity. Science *280*, 69–77.

Egebjerg, J., and Heinemann, S.F. (1993). Ca^{2+} permeability of unedited and edited versions of the kainate selective glutamate receptor GluR6. Proc. Natl. Acad. Sci. U.S.A. *90*, 755–759.

Feldmeyer, D., Kask, K., Brusa, R., Kornau, H.C., Kolhekar, R., Rozov, A., Burnashev, N., Jensen, V., Hvalby, O., Sprengel, R., *et al.* (1999). Neurological dysfunctions in mice expressing different levels of the Q/R site-unedited AMPAR subunit GluR-B. Nat. Neurosci. *2*, 57–64.

Gadsby, D.C., and Nakao, M. (1989). Steady-state current–voltage relationship of the Na/K pump in guinea pig ventricular myocytes. J. Gen. Physiol. *94*, 511–537.

Gadsby, D.C., Kimura, J., and Noma, A. (1985). Voltage dependence of Na/K pump current in isolated heart cells. Nature *315*, 63–65.

Gadsby, D.C., Rakowski, R.F., and De Weer, P. (1993). Extracellular access to the Na,K pump: pathway similar to ion channel. Science *260*, 100–103.

Gadsby, D.C., Bezanilla, F., Rakowski, R.F., De Weer, P., and Holmgren, M. (2012). The dynamic relationships between the three events that release individual Na^+ ions from the Na^+/K^+-ATPase. Nat. Commun. 3, 669.

Garrett, S., and Rosenthal, J.J. (2012). RNA editing underlies temperature adaptation in K^+ channels from polar octopuses. Science *335*, 848–851.

Gonzalez, C., Lopez-Rodriguez, A., Srikumar, D., Rosenthal, J.J., and Holmgren, M. (2011). Editing of human K(V)1.1 channel mRNAs disrupts binding of the N-terminus tip at the intracellular cavity. Nat. Commun. 2, 436.

Graveley, B.R., Brooks, A.N., Carlson, J.W., Duff, M.O., Landolin, J.M., Yang, L., Artieri, C.G., van Baren, M.J., Boley, N., Booth, B.W., *et al.* (2011). The developmental transcriptome of *Drosophila melanogaster*. Nature *471*, 473–479.

Gulbis, J.M., Zhou, M., Mann, S., and MacKinnon, R. (2000). Structure of the cytoplasmic beta subunit-T1 assembly of voltage-dependent K^+ channels. Science *289*, 123–127.

Herb, A., Higuchi, M., Sprengel, R., and Seeburg, P.H. (1996). Q/R site editing in kainate receptor GluR5 and GluR6 pre-mRNAs requires distant intronic sequences. Proc. Natl. Acad. Sci. U.S.A. 93, 1875–1880.

Hideyama, T., Yamashita, T., Suzuki, T., Tsuji, S., Higuchi, M., Seeburg, P.H., Takahashi, R., Misawa, H., and Kwak, S. (2010). Induced loss of ADAR2 engenders slow death of motor neurons from Q/R site-unedited GluR2. J. Neurosci. *30* 11917–11925.

Hideyama, T., Yamashita, T., Aizawa, H., Tsuji, S., Kakita, A., Takahashi, H., and Kwak, S. (2011). Profound down-regulation of the RNA editing enzyme ADAR2 in ALS spinal motor neurons. Neurobiol. Dis. *45*, 1121–1128.

Higuchi, M., Maas, S., Single, F.N., Hartner, J., Rozov, A., Burnashev, N., Feldmeyer, D., Sprengel, R., and Seeburg, P.H. (2000). Point mutation in an AMPA receptor gene rescues lethality in mice deficient in the RNA e enzyme ADAR2. Nature *406*, 78–81.

Hilgemann, D.W. (1994). Channel-like function of the Na,K pump probed at microsecond resolution in giant membrane patches. Science *263*, 1429–1432.

Hille, B. (2001). Ion Channels of Excitable Membranes, 3rd edn (Sinauer Associates, Inc., Suderland, MA).

Hoffman, D.A., Magee, J.C., Colbert, C.M., and Johnston, D. (1997). K^+ channel regulation of signal propagation in dendrites of hippocampal pyramidal neurons. Nature *387*, 869–875.

Hollmann, M., Hartley, M., and Heinemann, S. (1991). Ca^{2+} permeability of KA-AMPA—gated glutamate receptor channels depends on subunit composition. Science *252*, 851–853.

Holmgren, M., Smith, P.L., and Yellen, G. (1997). Trapping of organic blockers by closing of voltage-dependent K^+ channels: evidence for a trap door mechanism of activation gating. J. Gen. Physiol. *109*, 527–535.

Holmgren, M., Wagg, J., Bezanilla, F., Rakowski, R.F., De Weer, P., and Gadsby, D.C. (2000). Three distinct and sequential steps in the release of sodium ions by the Na^+/K^+-ATPase. Nature *403*, 898–901.

Hoopengardner, B., Bhalla, T., Staber, C., and Reenan, R. (2003). Nervous system targets of RNA editing identified by comparative genomics. Science *301*, 832–836.

Horsch, M., Seeburg, P.H., Adler, T., Aguilar-Pimentel, J.A., Becker, L., Calzada-Wack, J., Garrett, L., Gotz, A., Hans, W., Higuchi, M., *et al.* (2011). Requirement of the RNA editing enzyme ADAR2 for normal physiology in mice. J. Biol. Chem. *286*, 18614–18622.

Hoshi, T., Zagotta, W.N., and Aldrich, R.W. (1990). Biophysical and molecular mechanisms of Shaker potassium channel inactivation. Science *250*, 533–538.

Huang, H., Tan, B.Z., Shen, Y., Tao, J., Jiang, F., Sung, Y.Y., Ng, C.K., Raida, M., Kohr, G., Higuchi, M., *et al.* (2012). RNA editing of the IQ domain in Ca(v)1.3 channels modulates their Ca^{2+}-dependent inactivation. Neuron 73, 304–316.

Ingleby, L., Maloney, R., Jepson, J., Horn, R., and Reenan, R. (2009). Regulated RNA editing and functional epistasis in Shaker potassium channels. J. Gen. Physiol. *133*, 17–27.

Inouye, S.T., and Kawamura, H. (1979). Persistence of circadian rhythmicity in a mammalian hypothalamic 'island' containing the suprachiasmatic nucleus. Proc. Natl. Acad. Sci. U.S.A. *76*, 5962–5966.

Jones, A.K., Buckingham, S.D., Papadaki, M., Yokota, M., Sattelle, B.M., Matsuda, K., and Sattelle, D.B. (2009).

Splice-variant – and stage-specific RNA editing of the *Drosophila* GABA receptor modulates agonist potency. J. Neurosci. *29,* 4287–4292.

Kawahara, Y., Ito, K., Sun, H., Aizawa, H., Kanazawa, I., and Kwak, S. (2004). Glutamate receptors: RNA editing and death of motor neurons. Nature *427,* 801.

Keinanen, K., Wisden, W., Sommer, B., Werner, P., Herb, A., Verdoorn, T.A., Sakmann, B., and Seeburg, P.H. (1990). A family of AMPA-selective glutamate receptors. Science *249,* 556–560.

Kohler, M., Burnashev, N., Sakmann, B., and Seeburg, P.H. (1993). Determinants of Ca^{2+} permeability in both TM1 and TM2 of high affinity kainate receptor channels: diversity by RNA editing. Neuron *10,* 491–500.

Krestel, H.E., Shimshek, D.R., Jensen, V., Nevian, T., Kim, J., Geng, Y., Bast, T., Depaulis, A., Schonig, K., Schwenk, F., *et al.* (2004). A genetic switch for epilepsy in adult mice. J. Neurosci. *24* 10568–10578.

Lambolez, B., Ropert, N., Perrais, D., Rossier, J., and Hestrin, S. (1996). Correlation between kinetics and RNA splicing of alpha-amino-3-hydroxy-5-methylisoxazole-4-propionic acid receptors in neocortical neurons. Proc. Natl. Acad. Sci. U.S.A. *93,* 1797–1802.

Laube, B., Kuhse, J., and Betz, H. (1998). Evidence for a tetrameric structure of recombinant NMDA receptors. J. Neurosci. *18,* 2954–2961.

Li, J.B., Levanon, E.Y., Yoon, J.K., Aach, J., Xie, B., Leproust, E., Zhang, K., Gao, Y., and Church, G.M. (2009). Genome-wide identification of human RNA editing sites by parallel DNA capturing and sequencing. Science *324,* 1210–1213.

Li, M., Wang, I.X., Li, Y., Bruzel, A., Richards, A.L., Toung, J.M., and Cheung, V.G. (2011). Widespread RNA and DNA sequence differences in the human transcriptome. Science *333,* 53–58.

Liu, Y., Holmgren, M., Jurman, M.E., and Yellen, G. (1997). Gated access to the pore of a voltage-dependent K^+ channel. Neuron *19,* 175–184.

Lomeli, H., Mosbacher, J., Melcher, T., Hoger, T., Geiger, J.R., Kuner, T., Monyer, H., Higuchi, M., Bach, A., and Seeburg, P.H. (1994). Control of kinetic properties of AMPA receptor channels by nuclear RNA editing. Science *266,* 1709–1713.

Long, S.B., Campbell, E.B., and Mackinnon, R. (2005). Crystal structure of a mammalian voltage-dependent Shaker family K^+ channel. Science *309,* 897–903.

Maas, S., Melcher, T., Herb, A., Seeburg, P.H., Keller, W., Krause, S., Higuchi, M., and O'Connell, M.A. (1996). Structural requirements for RNA editing in glutamate receptor pre-mRNAs by recombinant double-stranded RNA adenosine deaminase. J. Biol. Chem. *271,* 12221–12226.

MacKinnon, R., Aldrich, R.W., and Lee, A.W. (1993). Functional stoichiometry of Shaker potassium channel inactivation. Science *262,* 757–759.

Mangoni, M.E., Couette, B., Marger, L., Bourinet, E., Striessnig, J., and Nargeot, J. (2006). Voltage-dependent calcium channels and cardiac pacemaker activity: from ionic currents to genes. Prog. Biophys. Mol. Biol. *90,* 38–63.

Monyer, H., Seeburg, P.H., and Wisden, W. (1991). Glutamate-operated channels: developmentally early and mature forms arise by alternative splicing. Neuron *6,* 799–810.

Morth, J.P., Pedersen, B.P., Toustrup-Jensen, M.S., Sorensen, T.L., Petersen, J., Andersen, J.P., Vilsen, B., and Nissen, P. (2007). Crystal structure of the sodium-potassium pump. Nature *450,* 1043–1049.

Nakagawa, T., Cheng, Y., Ramm, E., Sheng, M., and Walz, T. (2005). Structure and different conformational states of native AMPA receptor complexes. Nature *433,* 545–549.

Nakao, M., and Gadsby, D.C. (1989). [Na] and [K] dependence of the Na/K pump current–voltage relationship in guinea pig ventricular myocytes. J. Gen. Physiol. *94,* 539–565.

Ogawa, H., and Toyoshima, C. (2002). Homology modeling of the cation binding sites of Na^+K^+-ATPase. Proc. Natl. Acad. Sci. U.S.A. *99,* 15977–15982.

Ohlson, J., Pedersen, J.S., Haussler, D., and Ohman, M. (2007). Editing modifies the GABA(A) receptor subunit alpha3. RNA *13,* 698–703.

Palavicini, J.P., O'Connell, M.A., and Rosenthal, J.J. (2009). An extra double-stranded RNA binding domain confers high activity to a squid RNA editing enzyme. RNA *15,* 1208–1218.

Papazian, D.M., Schwarz, T.L., Tempel, B.L., Jan, Y.N., and Jan, L.Y. (1987). Cloning of genomic and complementary DNA from Shaker, a putative potassium channel gene from *Drosophila.* Science *237,* 749–753.

Paschen, W., Schmitt, J., Gissel, C., and Dux, E. (1997). Developmental changes of RNA editing of glutamate receptor subunits GluR5 and GluR6: *in vivo* versus *in vitro.* Brain Res. Dev. Brain Res. *98,* 271–280.

Patton, D.E., Silva, T., and Bezanilla, F. (1997). RNA editing generates a diverse array of transcripts encoding squid Kv2 K^+ channels with altered functional properties. Neuron *19,* 711–722.

Peng, Z., Cheng, Y., Tan, B.C., Kang, L., Tian, Z., Zhu, Y., Zhang, W., Liang, Y., Hu, X., Tan, X., *et al.* (2012). Comprehensive analysis of RNA-Seq data reveals extensive RNA editing in a human transcriptome. Nat. Biotechnol. *30,* 253–260.

Pennartz, C.M., de Jeu, M.T., Bos, N.P., Schaap, J., and Geurtsen, A.M. (2002). Diurnal modulation of pacemaker potentials and calcium current in the mammalian circadian clock. Nature *416,* 286–290.

Reppert, S.M., and Weaver, D.R. (2001). Molecular analysis of mammalian circadian rhythms. Annu. Rev. Physiol. *63,* 647–676.

Rettig, J., Heinemann, S.H., Wunder, F., Lorra, C., Parcej, D.N., Dolly, J.O., and Pongs, O. (1994). Inactivation properties of voltage-gated K^+ channels altered by presence of beta-subunit. Nature *369,* 289–294.

Rosenmund, C., Stern-Bach, Y., and Stevens, C.F. (1998). The tetrameric structure of a glutamate receptor channel. Science *280,* 1596–1599.

Rosenthal, J.J., and Bezanilla, F. (2002a). A comparison of propagated action potentials from tropical and temperate squid axons: different durations and conduction velocities correlate with ionic conductance levels. J. Experi. Biol. *205*, 1819–1830.

Rosenthal, J.J., and Bezanilla, F. (2002b). Extensive editing of mRNAs for the squid delayed rectifier K^+ channel regulates subunit tetramerization. Neuron *34*, 743–757.

Rosenthal, J.J., Vickery, R.G., and Gilly, W.F. (1996). Molecular identification of SqKv1A. A candidate for the delayed rectifier K channel in squid giant axon. J. Gen. Physiol. *108*, 207–219.

Rula, E.Y., Lagrange, A.H., Jacobs, M.M., Hu, N., Macdonald, R.L., and Emeson, R.B. (2008). Developmental modulation of GABA(A) receptor function by RNA editing. J. Neurosci. *28*, 6196–6201.

Ryan, M.Y., Maloney, R., Reenan, R., and Horn, R. (2008). Characterization of five RNA editing sites in Shab potassium channels. Channels (Austin) *2*, 202–209.

Sagar, A., and Rakowski, R.F. (1994). Access channel model for the voltage dependence of the forward-running Na+/K+ pump. J. Gen. Physiol. *103*, 869–893.

Schwarz, T.L., Tempel, B.L., Papazian, D.M., Jan, Y.N., and Jan, L.Y. (1988). Multiple potassium-channel components are produced by alternative splicing at the Shaker locus in *Drosophila*. Nature *331*, 137–142.

Shen, Y., Yu, D., Hiel, H., Liao, P., Yue, D.T., Fuchs, P.A., and Soong, T.W. (2006). Alternative splicing of the Ca(v)1.3 channel IQ domain, a molecular switch for Ca^{2+}-dependent inactivation within auditory hair cells. J. Neurosci. *26*, 10690–10699.

Shinoda, T., Ogawa, H., Cornelius, F., and Toyoshima, C. (2009). Crystal structure of the sodium-potassium pump at 2.4 A resolution. Nature *459*, 446–450.

Sobolevsky, A.I., Rosconi, M.P., and Gouaux, E. (2009). X-ray structure, symmetry and mechanism of an AMPA-subtype glutamate receptor. Nature *462*, 745–756.

Sommer, B., Keinanen, K., Verdoorn, T.A., Wisden, W., Burnashev, N., Herb, A., Kohler, M., Takagi, T., Sakmann, B., and Seeburg, P.H. (1990). Flip and flop: a cell-specific functional switch in glutamate-operated channels of the CNS. Science *249*, 1580–1585.

Sommer, B., Kohler, M., Sprengel, R., and Seeburg, P.H. (1991). RNA editing in brain controls a determinant of ion flow in glutamate-gated channels. Cell *67*, 11–19.

Stapleton, M., Carlson, J.W., and Celniker, S.E. (2006). RNA editing in *Drosophila melanogaster*: New targets and functional consequences. RNA *12*, 1922–1932.

Swanson, G.T., Feldmeyer, D., Kaneda, M., and Cull-Candy, S.G. (1996). Effect of RNA editing and subunit co-assembly single-channel properties of recombinant kainate receptors. J. Physiol. *492(Pt 1)*, 129–142.

Tadross, M.R., Ben Johny, M., and Yue, D.T. (2010). Molecular endpoints of Ca2+/calmodulin – and voltage-dependent inactivation of Ca(v)1.3 channels. J. Gen. Physiol. *135*, 197–215.

Taglialatela, M., Toro, L., and Stefani, E. (1992). Novel voltage clamp to record small, fast currents from ion channels expressed in Xenopus oocytes. Biophys. J. *61*, 78–82.

Tanabe, T., Takeshima, H., Mikami, A., Flockerzi, V., Takahashi, H., Kangawa, K., Kojima, M., Matsuo, H., Hirose, T., and Numa, S. (1987). Primary structure of the receptor for calcium channel blockers from skeletal muscle. Nature *328*, 313–318.

Tempel, B.L., Papazian, D.M., Schwarz, T.L., Jan, Y.N., and Jan, L.Y. (1987). Sequence of a probable potassium channel component encoded at Shaker locus of *Drosophila*. Science *237*, 770–775.

Tillotson, D. (1979). Inactivation of Ca conductance dependent on entry of Ca ions in molluscan neurons. Proc. Natl. Acad. Sci. U.S.A. *76*, 1497–1500.

Timpe, L.C., Schwarz, T.L., Tempel, B.L., Papazian, D.M., Jan, Y.N., and Jan, L.Y. (1988). Expression of functional potassium channels from Shaker cDNA in Xenopus oocytes. Nature *331*, 143–145.

Wilding, T.J., Zhou, Y., and Huettner, J.E. (2005). Q/R site editing controls kainate receptor inhibition by membrane fatty acids. J. Neurosci. *25*, 9470–9478.

Wilding, T.J., Fulling, E., Zhou, Y., and Huettner, J.E. (2008). Amino acid substitutions in the pore helix of GluR6 control inhibition by membrane fatty acids. J. Gen. Physiol. *132*, 85–99.

Wilding, T.J., Chen, K., and Huettner, J.E. (2010). Fatty acid modulation and polyamine block of GluK2 kainate receptors analyzed by scanning mutagenesis. J. Gen. Physiol. *136*, 339–352.

Yang, P.S., Alseikhan, B.A., Hiel, H., Grant, L., Mori, M.X., Yang, W., Fuchs, P.A., and Yue, D.T. (2006). Switching of Ca^{2+}-dependent inactivation of Ca(v)1.3 channels by calcium binding proteins of auditory hair cells. J. Neurosci. *26*, 10677–10689.

Zagotta, W.N., Hoshi, T., and Aldrich, R.W. (1990). Restoration of inactivation in mutants of Shaker potassium channels by a peptide derived from ShB. Science *250*, 568–571.

Zhou, M., Morais-Cabral, J.H., Mann, S., and MacKinnon, R. (2001). Potassium channel receptor site for the inactivation gate and quaternary amine inhibitors. Nature *411*, 657–661.

Mechanisms and Functions of RNA Editing in *Physarum polycephalum* 2

Jonatha M. Gott

Abstract

Mitochondrial RNAs in the acellular slime mould *Physarum polycephalum* are subject to the widest range of editing events observed thus far. Mitochondrial RNAs differ from the mitochondrial genome at over 1300 sites, and both coding (mRNAs) and non-coding RNAs (rRNAs and tRNAs) are affected. At least three distinct editing mechanisms are needed to account for the different forms of editing observed in the mitochondrial transcriptome: nucleotide insertions and deletions, C to U changes, and specific alterations at the 5′ end of tRNAs. Nucleotide insertions are co-transcriptional and require flanking regions of the template, but the exact signals that specify the site of nucleotide insertion and the identity of the nucleotide to be added remain an enigma. The rare instances of base changes and replacement of the first nucleotide of mitochondrial tRNAs are not directly linked to transcription and are likely to occur via processes related to those previously described in other mitochondrial editing systems.

Introduction

RNA editing and gene expression

The term RNA editing encompasses a range of local alterations within RNA molecules that could have been encoded at the DNA level. Many of these changes are highly specific and some are regulated in a developmental, hormonal, or tissue-specific manner (Powell *et al.*, 1987; Carrillo *et al.*, 2001). Editing has been observed in a wide spectrum of organisms, including mammals, viruses, plants, worms, flies, and unicellular eukaryotes and occurs via many different mechanisms (reviewed in Knoop, 2011). Editing affects most, if not all, classes of RNAs, including mRNAs, tRNAs, rRNAs, miRNAs, and their precursors; this list is likely to expand as more recently discovered RNAs (e.g. long non-coding RNAs) are characterized more fully. There are two general classes of editing events: base substitution and nucleotide insertion/deletion. The majority of base changes are due to deamination reactions (C to U and A to I) that occur on intact RNAs; these often result in codon alterations and can include the creation of start and stop signals within mRNAs (Blanc and Davidson, 2010; Farajollahi and Maas, 2010; Pullirsch and Jantsch, 2010). In contrast, the insertion or deletion of nucleotides creates frameshifts in mRNAs and changes in higher order structure in tRNAs and rRNAs (Hausmann *et al.*, 1999; Gott and Emeson, 2000; Aphasizhev and Aphasizheva, 2011). Nucleotide changes can range from a single programmed insertion, such as those used to access additional reading frames in viruses (Vidal *et al.*, 1990), to the addition of over 50% of the nucleotides in certain kinetoplastid mRNAs (Feagin *et al.*, 1988), resulting in the creation of entirely new open reading frames. Not surprisingly, many editing events have biologically significant effects, and deviations from normal editing patterns have been implicated in a number of human disorders (Maas *et al.*, 2006; Gott, 2011).

RNA editing in *Physarum polycephalum*

The most complex assortment of editing events thus far is found in the acellular slime mould, *Physarum polycephalum* (*Pp*), whose mitochondrial RNAs are extensively edited. Since the initial discovery of 54 single C insertions in the *atpA* mRNA by Miller and colleagues (Mahendran *et al.*, 1991), the list of editing types has continued to grow. Most of the ‘typical’ mitochondrial mRNAs, as well as the rRNAs and five encoded tRNAs are present in the *Pp* mitochondrial genome as 'cryptogenes' whose expression requires RNA editing (Takano *et al.*, 2001; Bundschuh *et al.*, 2011). Transcripts from a subset of these genes were characterized individually (Mahendran *et al.*, 1991; Gott *et al.*, 1993, 2005, 2010; Mahendran *et al.*, 1994; Antes *et al.*, 1998; Takano *et al.*, 2001; Beargie *et al.*, 2008; Bullerwell *et al.*, 2010), but even with a complete genome sequence (Takano *et al.*, 2001) not all genes that would be expected to be encoded by the mitochondrial genome could be identified. To surmount this problem, Bundschuh and colleagues used deep sequencing to characterize the entire *Pp* mitochondrial transcriptome in plasmodial cells (Bundschuh *et al.*, 2011), resulting in the identification of nine previously undiscovered genes and 775 additional editing sites. In total, *Pp* mitochondrial RNAs are altered at 1333 sites (Table 2.1). Of these, 94% involve C addition, but single U, A and G residues and multiple combinations of dinucleotides (AA, UU, UA, GU/UG, CG/GC, CU/UC) are also inserted at specific sites. This range of insertional RNA editing events has only been observed in *Pp* and closely related species (Antes *et al.*, 1998; Horton and Landweber, 2000; Krishnan *et al.*, 2007; Hendrickson and Silliker, 2010; Traphagen *et al.*, 2010). A total of 1347 nt are added at 1324 positions, with the balance of editing sites involving C > G and U > G changes at the 5′ end of mitochondrial tRNAs (Gott *et al.*, 2010), C > U changes within the *coxI* mRNA (Gott *et al.*, 1993), and the deletion of three encoded As from the *nad2* mRNA (Gott *et al.*, 2005). The entire set of editing events in the *Pp* mitochondrial transcriptome can be accessed through REDBASE (http://bioserv.mps.ohio-state.edu/redbase/).

The *Pp* mitochondrial genome and transcriptome

Pp mitochondrial genome

The 62,862 bp *Pp* mitochondrial genome is AT rich (74.1%) and densely packed with genes on both strands (Takano *et al.*, 2001; Bundschuh *et al.*, 2011). Many genes are separated by only a few basepairs, while others overlap (Fig. 2.1), sometimes substantially. The genome encodes 20 open reading frames (ORFs) of 300 nt or longer (Takano *et al.*, 2001) and a shorter 292 bp ORF with homology to known *atpB* genes (Bundschuh *et al.*, 2011). The functions of the proteins encoded by the longer ORFs are unknown; none of these have counterparts in other sequenced genomes. Although most ORFs are not expressed in the plasmodial stage examined in RNAseq experiments (Bundschuh *et al.*, 2011), the fact that so many are maintained in the mitochondrial genome makes it likely that these ORFs are transcribed under conditions that remain to be examined. The mitochondrial genome also encodes 45 'cryptogenes', i.e. genomic regions whose transcripts require editing to produce functional RNA molecules (Table 2.1). These include genes encoding 37 mRNAs, 5 tRNAs, the large (LSU) and small (SSU) rRNAs, and a 5S-like RNA (Mahendran *et al.*, 1991, 1994; Gott *et al.*, 1993, 2005, 2010; Antes *et al.*, 1998; Takano *et al.*, 2001; Beargie *et al.*, 2008; Bullerwell *et al.*, 2010; Bundschuh *et al.*, 2011).

Pp mitochondrial transcriptome

There are at least 20 separate transcripts expressed in *Pp* mitochondria, many of which are polycistronic (Bundschuh *et al.*, 2011). Only ~60% of the mitochondrial genome is transcribed in plasmodia, including only two of the ORFs, *php15* (ORF14) and *php25* (provisionally annotated as *atpB*). While the remaining ORFs fall within the portion of the genome that is not transcribed under the conditions analysed, it is likely that they are expressed at some other point in the life cycle or under different growth conditions. Indeed, two transcripts derived from the region encompassing ORFs17, 18, and19 were detected in Northern blots of RNAs isolated from a different *Pp* strain (Takano *et al.*, 2001). Two of the mitochondrially encoded tRNAs ($tRNA^{met1}$ and $tRNA^{met2}$) are edited at their 5′ ends; one of these ($tRNA^{met1}$) also contains internal nucleotide insertions (Gott *et al.*, 2010). All other mitochondrial RNAs contain non-encoded nucleotides; these are efficiently edited, with all but one of the insertion sites fully edited in steady-state RNA pools (Bundschuh *et al.*, 2011). Editing at the C to U sites within the

Table 2.1 Distribution of editing sites in *Physarum polycephalum* mitochondrial RNAs

	Editing type												
Gene/no. sites	+C	+U	+G	+A	−A	+AA	+UU	+UA	+CU/UC	+GU/UG	+GC/CG	C>U	N to N
	1255	43	2	1	3	4	2	2	9	4	2	4	2
nad5	69	7	1										
nadG	48												
rpS2	56	1											
rpS12	23												
rpS7	26	2											
rpL2	36	2											
rpS19	13							1					
php15[a]													
cox1	59	1						1	1	1		4	
nad7	37	4					2		2	1			
cox2	33												
php22[b]	38												
nad2	58	5			3								
rpS16	15												
rpL19	20	1								1			
atp8	9												
nad4L	13												
atp6	33												
nad4	56	5									1		
nad3	7												
rpL14	18												
php23	21	1											
rpS14	11												
rpS8	19												
rpL6	21	1											
rpS13	21	1											
nad9	23												
rpS11	35												
php24	24								1				
rpS4	32												
tRNAglu	1	1											
tRNAmet1	2												1
23S rRNA	52		1			2			2	1			
17S rRNA	40	1				2			1				
5S rRNA	2												
tRNAmet2													1
tRNAlys	3												
tRNApro	2												
php25[a,c]													
atpA	54												
cox3	33												
nad6	18	1											
rpL16	20			1									
rpS3	66	1											
nad1	38	2											
cytb	31	6									1		
atp9	9												
intergenic[d]	3												
UTR[e]	7												

[a]unedited mRNA; [b]may be rpL11; [c]likely atpB; [d]includes 2 +C between php22 and nad2, 1 +C between tRNAmet2 and tRNAlys; [e]includes 1 +C in php22 5′UTR, 2 +Cs in nad3 5′UTR, and 4 +C in atp9 3′UTR.

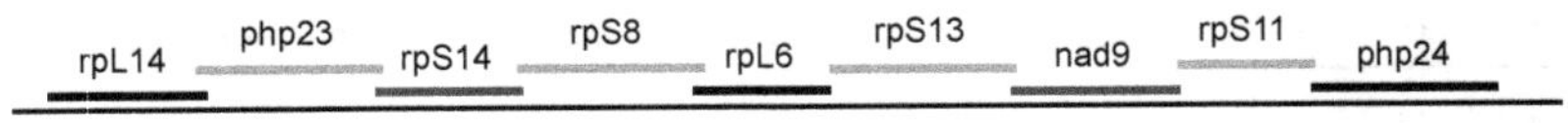

sequences surrounding translational start and stop codons in the rpL14->php24 transcript

```
5' UTR / rpL14 start   TTATTTTAATAATTTATTTTTTATAAAACCATAAAAATATGCGATcTGGTACAGTTGTTAAAGTAGCcGA
rpL14 / php23          TATATTCGTcTTATGTCTTTGGGTACcATCGCTTTATAATGAAATCTCTAGTcTTTTCTAAATTTAATAA
php23 / rpS14          ATCCTTTATTTTTTCATAAAACcGCTTATTTAGACTAAATGATcTCCATATTACAAAcATCAATAAAAGA
rpS14 / prS8           c...42 nt...CTGGATTACGTAACTCATCCTGGTAAATGACAGCCcGTTTTTCTGCTATGATcAGTAT
rpS8 / rpL6            AATAGGTTATCTGGCAAACTAcTAGCTGAAATTTTTATATGATcATTGCCTCTTCTTATCcTTTATTACA
rpL6 / rpS13           ATAAAATAAAACAACGTAAGAAAcGTAAATAAAAATTAATGATcGATACTTCTTATTTATCACTTCAACCT
rpS13 / nad9           c...48 nt...AAACATCAAAGACTTAAATCATAATTATGTTTATcAATAAAGATCAATcATTTTATTTA
nad9/ rpS11            AAGACCATTCGGAGTAAGATcGAAGTAATTTTTTGATAATGCTTAAAACTTTTGTcAGAGGTAGACTTTAT
rpS11 / php24          GCcTTACTAGAAAGAAAAAGGTTCGCCGTTTATAGAATATGACTCCAGcACTTCAAAAGACATTAcTAGAT
php24 stop / 3' UTR    c...59 nt...AATTTAAGAGACTTTATTAATTCTAATCTTAATAATTATTACTTTAAACATATGTATATA
```

Figure 2.1 Spacing of *Pp* mitochondrial genes. Top: Gene order on the *rpL14-php24* transcript. Bottom: Sequences surrounding the start and stop codons of the genes encoding ribosomal proteins *L14, S14, S8, L6, S13* and *S11*, the NADH dehydrogenase subunit 9 (*nad9*), and two hypothetical proteins (*php23* and *php24*). Gene names are given at left. Start codons (**ATG**, in bold) are aligned; stop codons are underlined. Inserted C residues (present only in the RNA) are shown in lower case with shaded backgrounds.

cox1 mRNA is slightly less efficient, but is still in the range of 95–98% (Bundschuh *et al.*, 2011).

Characteristics of edited transcripts

RNA editing in *Pp* mitochondria is extraordinarily precise and highly accurate; non-encoded nucleotides are seen only at the correct site and incorporation of the wrong nucleotide is virtually undetectable in steady-state RNA pools (Gott *et al.*, 1993; Rundquist and Gott, 1995; Byrne *et al.*, 2002; Byrne and Gott, 2004; Bundschuh *et al.*, 2011). The frequency of insertions is quite high, with inserted nucleotides (nts) making up ~4% of the residues in mRNAs and ~2% of non-coding RNAs (rRNAs and tRNAs). Editing occurs almost exclusively within coding regions and structural RNAs; only 10 of the 1255 C insertions are extragenetic. Curiously, C insertion sites frequently occur close to translation initiation sites, but are generally farther from translational stop codons, except in cases of closely spaced (or overlapping) genes on the same transcript (Fig. 2.2). Although not regularly spaced, insertion sites are distributed fairly evenly within an RNA, with a minimal distance of 9 nt between adjacent C insertion sites. As discussed below, this constraint is likely to be mechanistically significant. In contrast, the insertion of nucleotides other than C is much more sporadic, with dinucleotide insertions clustered within only nine RNAs.

Although deep sequencing has only been carried out on transcripts isolated from plasmodia, an analysis of total RNA from cells at various points in the life cycle indicated that the *cox1* mRNA is fully edited at each developmental stage (Rundquist and Gott, 1995). Since editing is required to create mRNAs encoding nearly all of the conserved mitochondrial proteins, it appears that this process is essential for mitochondrial function throughout development and is unlikely to be regulated during the life cycle.

Phylogenetic distribution

The unusual editing patterns described above appear to be limited to *Physarum* and closely related myxomycetes. Interestingly, the different forms of editing seem to have different evolutionary histories (Horton and Landweber, 2000). Comparison of *Pp* editing sites within a conserved ~1200 nt region of the *cox1* mRNA with those of *Didymium nigripes, Stemonitis flavogenita, Arcyria cinerea, and Clastoderma debaryanum* revealed an uneven distribution of editing types across species (Horton and Landweber, 2000). A small number of added Us (1–4) were observed in all five species, but the frequency of other types of insertions varied widely. *P. polycephalum, D. nigripes,* and *S. flavogenita* each contain large numbers (34–40) of C insertions as well as three types of dinucleotide

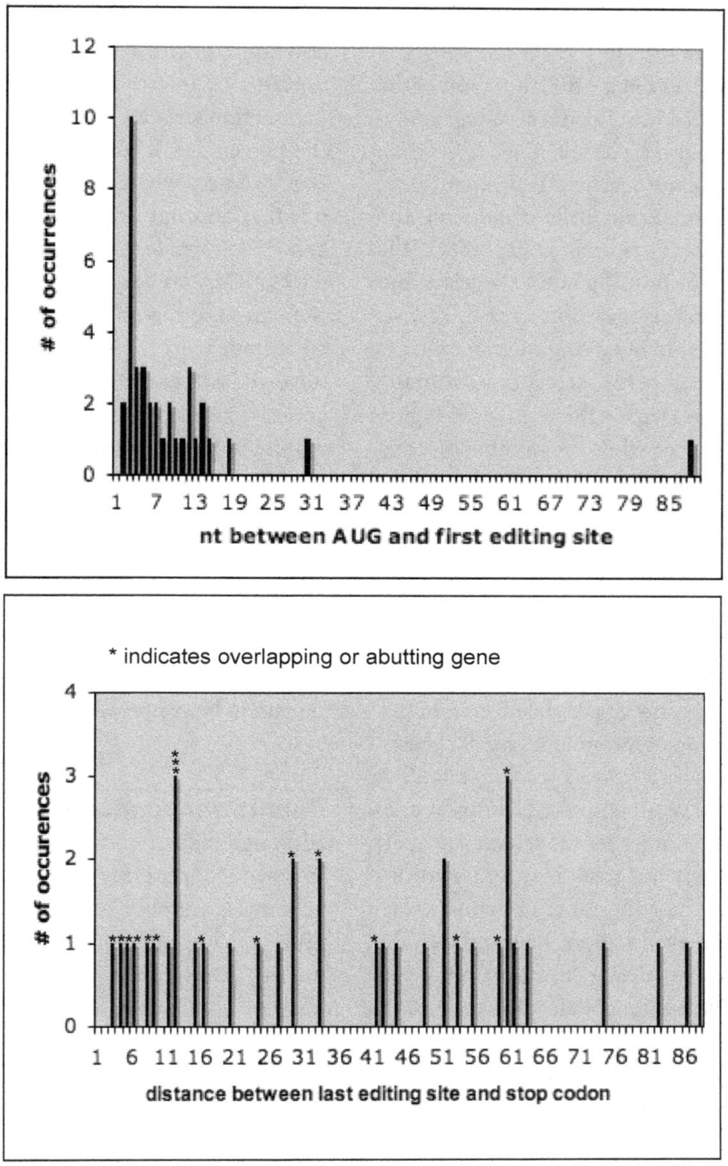

Figure 2.2 Distribution of editing sites relative to translation signals. Top: Distance in nucleotides (nt) between the start codon and the first editing site within mRNAs in *Pp* mitochondria. All but the *nad3* mRNA (29 nt) and the *atpA* mRNA (86 nt) have editing sites that fall within the first 20 nucleotides following the initiator AUG. Bottom: Distance in nucleotides between the last editing site and the stop codon within each mRNA in *Pp* mitochondria. This distance tends to be much shorter for closely spaced genes on the same transcript (indicated by asterisks, *).

insertions in this region of the *cox1* mRNA, while *A. cinerea* has only a single C insertion and no added dinucleotides, and *C. debaryanum* lacks both C and dinucleotide insertions. Curiously, the C to U changes have a different distribution, being present in *P. polycephalum*, *D. nigripes*, and *A. cinerea*, but absent in *S. flavogenita* and *C. debaryanum*. These findings suggest that U insertion was the ancestral activity and that the ability to add C and dinucleotides arose later in evolution (Horton and Landweber, 2000).

Miller and colleagues have examined the

distribution of editing sites in mitochondrially encoded tRNAs in *Pp* and *Didymium nigripes* (Antes *et al.*, 1998) and in a ~500 nt region of the small ribosomal RNA (SSU) from seven myxomycete species: *Physarum polycephalum, Physarum didermoides, Didymium nigripes, Didymium iridis, Stemonitis flavogenita, Echinostelium minutum,* and *Lycogala epidendrum* (Krishnan et al., 2007). The tRNAs are edited in both *Pp* and *D. nigripes*, but the sites of nucleotide insertion vary. Likewise, the overall distribution of editing sites is similar in the rRNAs examined, with each species containing 8–10 editing sites within the sequenced region of the rRNA. However, while the number of insertions is generally conserved, the exact location of editing sites is variable. The same 10 sites are edited in *P. polycephalum, D. nigripes,* and *D. iridis* and about half of these sites are shared by *S. flavogenita* and *P. didermoides*, but *L. epidendrum* has a completely non-overlapping set of editing sites. *S. flavogenita* also contains one instance of a U deletion, the only report of a U deletion in myxomycetes. C insertions predominate in all species, but five of the seven rRNAs have an inserted AA and six of seven have an added CU, both at conserved locations. Editing sites are at least 9 nt apart in all cases and there is a weak bias for insertions following a purine-U (48% of C insertion sites). Interestingly, conserved regions have close to the maximum density of editing sites (given the 9 nt constraint), while regions that diverge rapidly have fewer insertions. This has been attributed to the strong selective pressure on conserved regions important for function (Krishnan *et al.*, 2007).

The extensive characterization of *Didymium iridis* mitochondrial transcripts by Silliker and colleagues has revealed that *D. iridis* utilizes the entire spectrum of editing events observed in *P. polycephalum* (Hendrickson and Silliker, 2010; Traphagen *et al.*, 2010), including the first report of an A insertion for any myxomycete (Hendrickson and Silliker, 2010). An A was found at the same position upon deep sequencing of *Pp* RNAs and confirmed via RT-PCR (Bundschuh *et al.*, 2011). Transcripts from 16 of the 17 *D. iridis* mitochondrial genes examined contain abundant C insertions, infrequent U and dinucleotide insertions, and three C to U changes (one in *cox2*, two in *cox1*) with a similar frequency, spacing, and codon position. The context of C insertions is also similar, with 65% of C insertions following a purine-U. Interestingly, although ~77% of the C insertion sites are shared with *P. polycephalum*, the percentage is highly variable between genes. For instance, whereas all of the sites within the *nad4L* transcript are common to both *D. iridis* and *P. polycephalum*, only ~26% of sites within the *atp1* mRNA are identical. All dinucleotide insertion sites are shared between the two organisms, consistent with the extremely high conservation observed within rRNAs (Krishnan *et al.*, 2007). In contrast, none of the C to U sites are in common, although the final mRNA sequences are conserved at these positions. This difference is even more pronounced when partial sequences from the *cox1* mRNAs from *D. nigripes* and *A. cinerea* (Horton and Landweber, 2000) are included, with only 1 of 12 C to U changes in common. Editing sites within another seven *D. iridis* mitochondrial transcripts have recently been predicted, but remain to be confirmed (Chen *et al.*, 2012).

Bioinformatic studies

Although many typical mitochondrial genes were mapped to portions of the *Pp* mitochondrial genome, a number of the genes expected to be present were undetectable using standard gene finding programmes (Takano *et al.*, 2001). To identify possible cryptogenes, special algorithms were developed to map additional genes whose transcripts require editing. Initial studies utilized a position specific scoring matrix (PSSM) based on alignment of known mitochondrial proteins (Bundschuh, 2004, 2007). This algorithm (predictor of insertional editing, or PIE) predicted the existence of four additional cryptogenes; characterization of the transcripts generated from these regions of the genome validated this approach and, in the process, led to the discovery of deletion editing within the *nad2* mRNA (Gott *et al.*, 2005). A more recent application of a modified Smith–Waterman algorithm that takes into account additional characteristics of C insertion editing nearly doubled the number of predicted cryptogenes (Beargie *et al.*, 2008). The accuracy of these latter predictions was confirmed by experimental verification of two selected mRNAs (Beargie *et al.*, 2008) and subsequent deep sequencing (Bundschuh *et al.*, 2011). These computational tools have also proved useful in comparative studies of editing patterns in *D. iridis* and *P. polycephalum* (Chen *et al.*, 2012).

A second major motivation for bioinformatics studies has been the identification of editing signals in *Pp* mitochondria. Numerous attempts have been made to identify common sequence elements, secondary structures or motifs used to demark editing sites (Mahendran *et al.*, 1991; Miller *et al.*, 1993; Gott *et al.*, 2005; Liu and Bundschuh, 2005; Krishnan *et al.*, 2007; Bundschuh *et al.*, 2011; Chen *et al.*, 2012, and unpublished collaborations). A high frequency (~70%) of a purine-U immediately upstream of C insertion sites (at positions −2/−1) was noted initially by Miller and colleagues (Mahendran *et al.*, 1991; Miller *et al.*, 1993), and this bias is still apparent now that all editing sites are known, albeit at a somewhat lower level (~59%) (Bundschuh *et al.*, 2011). This extremely limited motif has insufficient information content to account for the location of editing sites yet, somewhat surprisingly, these two positions are the only ones in the vicinity of editing sites that deviate from expected base frequencies, even when the vastly expanded transcriptome data set is analysed (Bundschuh, 2011). A similar conclusion was reached based on comparisons between myxomycetes (Chen *et al.*, 2012). However, as discussed below (see 'Localization of template elements required for editing'), it is difficult to reconcile the lack of sequence information surrounding editing sites with the results of biochemical studies (Rhee *et al.*, 2009). A search for more subtle patterns based on the frequency of nucleotide pairs (such as those indicative of RNA secondary structure) yielded only weak correlations at positions −1 and −2, with an under-representation of identical bases immediately upstream of editing sites (Gott *et al.*, 2005). Thus, it still remains unclear where the information that specifies editing sites resides.

The precise site of insertion is ambiguous for the ~30% of added Cs inserted next to an encoded C and most of the non-C insertions (see 'Potential editing signals', below). It is therefore impossible to assign the exact codon position for roughly one third of *Pp* editing sites. When the analysis is limited to the ~800 unambiguous C insertions within protein coding genes, nearly half (49%) of the added Cs are found at the third codon position, with a significant under-representation (18%) at the second position (Bundschuh *et al.*, 2011). The observed codon bias has been attributed to the selection of random mutations at the protein level (Liu and Bundschuh, 2005); this model is supported by data from both *P. polycephalum* and *D. iridis* (Chen *et al.*, 2012). The Ile codon AUC is the codon most frequently created by editing in *Pp* (not surprising given the AU richness of the *Pp* mitochondrial genome and the bias for C insertion following a purine-U), but all C containing codons are represented at least once (Bundschuh *et al.*, 2011). A preferential creation of codons for hydrophobic amino acids has been noted by Silliker and colleagues in *D. iridis*, with AUC (Ile) again being the most frequently edited codon (Hendrickson and Silliker, 2010; Traphagen *et al.*, 2010).

The existence of multiple editing mechanisms in *Pp* mitochondria

Multiple editing mechanisms are involved in the maturation of *Pp* mitochondrial transcripts. As described in more detail below, the insertion of non-encoded nucleotides (and, likely the deletion of encoded nucleotides) is tightly coupled to transcription. In contrast, the C to U changes and editing of the 5′ end of tRNAs are clearly post-transcriptional; these last two forms of editing are quite distinct from one another, requiring different activities. Thus, minimally, there are at least three separate editing mechanisms operating in these organelles.

C to U editing

Little experimental work has been done on the mechanism by which Cs are converted to Us in *Pp* mitochondrial RNAs. There are only four C to U changes in the edited transcriptome (Bundschuh *et al.*, 2011), all located within the *coxI* mRNA. Whereas run-on transcripts made in isolated mitochondria are virtually completely edited at nucleotide insertion sites, these transcripts contain C rather than U (Visomirski-Robic and Gott,

1995). Pulse-chase experiments indicate that the *nascent* transcripts remain unedited at these positions even when the RNA polymerase is far downstream (Gott and Visomirski-Robic, 1998), suggesting that this process is entirely divorced from transcription. Conversion of C to U is likely to proceed via a deamination reaction analogous to those seen in plant mitochondria, but this remains to be tested directly.

The specific deamination of C to U was first reported in the mammalian apoB mRNA (Chen *et al.*, 1987; Powell *et al.*, 1987) and subsequently found to be abundant in mRNAs in plant mitochondria and plastids (reviewed in Chateigner-Boutin and Small, 2010). Metazoan C to U changes require the protein factors Apobec-1 and ACF1 and are primarily directed by an 11 nt 'mooring sequence' downstream of the editing site (reviewed in Blanc and Davidson, 2010). In contrast, C to U changes in plant organelles require pentatricopeptide repeat (PPR) proteins, with the sequences that direct editing lying primarily upstream of editing sites (see Zehrmann *et al.*, 2011).

The four C to U changes in the *Pp* coxI mRNA all fall within a ~50 nucleotide segment of the RNA (Gott *et al.*, 1993). Three are tightly grouped, falling within a four nucleotide stretch, with the fourth site 44 nucleotides downstream. There are no obvious sequence motifs or mooring sequence-like elements in the region surrounding the C to U sites, beyond being flanked by encoded Us. Genes encoding a number of PPR proteins have recently been identified in the *Pp* genome and cloning is under way. These proteins are likely candidates for involvement in C to U editing, but this remains to be tested. This area should prove to be fertile ground for future studies.

Editing at the 5′ end of mitochondrial tRNAs

Two of the five mitochondrially encoded tRNAs, $tRNA^{met1}$ and $tRNA^{met2}$, are edited at their 5′ ends (Gott *et al.*, 2010). In both cases the encoded nucleotide at the 5′ end is replaced by a non-encoded G, forming a Watson–Crick basepair with the C on the opposite side of the acceptor stem. Labelling studies and the lack of G insertions in nascent transcripts (Gott *et al.*, 2010) suggest that this occurs via a mechanism similar to that characterized initially in *Acanthamoeba castellanii* (Price and Grey, 1999) and subsequently in *Spizellomyces punctatus* (Bullerwell and Grey, 2005). This form of editing involves the removal of the first 1–3 nucleotides from the 5′ end of tRNAs followed by re-synthesis in a 3′ to 5′ direction to correct mismatches within the acceptor stem. This reverse polymerization reaction has many parallels to the G_{-1} addition reaction catalysed by $tRNA^{His}$ guanylyltransferase (Thg1), an enzyme found in most eukaryotes (Gu *et al.*, 2003). Thg1 adds a non-encoded G to the 5′ end of cytoplasmic $tRNA^{His}$ (Cooley *et al.*, 1982), providing an important identity element for the histidine tRNA synthetase (Himeno *et al.*, 1989; Nameki *et al.*, 1995). Database searches for proteins related to Thg1 uncovered an entire family of Thg1-like proteins (TLPs) (Heinemann *et al.*, 2011; Jackman *et al.*, 2012), including two TLPs in the *Pp* genome (Jackman *et al.*, 2012). TLPs from Bacteria and Archaea have been demonstrated to be capable of Watson–Crick templated, 3′ to 5′ reverse polymerization on a variety of truncated tRNA templates (Heinemann *et al.*, 2009, 2010; Abad *et al.*, 2010, 2011; Rao *et al.*, 2011), suggesting roles for this family of enzymes in tRNA 5′ repair and, potentially, RNA editing. Consistent with their proposed role in tRNA 5′ editing, TLPs from the cellular slime mould *Dictyostelium discoideum* have recently been shown to fill in the 5′ end of truncated tRNA editing substrates using the 3′ portion of the acceptor stem as template (Abad *et al.*, 2011). Thus, the TLPs identified in the *Pp* genome are strong candidates for the *Pp* mitochondrial tRNA 5′ editing activity.

Internal insertion of non-encoded nucleotides

The bulk of mechanistic studies on *Pp* editing have focused on the unique form of nucleotide insertion that is currently only known to exist in a small group of closely related myxomycetes. Nucleotide insertions were initially assumed by many to be post-transcriptional additions based on precedents in kinetoplasts, but the pattern of insertions is quite different, hinting that the two processes differ mechanistically. Unlike what had been observed in kinetoplasts, cDNA clones derived from *Pp* mitochondrial RNA are completely edited, with a given editing site always having the same 'extra' nucleotide (Gott *et al.*, 1993; Wang *et al.*, 1999). A handful of cDNAs that contain single nucleotide insertions but lack added dinucleotides have been reported (Wang *et al.*, 1999). However, because these were only detected upon two rounds of PCR amplification separated and followed by a negative selection step, such RNAs must be exceptionally rare in steady state RNA pools. Despite extensive searching, no convincing evidence for a pool of potential editing intermediates could be found, even when primers specific for edited or unedited sequences were used for reverse transcription and polymerase chain reaction (PCR) (Gott *et al.*, 1993; Rundquist and Gott, 1995; unpublished data). Likewise, primer extension sequencing of total mitochondrial RNA yields only a single sequence with end-labelled primers (see, for example, Gott *et al.*, 1993; Rundquist and Gott, 1995; Gott *et al.*, 2005), except for the region encompassing the single partially edited site identified by RNAseq (Bundschuh *et al.*, 2011). The lack of unedited or partially edited transcripts strongly suggested that nucleotide insertion in *Pp* mitochondria is tightly coupled to transcription and occurs via a different mechanism than that used in trypanosomes. The finding that nascent tRNAs and rRNAs are edited by nucleotide insertion also appeared to rule out any mechanism involving direct interactions between the transcription and translation machineries (Antes *et al.*, 1998; Byrne and Gott, 2002).

Mechanistic studies of insertion editing

Important mechanistic insights from an *in organello* transcription/editing system

Based on the hypothesis that editing and transcription are coupled in some way, Visomirski-Robic and Gott (1995) developed an isolated mitochondrial system to examine editing mechanisms. To distinguish newly made transcripts from the fully edited endogenous RNAs, run-on transcripts were synthesized in the presence of radiolabelled NTPs. Direct analysis of labelled RNAs synthesized in isolated mitochondria revealed that these run-on transcripts were accurately and efficiently edited, leading to the conclusion that insertion of non-templated nucleotides occurs on nascent transcripts (Visomirski-Robic and Gott, 1995).

A key finding using this *in vitro* approach was that when the concentration of CTP was kept very low during run-on transcription, RNAs lacking the added Cs (i.e. unedited RNAs) were made, indicating that edited RNAs are derived from the mitochondrial genome rather than an 'alternative' (edited) template and that transcription can proceed without editing under limiting CTP concentrations (Visomirski-Robic and Gott, 1997a). Follow-up experiments indicated that other cytidine nucleotides cannot substitute for CTP in the editing reaction. Inclusion of high levels of cytidine nucleotides that are not substrates for RNA polymerase (e.g. CDP, CMP, dCTP) failed to support C insertion at editing sites (JMG, unpublished data). These findings, coupled with nearest neighbour analyses demonstrating that [α^{32}P]CTP is incorporated at editing sites in isolated mitochondria (Visomirski-Robic and Gott, 1995, 1997a) led to the conclusion that cytidine triphosphate (CTP) is the substrate for C insertion.

RNA fingerprint analyses of labelled run-on transcripts demonstrated the total absence of unedited stretches of RNA, leading to the conclusions that (i) editing occurs with a 5′ to 3′ polarity (Visomirski-Robic and Gott, 1997a) and (ii) insertion of non-templated nucleotides occurs very close to the 3′ end of a nascent RNA (Visomirski-Robic and Gott, 1997a,b). Importantly, pulse-chase studies revealed that once unedited RNA is made, it is not subject to nucleotide insertion, even though downstream regions of the same transcript *are* edited if the concentration of CTP is raised during the chase reaction (Fig. 2.3). These results indicated that there is a limited window of opportunity in which editing can occur (Visomirski-Robic and Gott, 1997a). Taken together, these data confirmed that editing of *Pp* mitochondrial RNAs occurs via a completely different mechanism than that employed for uridine insertion and deletion in

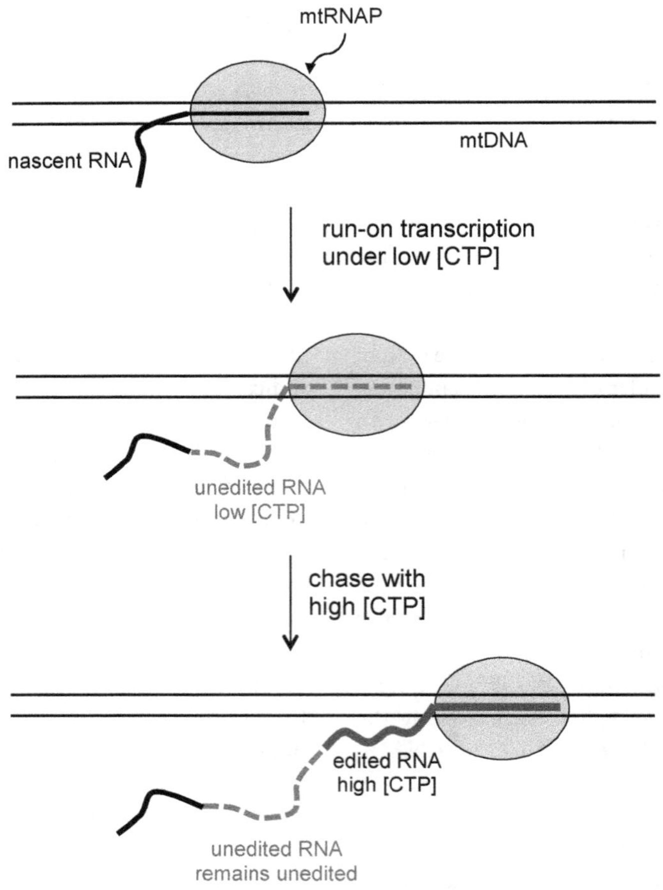

Figure 2.3 C insertion occurs within a very limited window. Schematic illustration of the findings of pulse-chase experiments described in Visomirski-Robic and Gott (1997a). Nascent transcripts extended in isolated mitochondria in the presence of low concentrations of CTP are largely unedited at C insertion sites within the newly synthesized RNA and remain unedited when the concentration of CTP is increased. In contrast, RNAs made during the chase reaction in the presence of high levels of CTP are fully edited. Nascent RNA, mitochondrial DNA (mtDNA), and the *Pp* mitochondrial RNA polymerase (mtRNAP) are pictured.

trypanosome kinetoplasts (Visomirski-Robic and Gott, 1997a,b).

Development of a soluble transcription/editing system

Although editing is highly accurate and efficient in isolated mitochondria (Visomirski-Robic and Gott, 1995), there are limitations to working with organelles. Chief among these are the presence of substantial pools of ATP and GTP and the inaccessibility of the transcription and editing machineries to manipulations. To develop a soluble transcription/editing system in which to analyse editing mechanisms, Cheng and Gott (2000) partially purified mitochondrial transcription elongation complexes (mtTECs) from mitochondrial lysates, using gel filtration chromatography to remove endogenous nuclease activities and nucleotide pools. Run-on transcripts made by mtTECs are edited to 40–60% at most sites, making this a useful system for studying editing mechanisms (Cheng and Gott, 2000).

Fundamental discoveries from *in vitro* transcription/editing studies

The availability of an *in vitro* transcription/editing system that is essentially free of endogenous nucleotides made it possible to make systematic changes in nucleotide concentrations and examine the effects on editing in many contexts. Importantly, the extent of editing at a given site is dependent upon relative nucleotide concentrations (Cheng *et al.*, 2001). Increasing the concentration of the nucleotide that is incorporated immediately downstream of a C insertion site favours transcription over editing, while dropping its concentration favours editing. For example, a high CTP–GTP ratio favours C insertion at sites followed by an encoded G, while the levels of C insertion at other sites remain unaffected. Likewise, high CTP–UTP ratios favour editing at sites followed by an encoded U and high CTP–ATP ratios favour editing at C insertion sites followed by A. These data are indicative of a direct competition between transcription and editing, i.e. that non-templated nucleotides are added to the 3′ end of nascent RNAs in *Pp* mitochondria (Cheng *et al.*, 2001).

Interestingly, the pattern of nucleotide concentration effects was strikingly different from that observed during paramyxoviral editing (Vidal *et al.*, 1990), strongly suggesting that these two forms of co-transcriptional editing occur via substantially different mechanisms. Editing of paramyxoviral transcripts occurs at a single, 'slippery' site, leading to the production of two (or three) different proteins from a single gene. Added nucleotides are 'pseudo-templated', i.e. the same region of the template is read more than once by the viral polymerase, as a result of stuttering and realignment within a homopolymer tract (Jacques and Kolakofsky, 1991). In contrast, editing in *Physarum* mitochondrial RNAs occurs in a variety of contexts, the majority of which are not compatible with a stuttering mechanism. Thus, the available data points to an entirely novel RNA editing mechanism in *Pp* mitochondria.

General model for insertion editing

Our current model for insertion editing (Rhee *et al.*, 2009) is shown in Fig. 2.4, which illustrates the basic steps involved in the co-transcriptional addition of non-encoded nucleotides to *Pp* mitochondrial RNAs. The mitochondrial RNA polymerase (mtRNAP) normally makes a faithful copy of the mitochondrial DNA (mtDNA) template. However, upon reaching an editing site, normal transcription is interrupted to allow for the selection and incorporation of the nucleotide(s) to be inserted at that site. It is unknown what activities are required for these steps, but a number of possibilities are discussed in 'Editing machinery' below. Templated transcription then resumes, initially from an unpaired 3′ end, with the extra nucleotide(s) accommodated within the RNA–DNA hybrid until emerging from the

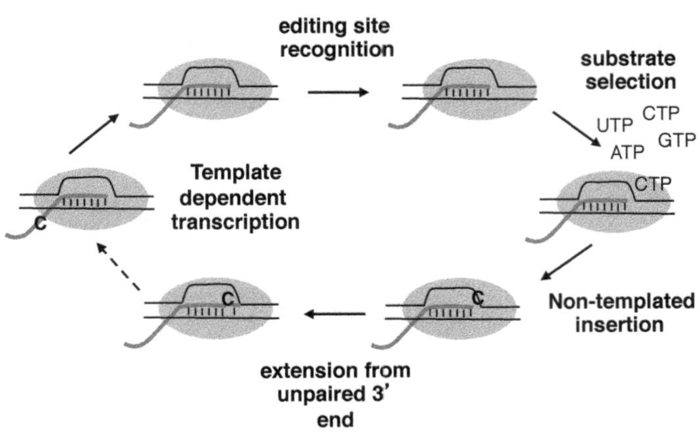

Figure 2.4 Model of the insertion editing cycle in *Pp* mitochondria. Schematic illustration of the hypothetical steps in the insertion editing process, including template dependent transcription by the *Pp* mitochondrial RNA polymerase (shaded ovals), insertion site recognition, selection of the residue to be added, non-templated nucleotide insertion, and templated extension from the unpaired 3′ end prior to resumption of templated transcription. Nascent RNA, mitochondrial DNA, and the *Pp* mitochondrial RNA polymerase are indicated as in Fig. 2.3.

mtRNAP. Note that this cycle must be repeated many times during the course of RNA synthesis as, on average, editing contributes 1 in every 25 nucleotides within mRNAs and 1 in every 40 nucleotides in rRNAs and tRNAs (Miller *et al.*, 1993; Bundschuh *et al.*, 2011).

Key findings from the development and use of 'chimeric' templates

The identification of the specific sequences required for editing requires defined template changes in the vicinity of editing sites. However, the edited RNAs synthesized by isolated mtTECs are run-on transcripts made by complexes assembled *in vivo*. Thus, as in isolated mitochondria, the template in mtTECs is the entire mitochondrial genome. Unfortunately, attempts to achieve editing from exogenous templates have thus far been unsuccessful, necessitating the development of alternative approaches to making template alterations.

The DNA in mtTECs is readily digested by restriction endonucleases and run-off RNAs made from these templates are edited to the same extent as those from the intact genome (A.E. Majewski, unpublished data; Byrne and Gott, 2002), indicating that linear templates support editing. By digesting mtDNA at restriction sites at various distances from editing sites, it was determined that 14 basepairs of downstream DNA is sufficient to support accurate editing (A.E. Majewski, unpublished). However, uncertainties regarding the stability of transcription complexes with even smaller stretches of DNA downstream of editing sites made it difficult to interpret the results when the DNA template was cleaved at positions very close to editing sites. To circumvent these problems, Byrne and colleagues (Byrne and Gott, 2002; Byrne, 2004; Byrne *et al.*, 2007) developed methods that allowed the creation of chimeric templates. Reasoning that mtTEC preparations contain the components necessary for editing and could potentially be used as the source of the mitochondrial RNA polymerase and any editing factors, chimeric templates were created by digesting the DNA present in mtTECs with restriction endonucleases, then ligating the resulting fragments to exogenously supplied DNA fragments (Fig. 2.5). The added DNAs are then transcribed by the endogenous RNA polymerase as it passes from the native template to the ligated fragment. Run-on transcripts are directly assayed for RNA editing using S1 nuclease protection and RNase T1 analysis (for labelled RNAs) or used as substrates for RT-PCR and subsequent restriction enzyme analysis (using enzymes whose recognition sites are created by C insertion) or cloning and sequencing of individual cDNAs (Byrne

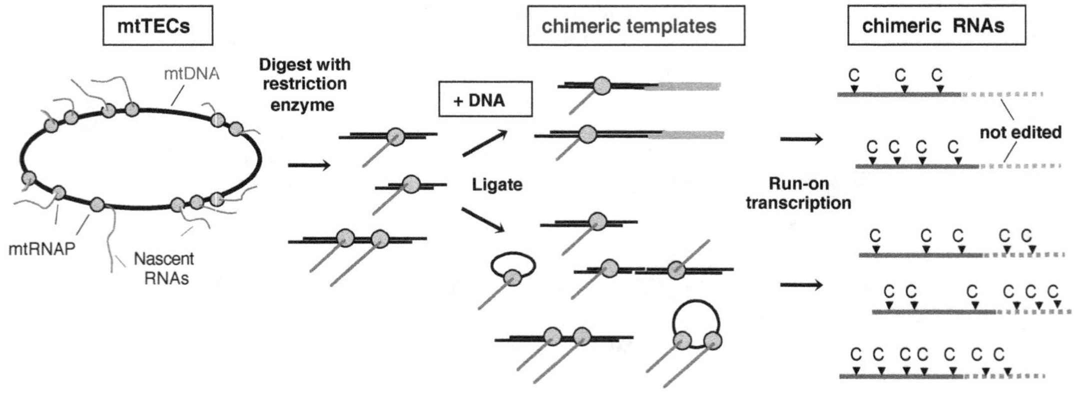

Figure 2.5 Synthesis of chimeric templates and RNAs. Chimeric templates are formed by digestion of the DNA (mtDNA) within mitochondrial transcription elongation complexes (mtTECs) with restriction endonucleases, followed by ligation in the presence (+DNA) or absence of exogenous DNA fragments (in light grey). Upon addition of nucleotides, the RNA polymerases associated with the mtDNA transcribe the mixture of templates, producing chimeric RNAs. Dotted lines indicate the portion of the run-on RNAs that are synthesized from DNA in an unnatural context (i.e. from a rearranged template). The presence or absence of added Cs is indicated for each species of chimeric RNA.

and Gott, 2002; Byrne, 2004; Byrne *et al.*, 2007). Surprisingly, these analyses demonstrated that chimeric transcripts are edited *only* in regions synthesized from the endogenous mitochondrial DNA, even in cases where the gene sequence is precisely reconstructed (Byrne and Gott, 2002) (Fig. 2.5). This was true irrespective of whether the added DNA was derived from plasmids, PCR products or phenol-extracted mitochondrial DNA (Byrne and Gott, 2002).

When the efficiency of chimeric template formation was assessed by Southern blotting, Byrne and Gott (2002) noted the presence of circularized DNA molecules created by ligation of the ends of restriction fragments derived from mtTEC DNA. These ligation events led to the juxtaposition of sequences that are normally distant from one another, creating templates in novel configurations. RNAs produced from these circular templates were analysed via sequencing of RT-PCR products generated using primers specific for the sequences upstream and downstream of ligation junctions in these unnatural contexts. Importantly, transcripts derived from rearranged mtTEC fragments (which have endogenous DNA both upstream and downstream of the hybrid junction) are edited throughout, despite the fact that sequences from different genes are artificially joined in these constructs. These results demonstrated that the lack of editing from the exogenous DNA cassettes is not due to the cleavage and ligation steps and led to a number of conclusions, including (i) any *cis*-acting editing determinants must be within 20 bp of the C insertion sites; (ii) insertion sites are edited independently; and (iii) template sequences alone are insufficient to specify nucleotide insertion, i.e. template-associated factors are required for RNA editing in *Pp* mitochondria (Byrne and Gott, 2002).

Localization of template elements required for editing

These findings were exploited further by Rhee *et al.* (2009) to define the location of sequences required for the identification of editing sites and nucleotide insertion. A series of chimeric templates using restriction enzymes that cut close to sites of C insertion were made, resulting in sequence changes upstream or downstream of specific editing sites. These studies indicated that the template sequences required for nucleotide addition lie within ~9 bp upstream and 9–10 bp downstream of editing sites (Fig. 2.6); changes outside this 18-bp 'critical region' did not affect the accuracy or efficiency of editing (Rhee *et al.*, 2009). This finding is consistent with the known distribution of editing sites, as single C insertion sites are no closer than nine nucleotides apart. In addition, editing patterns within RNAs generated from chimeric templates strongly suggested that upstream and downstream elements affect different steps in the editing cycle, with sequences upstream of editing sites contributing to the selection and/or insertion of added nucleotides and downstream sequences likely playing a role in editing site recognition and templated extension from the added nucleotide (Fig. 2.6). Curiously, however, although the experimental data are clear-cut (Rhee *et al.*, 2009),

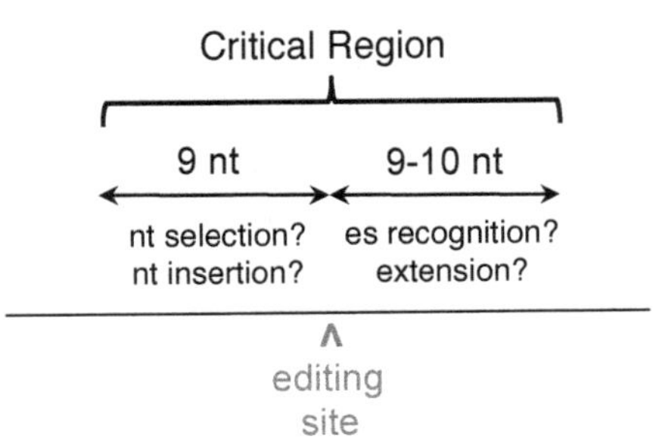

Figure 2.6 Editing signals are local. Regions of the mitochondrial DNA template that are critical for the insertion of non-encoded nucleotides fall within 9–10 nucleotides upstream and downstream of C insertion sites. Inferred roles for each (Rhee *et al.*, 2009) are indicated. nt: nucleotide; es: editing site.

genome-wide bioinformatics studies indicate that this region appears to contain insufficient information at the sequence level to define an editing site (Bundschuh *et al.*, 2011; Chen *et al.*, 2012). The reason for this discrepancy is not currently understood.

Mechanistic clues from misediting events *in vitro*

Sequence analysis of RNAs made *in vitro* indicated that RNAs synthesized from both untreated and rearranged mtTECs were misedited at a low level (~5%) (Byrne *et al.*, 2002). Mistakes included misincorporation of nucleotides at C insertion sites, omission of encoded nucleotides adjacent to editing sites, and larger deletions which lack all templated nucleotides between adjoining editing sites. These findings contrasted with the accuracy of editing previously reported in steady state RNA (Miller *et al.*, 1993; Rundquist and Gott, 1995; Byrne *et al.*, 2002), prompting the question of whether the observed fidelity of editing *in vivo* was dependent upon a proofreading or surveillance mechanism. Analyses of RNAs associated with mtTECs prior to run-on transcription (i.e. nascent transcripts made *in vivo*) indicated that editing of nascent transcripts is a highly accurate and efficient process *in vivo*, with no detectable misediting (Byrne *et al.*, 2002).

The patterns of sporadic *in vitro* misediting are revealing in that while the wrong nucleotide can be added at a given editing site, the insertion of an extra nucleotide has not been observed at non-editing sites. This argues that editing site recognition and nucleotide addition are separate events. In addition, because misediting occurs only on templates that support editing, these results strongly suggest that the *recognition* of insertion sites involves features of the mitochondrial template. This hypothesis is strengthened by the finding that the transcription/editing machinery can 'jump' from one editing site to the next, without transcribing the intervening templated nucleotides, potentially via an intrastrand transfer or 'looping out' mechanism. These results suggest that the polymerase may interact directly with editing site determinants associated with the template (Byrne *et al.*, 2002).

Insights into the process of dinucleotide insertion

Early work indicated that inserted C residues are added to the end of growing transcripts (Cheng *et al.*, 2001) and that dinucleotides are also added to nascent RNAs (Visomirski-Robic and Gott, 1995, 1997a,b). Phylogenetic analyses by Horton and Landweber (2000) indicate that the ability to insert single Us, single Cs, and dinucleotides arose at different times in evolution. Thus, although the simplest possibility is that all types of nucleotide insertions utilize similar mechanisms, it is also possible that different mechanisms are used in each case (Wang *et al.*, 1999). The patterns of partial editing and misediting at single and dinucleotide insertion sites suggest that single and dinucleotide insertions occur by similar mechanisms, but that there may be separate factors for dinucleotide insertions, a subset of which may be less stably bound to the editing template (Byrne and Gott, 2004). Under limiting nucleotide conditions, single nucleotides are added at dinucleotide insertion sites in isolated mitochondria (Visomirski-Robic and Gott, 1997a), and both the addition of single nucleotides at dinucleotide sites and the insertion of dinucleotides at single nucleotide sites is occasionally observed during run-on transcription in mtTECs (Byrne and Gott, 2004). These data are consistent with sequential addition at dinucleotide sites.

Potential editing signals

Potential sources of editing information

One of the most compelling questions in the field of *Physarum* editing is 'Where's the information?', i.e. what specifies the *site* of nucleotide insertion and the *identity* of the nucleotide(s) added at that site? In other instances of internal insertion editing, nucleotide additions are directed by RNA sequences acting *in cis* and/or *in trans*. There are also ample precedents for the use of RNAs to target processes other than editing, including snRNAs (splicing; Dieci *et al.*, 2009), snoRNAs (rRNA modification and cleavage; Kiss *et al.*, 2010), microRNAs (miRNAs, translational regulation; Bartel, 2009), RNase P RNA (tRNA processing; Ellis and Brown, 2009), and telomerase RNA (maintenance of chromosomal ends; Blackburn and Collins, 2011). Although each interaction involves at least some basepairing, these RNAs bind in a range of configurations that in many cases would have been difficult to predict *a priori*.

The widespread insertion and deletion of uridines in kinetoplastid mRNAs are directed by *trans*-acting guide RNAs (gRNAs) (Blum *et al.*, 1990), which basepair to sequences immediately downstream of the sites to be edited. However, no sequences that could encode similar guide-like RNAs were detected by Takano and colleagues in their characterization of the *Pp* mitochondrial genome (Takano *et al.*, 2001). The likelihood of potential RNA templates was also investigated using labelling strategies that would detect abundant RNAs of a discreet size or species having characteristics similar to gRNAs or microRNAs (Bullerwell *et al.*, 2010), but no such RNAs were detected (Jonatha M. Gott, unpublished). More tellingly, the RNA-seq library made by Bundschuh *et al.* (2011) was designed to maximize the chances of finding potential templates via deep sequencing. Despite the use of non-size selected, total mitochondrial RNA and directional cloning during library construction coupled with extensive bioinformatic analysis, no candidate RNAs were found (Bundschuh *et al.*, 2011). Thus, it is unlikely that *trans*-acting RNA templates direct editing in *Pp* mitochondria.

Cis-acting RNA elements involved in editing could potentially be located either upstream or downstream of editing sites. A role for downstream RNA does not seem plausible based on the absence of unedited stretches of RNA in fingerprinting experiments (Visomirski-Robic and Gott, 1995), the inability to add non-encoded nucleotides to unedited regions of RNA once made (Visomirski-Robic and Gott, 1997a), and the competition between transcription and editing (Cheng *et al.*, 2001), arguing that nucleotides are added prior to the synthesis of the downstream portion of the transcript. The addition of one or more purines within the P mRNA of paramyxoviruses (Hausmann *et al.*, 1999) and Ebola virus (Volchkov *et al.*, 1995; Sanchez *et al.*, 1996) is also co-transcriptional and is templated by an upstream homopolymer tract within the RNA template. However, very few of the editing sites in *Pp* mitochondria occur in the context of a homopolymer run, and the different effects of relative nucleotide concentration on editing in the two systems (described above) are not consistent with a similar slippage model in *Pp* mitochondria (Cheng *et al.*, 2001), at least at C insertion sites.

Regulatory events that occur during transcription elongation often depend upon upstream elements within the growing RNA chain. RNA hairpins, for example, frequently modulate elongation, and can be located either distal to the site of action (e.g. λ nut site in antitermination) (Whalen and Das, 1990) or proximal (e.g. pause hairpins in attenuation and termination) (Gutierrez-Preciado *et al.*, 2009; Santangelo and Artsimovitch, 2011). No common RNA secondary structural elements have been identified upstream of editing sites using RNA folding programmes, but this does not, of course, preclude RNA involvement. However, two lines of evidence strongly argue against a role for distal RNA in the insertion of non-encoded nucleotides. First, oligonucleotide-directed RNaseH cleavage has been used to test directly whether upstream RNA influences editing. Removal of the accessible portion of nascent transcripts prior to run-on transcription has no effect on non-encoded nucleotide insertion (Fig. 2.7) (A. Majewski and Jonatha M. Gott, unpublished data), indicating that distal RNA elements are not required for this process. Second, experiments with chimeric templates have demonstrated that complete replacement of RNA sequences more than nine nucleotides upstream of C insertion sites does not affect editing (Rhee *et al.*, 2009). Thus, if nascent RNA plays a role in editing, its effects are likely limited to the region within or immediately adjacent to the RNA–DNA hybrid.

Although bioinformatics studies have not detected any motif or consensus sequence surrounding C insertion sites (see above), the context of other insertion sites is far from random. The specific insertion of single nucleotides other than C is too rare to allow for meaningful statistical analysis, but their contexts are nevertheless interesting. Whereas only 30% of the 1255 single C insertions are added next to an encoded C, the sole A insertion is added in the context of three

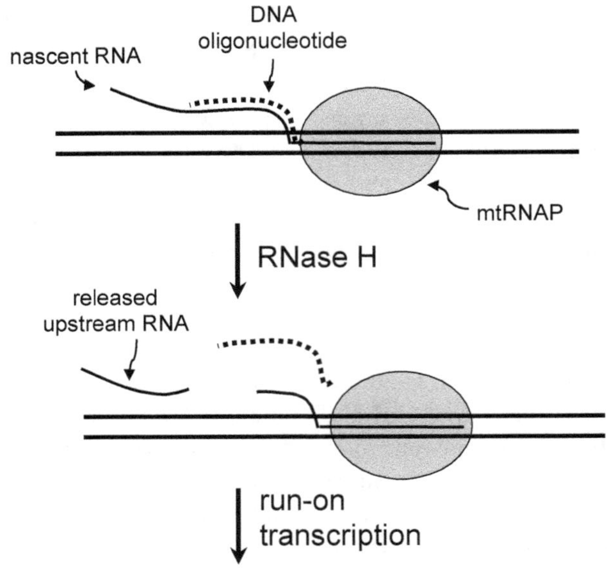

Figure 2.7 Strategy for removal of upstream RNA. Removal of upstream RNA from elongation transcription complexes via oligonucleotide-directed RNaseH cleavage prior to run-on transcription.

encoded As, the two single G insertions are added next to one or two encoded Gs, and 60% of the inserted single Us (26 of 43) are adjacent to encoded Us (Bundschuh *et al.*, 2011). The sequences surrounding dinucleotide insertion sites are even more striking in that in most cases the order of nucleotide addition is not even clear (Table 2.2) (Byrne and Gott, 2004; Bundschuh *et al.*, 2011). An extreme example is provided by the addition of an adjacent G and U to the *coxI* mRNA, which results in a sequence change from the encoded GUUGUUA to GUUGUGUUA. This could potentially be accomplished in a variety of ways, including GUugUGUUA, GUUguGUUA, GUUGugUUA, or GUUGUguUA (where the inserted nucleotides are shown in lower case), or even GuUgUGUUA or GUUGUgUuA. In this case, the actual site of dinucleotide insertion is almost certainly GUUGugUUA based on two misediting events in which only a single nucleotide (+U or +G) is added at this site during run-on transcription in mtTECs (Byrne *et al.*, 2002). Only in the case of the UA insertion within the *Pp coxI* mRNA is the insertion order unambiguous (GUUCA to GUUUACA), but even here the added nucleotides could potentially be added one nucleotide apart based on sequence context alone. Once again, misediting patterns support the addition of adjacent nucleotides at this site (GUUuaC), and the same is true for other dinucleotide sites as well (Byrne and Gott, 2004; Rhee *et al.*, 2009). The sequence bias seen at non-C insertion sites may potentially be mechanistically significant, as discussed below.

Another possible source of editing information is the presence of epigenetic marks within the mitochondrial genome. If, for example, a particular modified base was used to demark *Physarum* editing sites, this base would be expected to comprise ~2% of the genome (1333 sites in 62,862 bp), which is well within the detection limits of mass spectrometry. Preliminary analysis of CsCl gradient-purified DNAs did not detect modified bases anywhere approaching this level in *Pp* mitochondrial DNA (P. Crain, personal communication). However, because contamination by even a small amount of genomic DNA could have masked signal from the mitochondrial DNA, the mass spectrometry results cannot be considered absolutely definitive. The finding that deproteinized mitochondrial DNA fragments fail to support editing upon ligation to digested mtTEC DNA also strongly suggests that editing sites are unlikely to be 'flagged' by covalent modifications (Byrne

Table 2.2 Context of dinucleotide insertions

Added dinucleotide	Gene location	Edited sequence context (*P. polycephalum*)	Edited sequence context (*D. iridis*)	Edited sequence context (*D. nigripes*)	Edited sequence context (*S. flavogenita*)
UC/CU	cox1	CUC	CUC	CUC	CUC
	nad7	UCUCUU	UCUCU		
	nad7	UUUCUCU	UUUCUCU		
	LSU	CUC			
	LSU	CUCU			
	SSU	CUCC		CUCC	CUCC
	cytb	CCUC	CCUC		
	cytb	CUCUU	CUCU		
	php24	CUCC			
UG/GU	cox1	UUGUGUU	UUGUGUU	UUGUGUU	UUGUGUU
	nad7	GUGUU	GUGUUUU		
	LSU	GGUGU			
	rpL19	GUGU			
CG/GC	nad4	GGGCGC			
	cytb	GCGCC	GCGCC		
AA	LSU	AAA			
	LSU	AAA			
	SSU	AAA		AAA	AAA
	SSU	AAA			
UA	cox1	UU**UACA**UUUU	UU**UAUA**UUUU	UU**UACA**UUUU	UU**UAUA**UCUU
	rpS19	UAUA			
UU	nad7	UUU	UUU		
	nad7	UUUU	UUUU		

and Gott, 2002), but it could be argued that modifications might be necessary but not sufficient. Thus, although current data do not support a role for modified nucleotides in defining editing sites, this possibility cannot be conclusively ruled out.

Editing machinery

The second major outstanding question in the field of *Pp* insertion RNA editing is the composition of the editing apparatus. All existing evidence indicates that the addition of 'non-templated' nucleotides occurs as the RNAs are being synthesized (Visomirski-Robic and Gott, 1995, 1997a,b; Cheng *et al.*, 2001; Byrne and Gott, 2002; Byrne *et al.*, 2002), implicating the *Physarum* mitochondrial RNA polymerase (mtRNAP) in the editing process. At the very least, the polymerase must be involved in the initiation and resolution steps of editing, as templated transcription elongation must stop at an editing site and then resume from an extra, non-paired nucleotide in a template-directed manner. However, it is also clear that the mtRNAP is unlikely the whole story and that one or more additional factors are required.

The mitochondrial RNA polymerase

The cDNA encoding the *Pp* mtRNAP has been cloned independently in two laboratories (Miller *et al.*, 2006; Gott and Rhee, 2007). Like other mitochondrial RNA polymerases, the *Pp* mtRNAP is homologous to the well-studied RNA polymerase from bacteriophage T7 and contains the same highly conserved motifs, including the residues important for catalysis (Gott and Rhee, 2007). Recombinant versions of the *Pp* mtRNAP have been expressed in *E. coli* as either a maltose-binding protein (MBP) fusion protein (Miller *et al.*, 2006) or with a poly-histidine tag at its N terminus (Gott and Rhee, 2007). Neither recombinant enzyme adds extra nucleotides at internal sites when transcribing cloned *Pp* mitochondrial genes (Miller *et al.*, 2006; Gott and Rhee, 2007; and unpublished data). However, the MBP–mtRNAP fusion protein has been shown to add extra nucleotides to the 3′ end of RNAs in a non-specific manner, even in the absence of a DNA template (Miller and Miller, 2008), an activity that may be related to editing. Unlike the transcription reaction, this activity, which was also observed in parallel reactions with the bacteriophage T7 RNAP, does not require the presence of all four ribonucleoside triphosphates (rNTPs). The mtRNAP does not appear to discriminate between rNTPs, as all four ribonucleotides can be used as substrates for this end addition reaction (Miller and Miller, 2008).

Studies with the native mitochondrial polymerase also suggest that the polymerase alone cannot carry out the editing reaction. *In vitro* transcription using fractionated mitochondrial extracts highly enriched for the *Pp* mtRNAP results in the synthesis of totally unedited transcripts (A. Majewski, E. Byrne, and Jonatha M. Gott, unpublished data). Likewise, exogenously supplied *Pp* sequences fail to support editing in the context of chimeric templates transcribed by the native enzyme (Fig. 2.5). During run-on transcription in mtTECs, lengthening the dwell time at editing sites increases the extent of editing at those sites *in vitro* (Cheng *et al.*, 2001). It is possible, therefore, that the rate of transcription elongation on naked DNA present in chimeric templates is simply too fast for the polymerase and/or associated factors to respond to editing signals. To test this hypothesis, conditions that slow transcription (extremely low nucleotide concentrations and low temperatures), were tested to determine whether they would allow editing sites to be recognized on naked DNA. No editing was observed from deproteinized DNA templates under any condition tested (A. Rhee and Jonatha M. Gott, unpublished data), again strongly implying that template-associated factors are required for *Physarum* editing. Moreover, in preliminary mitochondrial transformation experiments in which DNA is introduced into isolated mitochondria via electroporation, transcripts synthesized from the exogenously supplied DNA are unedited (Jonatha M. Gott, unpublished data). Thus, the available data suggest that, on its own, the *Pp* mtRNAP is unable to catalyse the insertion of non-encoded nucleotides into nascent transcripts, pointing to the existence of additional editing factors.

Potential editing factors

Additional factors are almost certainly involved at one or more steps of the editing cycle. Possible roles include the identification of editing sites and/or the specification of the nucleotide(s) to be added, or actual catalysis. Experiments with chimeric templates demonstrated that sequence alone is not sufficient to support editing and that editing signals are not recognized on deproteinized DNA (Byrne and Gott, 2002; Byrne *et al.*, 2002), even under conditions in which transcription is slowed (A. Rhee and Jonatha M. Gott, unpublished data). Dinucleotide insertions are also added cotranscriptionally (Visomirski-Robic and Gott, 1997a), presumably involving at least some of the same basic machinery as the single nucleotide insertions. *In vitro* experiments suggest that dinucleotides are added sequentially (Visomirski-Robic and Gott, 1997a; Byrne and Gott, 2004) and that separate or additional factors may be required at these sites (Byrne and Gott, 2004).

The most likely candidates for *trans*-acting editing factors are proteins and small RNAs, but their identity remains a mystery. As discussed above, a role for *trans*-acting RNAs is unlikely, based on labelling, deep sequencing, and genomic studies. Early attempts to reconstitute editing by adding mitochondrial fractions to *in vitro* transcription reactions were unsuccessful, but extensive complementation experiments have yet to be attempted with more highly purified fractions. We have also been unable to separate the transcription and editing activities of mtTECs using an array of detergents and salts (Y.-W. Cheng and Jonatha M. Gott, unpublished data), suggesting that the mtRNAP is released from the mtDNA before or concurrently with any editing factors.

Any interactions may be transient or non-specific; no 'footprints' are detectable in DNaseI protection studies (L. Tsujikawa and Jonatha M. Gott, unpublished data) and the mitochondrial DNA in editing-competent mtTECs is freely accessible to restriction enzymes (Byrne and Gott, 2002; Byrne, 2004; Byrne *et al.*, 2007), whose cleavage is virtually complete even at sites where editing and restriction sites overlap (Rhee *et al.*, 2009). A yeast two-hybrid screen using the *Pp* mtRNAP as bait failed to identify possible interacting proteins, but this could potentially be due to issues with library construction, protein localization or folding, choice of bait construct, or non-specific DNA binding by the polymerase, rather than a lack of interacting partners. Efforts are under way to identify proteins closely associated with active transcription complexes and/or the mitochondrial RNAP in hopes of identifying likely candidates for further study.

Possible models for insertion RNA editing in *Pp* mitochondria

Editing site identity

Models are needed to explain both how editing sites are recognized and how the nucleotide/s to be incorporated is/are specified. Data from chimeric template experiments (Rhee *et al.*, 2009) indicate that the template sequences required for editing lie within a very limited region surrounding insertion sites. Based on misediting events that occur when changes are made within this critical region, Rhee *et al.* (2009) inferred that template sequences upstream of C insertion sites are likely to be involved in nucleotide selection and/or insertion, while downstream sequences are more likely to be involved in editing site recognition. However, statistical analyses indicate that these regions do not contain sufficient information to uniquely specify editing sites (Bundschuh *et al.*, 2011; Chen *et al.*, 2012). Other potential means of identifying editing sites have been put forward by Bundschuh and colleagues (Chen *et al.*, 2012). These include (i) an editing apparatus that does not recognize the same positions within the critical region at every site, (ii) use of the mature mRNA as the editing template, and (iii) a two-step recognition process. They consider the first option unlikely based on statistical arguments and find the second difficult to reconcile with the co-transcriptional nature of the editing process. The third scenario invokes distal recognition elements that direct the editing machinery to flag insertion sites. In this 'marker model', it is these tags that are recognized during transcription and editing. The nature of these 'marks' is unclear, however. Another alternative is that there is a common signal in the vicinity of editing sites that could be generated by a range of sequences. Possibilities include sequences that cause changes in local DNA structure (e.g. bends) and/or polymerase pausing. While this model is attractive, such signals would still need to function in conjunction with additional editing factors given that sequence alone is insufficient to support editing in the context of chimeric templates (Byrne and Gott, 2002).

A number of potential models can be invoked to explain the specificity of nucleotide selection. There may be site-specific factors that act at only one or a few sites, as observed in plant organelles (Zehrmann *et al.*, 2011) or editing type-specific factors, i.e. a single factor for all GC insertions. It is also possible that C insertion is the default mode of editing, and that factors involved in nucleotide selection are only required at non-C insertion sites. In this scenario, the requirement for some feature of the native template (Byrne and Gott, 2002) would be attributed to a role in editing site recognition or the physical insertion of the added nucleotide. Another appealing possibility is that the information is contributed by a combination of nucleic acid and protein functional groups in a manner similar to the class I CCA-adding enzymes (Shi *et al.*, 1998; Xiong and Steitz, 2004).

Based on the sequence bias seen at non-C insertion sites (Table 2.2), it is possible that insertions at these sites are completely or partially pseudo-templated via slippage of the RNA–DNA hybrid in a manner similar to that observed in paramyxoviruses (Vidal *et al.*, 1990). Alternatively, the surrounding sequences could contribute to the efficiency of templated extension *after* editing has occurred; extension of an RNA having two unpaired nucleotides at its 3′ end might be particularly problematic. In this scenario, slippage of the RNA–DNA hybrid once editing has occurred would allow the mitochondrial RNA polymerase to extend the RNA from a paired, rather than an unpaired 3′ end. Of particular interest in this context is that while the non-encoded UA in the *cox1* mRNAs from *P. polycephalum* and *D. nigripes* is added next to an encoded CA, it is inserted next to an encoded UA in *D. iridis* and *S. flavogenita* (Table 2.2). This raises the remarkable possibility that the *cox1* mRNA UA insertions are actually the result of a CA insertion followed by deamination of the C to a U. This may actually be the case in *Pp* mitochondria, based on the findings that approximately half of the *in vivo* transcripts associated with transcription complexes contain a CA at this position and that *cox1* transcripts made by mtTECs *in vitro* contain a CA at this site (Byrne *et al.*, 2002). Even if non-C insertions are pseudo-templated, this does not get around the problem of how sites are initially recognized, however. These (and other) potential mechanisms await further study.

Potential modes of nucleotide insertion

Insertion editing in *Pp* mitochondria occurs in the context of the transcription elongation complex, but it is currently unknown whether the actual insertion of non-encoded nucleotides is carried out by the mtRNAP or a separate, editing enzyme. Paramyxovirus editing occurs via a polymerase stuttering mechanism, involving slippage of the nascent RNA relative to the ribonucleoprotein complex that serves as the editing template (Jacques and Kolakofsky, 1991). DNA-dependent RNA polymerases can also add nucleotides in a non-templated fashion, generally at template ends (Melton *et al.*, 1984; Milligan *et al.*, 1987). The demonstrated ability of the *Pp* mtRNAP to add non-templated nucleotides to the 3′ end of RNAs is consistent with a potential role in nucleotide insertion (Miller and Miller, 2008). The finding that all required template elements fall within ~9 bp of editing sites (Rhee *et al.*, 2009), coupled with the minimal 9 nt spacing between C insertion sites observed in nature (Miller *et al.*, 1993; Bundschuh *et al.*, 2011), suggests that this constraint may be mechanistically significant. In this context it is interesting to note that the RNA–DNA hybrid within the *Pp* mtRNAP is likely 8 bp in length, based on X-ray crystal structures of T7 RNA polymerase elongation complexes (Tahirov *et al.*, 2002; Yin and Steitz, 2002). This distance constraint would be consistent with the involvement of the mtRNAP polymerase active site in non-encoded nucleotide addition.

If the *Pp* mtRNAP adds both templated and non-templated nucleotides, it may use partially or completely overlapping polymerization sites that adopt slightly different conformations, similar to what is observed with multisubunit RNA polymerases, which use a single active site for both polymerization and cleavage (Kettenberger *et al.*, 2003; Opalka *et al.*, 2003). However, there are also precedents for polymerases having independent active sites. DNA polymerases use two separate active sites for polymerization and 'editing' (proofreading). These two sites are roughly 30 Å apart (see Joyce and Steitz, 1994) and backtracking of the DNA polymerase and concomitant unwinding of a portion of the DNA helix are required for removal of incorrectly added nucleotides (see Johnson, 1993). Reverse transcriptase also has two separate catalytic domains for its polymerization and RNase H activities (see Kohlstaedt *et al.*, 1992). While the carboxy-terminus of the mtRNAP is highly similar to the T7 family of RNA polymerases, its amino-terminus does not align well with either the T7 enzyme or other mitochondrial RNA polymerases (note, however, that the N-terminal portions of mitochondrial RNA polymerases are not highly conserved). Thus, it is possible that the N-terminal domain of the *Pp* mtRNAP might also play a role in editing in a manner analogous to the enzymes mentioned above. Structural data on *Pp* mitochondrial elongation complexes are needed to address these issues.

If a separate enzyme is responsible for nucleotide insertion, the editing activity would need access to the 3′ end of the RNA. This would most likely require at least limited backtracking by the mtRNAP and partial unwinding of the RNA–DNA hybrid and/or a conformational change on the part of the RNA polymerase. Precedents for this type of activity include transcription elongation factors from both bacteria (GreA, GreB) and eukaryotes (TFIIS), which promote transcript cleavage within stalled or arrested elongation complexes. These proteins have long, finger-like projections that are thought to coordinate a Mg^{2+} ion, altering the geometry of the polymerase active site to allow for cleavage of mis-incorporated nucleotides (Kettenberger *et al.*, 2003; Opalka *et al.*, 2003). Oligonucleotide-directed RNase H experiments in which the upstream portion of the transcript is removed from *Pp* mtTECs prior to run-on transcription (Fig. 2.7) suggest that if backtracking is involved, the polymerase movement is likely to be limited to only a few nucleotides. In either case, it is likely that the *Pp* mtRNAP is closely associated with the editing activity. However, in the absence of an *in vitro* system that carries out accurate editing using a defined template, it has been difficult to test any of these models directly.

Summary and future perspectives

Important questions remain. Although non-encoded nucleotides are added to *Pp* mitochondrial RNAs in a highly specific fashion, the basis of this precision is still completely unknown. Our current hypothesis is that the mtRNAP is directly involved in nucleotide insertion, but that additional factor(s) are required at one or more steps of the editing cycle. Only template elements close to the point of insertion are necessary for editing and most of the upstream RNA within a transcription complex can be removed or replaced without affecting accuracy or efficiency of the process (Rhee *et al.*, 2009). Our data also indicate that there are template-associated *trans*-acting factors required for editing site recognition, that recognition of editing sites and nucleotide insertion are separable processes, and that there may be additional editing factors involved in specifying the nucleotide to be inserted or signalling resumption of template-directed transcription (Byrne and Gott, 2002, 2004; Byrne *et al.*, 2002). Taken together, the data make a compelling case that the factors that participate in the various steps of RNA editing in *Pp* mitochondria must be identified in order to understand this remarkable process. Because virtually nothing is known about the components of the editing machinery, a combination of biochemical, affinity, bioinformatics, and directed genetic approaches will be necessary to identify the *trans*-acting factors involved in editing and to generate the tools necessary for dissecting *Pp* editing mechanisms. Fortunately, the final assembly of the *Pp* (nuclear) genome project by the Washington University Genome Center is now becoming available, making a more global approach to the identification of editing factors feasible. This, along with the establishment of a means of introducing DNA into isolated organelles (Jonatha M. Gott, unpublished data), should allow us to address the critical challenges within the field, the identification of editing factors and editing signals.

References

Abad, M.G., Rao, B.S., and Jackman, J.E. (2010). Template-dependent 3′–5′ nucleotide addition is a shared feature of tRNAHis guanylyltransferase enzymes from multiple domains of life. Proc. Natl. Acad. Sci. U.S.A. *107*, 674–679.

Abad, M.G., Long, Y., Willcox, A., Gott, J.M., Gray, M.W., and Jackman, J.E. (2011). A role for tRNA(His) guanylyltransferase (Thg1)-like proteins from *Dictyostelium discoideum* in mitochondrial 5′-tRNA editing. RNA *17*, 613–623.

Antes, T., Costandy, H., Mahendran, R., Spottswood, M., and Miller, D. (1998). Insertional editing of mitochondrial tRNAs of *Physarum polycephalum* and *Didymium nigripes*. Mol. Cell. Biol. *18*, 7521–7527.

Aphasizhev, R., and Aphasizheva, I. (2011). Uridine insertion/deletion editing in trypanosomes: a playground for RNA-guided information transfer. Wiley Interdiscip. Rev. RNA 2, 669–685.

Bartel, D.P. (2009). MicroRNAs: target recognition and regulatory functions. Cell *136*, 215–233.

Beargie, C., Liu, T., Corriveau, M., Lee, H.Y., Gott, J., and Bundschuh, R. (2008). Genome annotation in the presence of insertional RNA editing. Bioinformatics *24*, 2571–2578.

Blackburn, E.H., and Collins, K. (2011). Telomerase: an RNP enzyme synthesizes DNA. Cold Spring Harb. Perspect. Biol. 3, pii: a003558.

Blanc, V., and Davidson, N.O. (2010). Apobec-1 mediated RNA editing. Wiley Interdiscip. Rev. Syst. Biol. Med. *2*, 594–602.

Blum, B., Bakalara, N., and Simpson, L. (1990). A model for RNA editing in kinetoplastid mitochondria: 'guide' RNA molecules transcribed from maxicircle DNA provide the edited information. Cell *60*, 189–198.

Bullerwell, C.E., and Gray, M.W. (2005). *In vitro* characterization of a tRNA editing activity in the mitochondria of *Spizellomyces punctatus*, a Chytridiomycete fungus. J. Biol. Chem. *280*, 2463–2470.

Bullerwell, C.E., Burger, G., Gott, J.M., Kourennaia, O., Schnare, M.N., and Gray, M.W. (2010). Abundant 5S rRNA-like transcripts encoded by the mitochondrial genome in amoebozoa. Eukaryot. Cell 9, 762–773.

Bundschuh, R. (2004). Computational prediction of RNA editing sites. Bioinformatics *20*, 3214–3220.

Bundschuh, R. (2007). Computational approaches to insertional RNA editing. Methods Enzymol. *424*, 173–195.

Bundschuh, R., Altmuller, J., Becker, C., Nurnberg, P., and Gott, J.M. (2011). Complete characterization of the edited transcriptome of the mitochondrion of *Physarum polycephalum* using deep sequencing of RNA. Nucleic Acids Res. *39*, 6044–6055.

Byrne, E.M. (2004). Chimeric templates and assays used to study *Physarum* cotranscriptional insertional editing *in vitro*. Methods Mol. Biol. *265*, 293–314.

Byrne, E.M., and Gott, J.M. (2002). Cotranscriptional editing of *Physarum* mitochondrial RNA requires local features of the native template. RNA *8*, 1174–1185.

Byrne, E.M., and Gott, J.M. (2004). Unexpectedly complex editing patterns at dinucleotide insertion sites in *Physarum* mitochondria. Mol. Cell. Biol. *24*, 7821–7828.

Byrne, E.M., Stout, A., and Gott, J.M. (2002). Editing site recognition and nucleotide insertion are separable processes in *Physarum* mitochondria. EMBO J. *21*, 6154–6161.

Byrne, E.M., Visomirski-Robic, L., Cheng, Y.W., Rhee, A.C., and Gott, J.M. (2007). RNA editing in *Physarum* mitochondria: assays and biochemical approaches. Methods Enzymol. *424*, 143–172.

Carrillo, R., Thiemann, O.H., Alfonzo, J.D., and Simpson, L. (2001). Uridine insertion/deletion RNA editing in *Leishmania tarentolae* mitochondria shows cell cycle dependence. Mol. Biochem. Parasitol. *113*, 175–181.

Chateigner-Boutin, A.-L., and Small, I. (2010). Plant RNA editing. RNA Biol. *7*, 213–219.

Chen, C., Frankhouser, D., and Bundschuh, R. (2012). Comparison of insertional RNA editing in Myxomycetes. PLoS Comput. Biol. *8*, e1002400.

Chen, S., Habib, G., Yang, C., Gu, Z., Lee, B., Weng, S., Silberman, S., Cai, S., Deslypere, J., and Rosseneu, M. (1987). Apolipoprotein B-48 is the product of a messenger RNA with an organ-specific in-frame stop codon. Science *238*, 363–366.

Cheng, Y.W., and Gott, J.M. (2000). Transcription and RNA editing in a soluble *in vitro* system from *Physarum* mitochondria. Nucleic Acids Res. *28*, 3695–3701.

Cheng, Y.W., Visomirski-Robic, L.M., and Gott, J.M. (2001). Non-templated addition of nucleotides to the 3′ end of nascent RNA during RNA editing in *Physarum*. EMBO J. *20*, 1405–1414.

Cooley, L., Appel, B., and Soll, D. (1982). Post-transcriptional nucleotide addition is responsible for the formation of the 5′ terminus of histidine tRNA. Proc. Natl. Acad. Sci. U.S.A. *79*, 6475–6479.

Dieci, G., Preti, M., and Montanini, B. (2009). Eukaryotic snoRNAs: a paradigm for gene expression flexibility. Genomics *94*, 83–88.

Ellis, J.C., and Brown, J.W. (2009). The RNase P family. RNA Biol. *6*, 362–369.

Farajollahi, S., and Maas, S. (2010). Molecular diversity through RNA editing: a balancing act. Trends Genet. *26*, 221–230.

Feagin, J.E., Abraham, J.M., and Stuart, K. (1988). Extensive editing of the cytochrome c oxidase III transcript in *Trypanosoma brucei*. Cell *53*, 413–422.

Gott, J. (2011). RNA editing and human disorders. In Encyclopedia of Life Science (John Wiley & Sons, Chichester).

Gott, J.M., and Emeson, R.B. (2000). Functions and mechanisms of RNA editing. Annu. Rev. Genet. *34*, 499–531.

Gott, J.M., and Rhee, A.C. (2007). Insertion/deletion editing in *Physarum polycephalum*. In RNA Editing, Goringer, H.U., ed. (Springer-Verlag, Berlin), pp. 85–104.

Gott, J.M., and Visomirski-Robic, L.M. (1998). RNA editing in *Physarum* mitochondria. In Modification and Editing of RNA, Grosjean, H., and Benne, R., eds. (ASM Press, Washington, DC), pp. 395–411.

Gott, J.M., Visomirski, L.M., and Hunter, J.L. (1993). Substitutional and insertional RNA editing of the cytochrome c oxidase subunit 1 mRNA of *Physarum polycephalum*. J. Biol. Chem. *268*, 25483–25486.

Gott, J.M., Parimi, N., and Bundschuh, R. (2005). Discovery of new genes and deletion editing in *Physarum* mitochondria enabled by a novel algorithm for finding edited mRNAs. Nucleic Acids Res. *33*, 5063–5072.

Gott, J.M., Somerlot, B.H., and Gray, M.W. (2010). Two forms of RNA editing are required for tRNA maturation in *Physarum* mitochondria. RNA *16*, 482–488.

Gu, W., Jackman, J.E., Lohan, A.J., Gray, M.W., and Phizicky, E.M. (2003). tRNAHis maturation: an essential yeast protein catalyzes addition of a guanine nucleotide to the 5′ end of tRNAHis. Genes Dev. *17*, 2889–2901.

Gutierrez-Preciado, A., Henkin, T.M., Grundy, F.J., Yanofsky, C., and Merino, E. (2009). Biochemical features and functional implications of the RNA-based T-box regulatory mechanism. Microbiol. Mol. Biol. Rev. *73*, 36–61.

Hausmann, S., Garcin, D., Delenda, C., and Kolakofsky, D. (1999). The versatility of paramyxovirus RNA polymerase stuttering. J. Virol. *73*, 5568–5576.

Heinemann, I.U., O'Donoghue, P., Madinger, C., Benner, J., Randau, L., Noren, C.J., and Soll, D. (2009). The appearance of pyrrolysine in tRNAHis guanylyltransferase by neutral evolution. Proc. Natl. Acad. Sci. U.S.A. *106*, 21103–21108.

Heinemann, I.U., Randau, L., Tomko, R.J., Jr., and Soll, D. (2010). 3′–5′ tRNAHis guanylyltransferase in bacteria. FEBS Lett. *584*, 3567–3572.

Heinemann, I.U., Nakamura, A., O'Donoghue, P., Eiler, D., and Soll, D. (2011). tRNAHis-guanylyltransferase establishes tRNAHis identity. Nucleic Acids Res. *40*, 333–344.

Hendrickson, P.G., and Silliker, M.E. (2010). RNA editing in six mitochondrial ribosomal protein genes of *Didymium iridis*. Curr. Genet. *56*, 203–213.

Himeno, H., Hasegawa, T., Ueda, T., Watanabe, K., Miura, K., and Shimizu, M. (1989). Role of the extra G–C pair at the end of the acceptor stem of tRNA(His) in aminoacylation. Nucleic Acids Res. *17*, 7855–7863.

Horton, T.L., and Landweber, L.F. (2000). Evolution of four types of RNA editing in myxomycetes. RNA *6*, 1339–1346.

Jackman, J.E., Gott, J.M., and Gray, M.W. (2012). Doing it in reverse: 3′-to-5′ polymerization by the Thg1 superfamily. RNA *18*, 886–899.

Jacques, J.P., and Kolakofsky, D. (1991). Pseudo-templated transcription in prokaryotic and eukaryotic organisms. Genes Dev. *5*, 707–713.

Johnson, K.A. (1993). Conformational coupling in DNA polymerase fidelity. Annu. Rev. Biochem. *62*, 685–713.

Joyce, C.M., and Steitz, T.A. (1994). Function and structure relationships in DNA polymerases. Annu. Rev. Biochem. *63*, 777–822.

Kettenberger, H., Armache, K.J., and Cramer, P. (2003). Architecture of the RNA polymerase II–TFIIS complex and implications for mRNA cleavage. Cell *114*, 347–357.

Kiss, T., Fayet-Lebaron, E., and Jady, B.E. (2010). Box H/ACA small ribonucleoproteins. Mol. Cell 37, 597–606.

Knoop, V. (2011). When you cannot trust the DNA: RNA editing changes transcript sequences. Cell. Mol. Life Sci. *68*, 567–586.

Kohlstaedt, L.A., Wang, J., Friedman, J.M., Rice, P.A., and Steitz, T.A. (1992). Crystal structure at 3.5 A resolution of HIV-1 reverse transcriptase complexed with an inhibitor. Science *256*, 1783–1790.

Krishnan, U., Barsamian, A., and Miller, D.L. (2007). Evolution of RNA editing sites in the mitochondrial small subunit rRNA of the Myxomycota. Methods Enzymol. *424*, 197–220.

Liu, T., and Bundschuh, R. (2005). Model for codon position bias in RNA editing. Phys. Rev. Lett. *95*, 088101.

Maas, S., Kawahara, Y., Tamburro, K.M., and Nishikura, K. (2006). A-to-I RNA editing and human disease. RNA Biol. 3, 1–9.

Mahendran, R., Spottswood, M.R., and Miller, D.L. (1991). RNA editing by cytidine insertion in mitochondria of *Physarum polycephalum*. Nature *349*, 434–438.

Mahendran, R., Spottswood, M.S., Ghate, A., Ling, M.L., Jeng, K., and Miller, D.L. (1994). Editing of the mitochondrial small subunit rRNA in *Physarum polycephalum*. EMBO J. *13*, 232–240.

Melton, D.A., Krieg, P.A., Rebagliati, M.R., Maniatis, T., Zinn, K., and Green, M.R. (1984). Efficient *in vitro* synthesis of biologically active RNA and RNA hybridization probes from plasmids containing a bacteriophage SP6 promoter. Nucleic Acids Res. *12*, 7035–7056.

Miller, D., Mahendran, R., Spottswood, M., Costandy, H., Wang, S., Ling, M.L., and Yang, N. (1993). Insertional editing in mitochondria of *Physarum*. Semin. Cell Biol. *4*, 261–266.

Miller, M.L., and Miller, D.L. (2008). Non-DNA-templated addition of nucleotides to the 3′ end of RNAs by the mitochondrial RNA polymerase of *Physarum polycephalum*. Mol. Cell. Biol. *28*, 5795–5802.

Miller, M.L., Antes, T.J., Qian, F., and Miller, D.L. (2006). Identification of a putative mitochondrial RNA polymerase from *Physarum polycephalum*: characterization, expression, purification, and transcription *in vitro*. Curr. Genet. *49*, 259–271.

Milligan, J.F., Groebe, D.R., Witherell, G.W., and Uhlenbeck, O.C. (1987). Oligoribonucleotide synthesis using T7 RNA polymerase and synthetic DNA templates. Nucleic Acids Res. *15*, 8783–8798.

Nameki, N., Asahara, H., Shimizu, M., Okada, N., and Himeno, H. (1995). Identity elements of *Saccharomyces cerevisiae* tRNA(His). Nucleic Acids Res. *23*, 389–394.

Opalka, N., Chlenov, M., Chacon, P., Rice, W.J., Wriggers, W., and Darst, S.A. (2003). Structure and function of the transcription elongation factor GreB bound to bacterial RNA polymerase. Cell *114*, 335–345.

Powell, L., Wallis, S., Pease, R., Edwards, Y., Knott, T., and Scott, J. (1987). A novel form of tissue-specific RNA processing produces apolipoprotein-B48 in intestine. Cell *50*, 831–840.

Price, D.H., and Gray, M.W. (1999). A novel nucleotide incorporation activity implicated in the editing of mitochondrial transfer RNAs in *Acanthamoeba castellanii*. RNA *5*, 302–317.

Pullirsch, D., and Jantsch, M.F. (2010). Proteome diversification by adenosine to inosine RNA editing. RNA Biol. 7, 205–212.

Rao, B.S., Maris, E.L., and Jackman, J.E. (2011). tRNA 5′-end repair activities of tRNAHis guanylyltransferase (Thg1)-like proteins from Bacteria and Archaea. Nucleic Acids Res. *39*, 1833–1842.

Rhee, A.C., Somerlot, B.H., Parimi, N., and Gott, J.M. (2009). Distinct roles for sequences upstream of and downstream from *Physarum* editing sites. RNA *15*, 1753–1765.

Rundquist, B.A., and Gott, J.M. (1995). RNA editing of the col mRNA throughout the life cycle of Physarum polycephalum. Mol. Gen. Genet. *247*, 306–311.

Sanchez, A., Trappier, S.G., Mahy, B.W., Peters, C.J., and Nichol, S.T. (1996). The virion glycoproteins of Ebola viruses are encoded in two reading frames and are expressed through transcriptional editing. Proc. Natl. Acad. Sci. U.S.A. *93*, 3602–3607.

Santangelo, T.J., and Artsimovitch, I. (2011). Termination and antitermination: RNA polymerase runs a stop sign. Nat. Rev. Microbiol. *9*, 319–329.

Shi, P.Y., Maizels, N., and Weiner, A.M. (1998). CCA addition by tRNA nucleotidyltransferase: polymerization without translocation? EMBO J. *17*, 3197–3206.

Tahirov, T.H., Temiakov, D., Anikin, M., Patlan, V., McAllister, W.T., Vassylyev, D.G., and Yokoyama, S. (2002). Structure of a T7 RNA polymerase elongation complex at 2.9 A resolution. Nature *420*, 43–50.

Takano, H., Abe, T., Sakurai, R., Moriyama, Y., Miyazawa, Y., Nozaki, H., Kawano, S., Sasaki, N., and Kuroiwa, T. (2001). The complete DNA sequence of the mitochondrial genome of *Physarum polycephalum*. Mol. Gen. Genet. *264*, 539–545.

Traphagen, S.J., Dimarco, M.J., and Silliker, M.E. (2010). RNA editing of 10 *Didymium iridis* mitochondrial genes and comparison with the homologous genes in *Physarum polycephalum*. RNA *16*, 828–838.

Vidal, S., Curran, J., and Kolakofsky, D. (1990). A stuttering model for paramyxovirus P mRNA editing. EMBO J. *9*, 2017–2022.

Visomirski-Robic, L.M., and Gott, J.M. (1995). Accurate and efficient insertional RNA editing in isolated *Physarum* mitochondria. RNA *1*, 681–691.

Visomirski-Robic, L.M., and Gott, J.M. (1997a). Insertional editing in isolated *Physarum* mitochondria is linked to RNA synthesis. RNA 3, 821–837.

Visomirski-Robic, L.M., and Gott, J.M. (1997b). Insertional editing of nascent mitochondrial RNAs in *Physarum*. Proc. Natl. Acad. Sci. U.S.A. *94*, 4324–4329.

Volchkov, V.E., Becker, S., Volchkova, V.A., Ternovoj, V.A., Kotov, A.N., Netesov, S.V., and Klenk, H.D. (1995). GP mRNA of Ebola virus is edited by the Ebola virus polymerase and by T7 and vaccinia virus polymerases. Virology *214*, 421–430.

Wang, S.S., Mahendran, R., and Miller, D.L. (1999). Editing of cytochrome b mRNA in *Physarum* mitochondria. J. Biol. Chem. *274*, 2725–2731.

Whalen, W.A., and Das, A. (1990). Action of an RNA site at a distance: role of the nut genetic signal in transcription antitermination by phage-lambda N gene product. New Biol. *2*, 975–991.

Xiong, Y., and Steitz, T.A. (2004). Mechanism of transfer RNA maturation by CCA-adding enzyme without using an oligonucleotide template. Nature *430*, 640–645.

Yin, Y.W., and Steitz, T.A. (2002). Structural basis for the transition from initiation to elongation transcription in T7 RNA polymerase. Science *298*, 1387–1395.

Zehrmann, A., Verbitskiy, D., Hartel, B., Brennicke, A., and Takenaka, M. (2011). PPR proteins network as site-specific RNA editing factors in plant organelles. RNA Biol. *8*, 67–70.

Transfer RNA Modification and Editing

3

Bhalchandra S. Rao and Jane E. Jackman

Abstract

Transfer RNAs (tRNAs) are critical players in gene expression due to their essential function as adaptor molecules. Yet, the true enigma in their biological function lies in their ability to consistently transport 22 distinct amino acids, as specified by their identity, to the decoding centre with extraordinarily high fidelity, yielding functional polypeptides. Previous research has indicated that mechanisms underlying the establishment of tRNA identity and high fidelity aminoacylation are tightly linked to recognition of an optimal three dimensional structure of tRNAs. Hence it is not surprising that diverse biochemical pathways of tRNA processing have evolved in all three domains of life that directly influence the functionality and structural integrity of tRNAs. The following chapter reviews tRNA modification and editing mechanisms that directly influence tRNA homeostasis in all three domains of life.

Introduction

The observation of modified nucleotides in tRNA is nearly as old as the discovery of tRNA molecules themselves. Not long after 'soluble RNA' (soon known as 'transfer RNA' to denote its ability to transfer labelled amino acids to protein in an ATP and GTP-dependent series of reactions) was discovered, it was observed that additional nucleotide components were present beyond the adenosine, guanosine, cytidine and uridine expected to constitute cellular RNA (Hoagland *et al.*, 1958). The first primary tRNA sequence determined was that of alanine-tRNA, for which Robert Holley was awarded the Nobel Prize in 1968, and which clearly demonstrated the presence of non-standard nucleotide species, such as dihydrouridine (D), *N*1-methyl guanosine (m^1G), *N*2-dimethyl guanosine (m^2G), inosine (I) and thymidine (rT) (Holley *et al.*, 1965). Rapid progress was made towards identification of new types of modified nucleotides in tRNA as more tRNA sequences became available. By the early 1970s, more than 50 modified nucleotides had already been observed in tRNA and now more than 100 nucleotide modifications of more than 60 distinct chemical types are known to occur in tRNA (Grosjean and Benne, 1998; Czerwoniec *et al.*, 2009; Grosjean, 2009; Juhling *et al.*, 2009). Moreover, many tRNA modifications are highly conserved, found at the same position in tRNAs from multiple organisms in multiple domains of life.

Although the identities of modified nucleotides in tRNA were largely known, progress towards identification of the enzymes responsible for formation of these modified species was relatively slower. Attempts to purify and characterize the enzymatic activities were hampered initially by difficulties in producing tRNA substrates for each activity (Mandel and Borek, 1963; Soll, 1971). Prior to the development of facile *in vitro* transcription techniques to produce unmodified *in vitro* tRNA transcripts of defined sequence, tRNA substrates for modification assays were purified from cells in which modification levels were decreased by limitation for essential cofactors by mutation or growth conditions. Thus, cells starved for methionine, the precursor for *S*-adenosylmethionine (SAM) could be forced to accumulate hypo-methylated tRNAs because of relatively low amounts of SAM, which serves as the methyl-group donor for many tRNA methylated nucleotides. However, a significant drawback to these *in vivo* methods is the simultaneous loss of multiple methylations on the tRNA, leading to difficulties in identifying specific methylation events catalysed by enzymes. A second factor that led to a delay in identifying enzymes associated with tRNA modification is the lack of an observable growth phenotype frequently associated with loss of the tRNA modification activity. Thus, development of genetic assays to identify mutants in tRNA-modifying enzymes was often not possible because of lack of identifiable growth defects associated with their loss. Notably, even for the few genes initially identified through genetic mutations, identification of the biochemical function of the gene in tRNA modification was not initially obvious. For example, the *GCD10/GCD14* (*TRM6/TRM61*) genes in *Saccharomyces cerevisiae* were first identified genetically due to their participation in general control of amino acid biosynthetic enzymes, and were only later shown to catalyse the essential *N*1-methyladenosine modification at position 58 of tRNA (Garcia-Barrio *et al.*, 1995; Anderson *et al.*, 1998, 2000).

Along with tRNA modifications, tRNA molecules can undergo multiple types of editing events to generate the mature tRNA molecules that function in translation. The distinction between tRNA modification and tRNA editing can be debated, since some nucleotide base changes, particularly deamination of adenosine to inosine or cytidine to uridine, are frequently classified as editing events in other cellular RNAs. For the purposes of this chapter, however, the phenomenon of tRNA editing will be defined as an event where the phosphate backbone of the initial precursor-tRNA transcript is cleaved (either during or subsequent to end-processing), and the genomically encoded nucleotides are replaced with nucleotides that are found in the final mature tRNA. The field of tRNA editing is relatively more recent, in part at least due to the fact that standard nucleotides are incorporated into tRNA as a result of tRNA editing. Thus these events are not obvious from sequencing of isolated tRNA species, but instead require knowledge of genome sequences to discern differences between mature tRNA sequences and the sequences encoded by the genome of the organism.

tRNA modification – genomic and proteomic approaches to identification of modification enzymes

The past three decades have seen an explosion in identification of tRNA modification enzymes. Before genome sequence data were available, a few isolated genes that encode modification enzymes were identified by painstaking mapping and purification techniques. These include the first known tRNA modifying enzyme, TrmA from *E. coli*, which catalyses formation of the highly conserved 5-methyluridine (also known as ribothymidine, rT) modification at position 54 of tRNA (Ny and Bjork, 1980; Ny *et al.*, 1988; Persson *et al.*, 1992). As the number of sequenced genomes increased, however, bioinformatics approaches to gene identification allowed for an extraordinarily rapid increase in the number of known tRNA modification enzymes (de Crecy-Lagard, 2007; de Crecy-Lagard *et al.*, 2007; Tkaczuk *et al.*, 2007). For this approach, identification of one family member, for example a methyltransferase involved in rRNA modification, is used to identify potential homologues that might catalyse a similar reaction with other RNA substrates, such as tRNA. This powerful approach has resulted in identification of numerous tRNA modification enzymes, from multiple yeast methyltransferases (such as Trm3, Trm4 and Trm5) up to the most recent new members of the methyltransferase enzyme family in Archaea (such as Trm14, TrmY and TrmK) (Persson *et al.*, 1997; Cavaille *et al.*, 1999; Motorin and Grosjean, 1999; Bjork *et al.*, 2001; Kalhor and Clarke, 2003; Armengaud *et al.*, 2004; Roovers *et al.*, 2004; Purushothaman *et al.*, 2005; Renalier *et al.*, 2005; Urbonavicius *et al.*, 2005; Kempenaers *et al.*, 2010; Menezes *et al.*, 2011; Chatterjee *et al.*, 2012).

An alternative approach was required, however, to identify modifying enzymes that possess little or no detectable sequence similarity to other well-studied enzyme families. To this end, the yeast biochemical genomic approach for identification of unknown genes based on known biochemical activities was also highly successful in identifying unknown tRNA modifying enzymes, including the dihydrouridine synthase enzyme family (Dus1, Dus2, Dus3, Dus4), several tRNA methyltransferases and the unusual 3′–5′ polymerase enzyme, Thg1 (Martzen *et al.*, 1999; Alexandrov *et al.*, 2002; Xing *et al.*, 2002; Gu *et al.*, 2003; Jackman *et al.*, 2003; Wilkinson *et al.*, 2007; Kotelawala *et al.*, 2008). The highly complementary nature of these proteomic and genomic approaches resulted in the identification of a large number of new tRNA modification genes over the past 20 years, with genes now identified for all examples of simple one or two-step catalysed modifications in yeast and *E. coli* (Table 3.1; Czerwoniec *et al.*, 2009). More recently, attention has turned towards identification of multisubunit enzyme complexes involved in formation of complex modifications such as wybutosine (yW), *N*6-threonyl adenosine (t^6A), and the various U_{34} wobble nucleotide modifications. Identification of the elusive enzymes that catalyse these complex modifications required more sophisticated proteomic and genetic approaches, and characterization of the biochemical activities associated with each of these relatively recently described enzymes is ongoing (Huang *et al.*, 2005; Kalhor *et al.*, 2005; Lu *et al.*, 2005; Esberg *et al.*, 2006; Noma *et al.*, 2006; Goto-Ito *et al.*, 2007; Suzuki *et al.*, 2007, 2009; El Yacoubi *et al.*, 2009; Umitsu *et al.*, 2009; de Crecy-Lagard *et al.*, 2010; Iyer *et al.*, 2010; Harris *et al.*, 2011; Kato *et al.*, 2011; Srinivasan *et al.*, 2011; Young and Bandarian, 2011).

A large majority of enzymes that have been identified using these varied approaches can be classified into relatively few groups on the basis of the related nature of the chemical reactions that they catalyse. Easily the largest group is the methyltransferases, reflecting the fact that methylated nucleotides (of all types, taken together) are the most frequent nucleotide modification and that methylations have been observed to occur at nearly every nitrogen atom on purine and pyrimidine rings, as well as the C5 carbon of pyrimidines and the 2′-hydroxyl of ribose (Fig. 3.1A). The most abundant nucleotide modification of a single type found in tRNA is pseudouridine (Ψ), and likewise, the pseudouridine synthases (Pus) constitute a second large class with members in all three domains of life and multiple enzymes encoded in most organisms (Fig. 3.1B). The deaminases and dihydrouridine synthases (Dus) families constitute relatively smaller groups of only two to four enzymes in each species where the modification is found. The deaminases include the Tad (tRNA adenosine deaminase) enzymes that catalyse the essential conversion of the wobble-decoding adenosine to inosine, as well as enzymes that catalyse conversion of cytidine to uridine in archaeal and mitochondrial tRNA.

The remaining known modification enzymes that do not readily fall into one of these general classes instead catalyse more specialized chemical reactions, frequently with a more limited number of tRNA substrates (Fig. 3.1C). A number of these modification enzymes, such as the acetyltransferase that catalyses *N*4-acetylcytidine formation (Tan1 in yeast) or the dimethylallyltransferase that catalyses *N*6-isopentenyladenosine (Mod5 in yeast) exist as single subunit enzymes in at least some species. Others, such as the enzymes that introduce the highly conserved wybutosine and queuosine bases (and their derivatives) exist as multi-subunit complexes and require multiple steps for their synthesis and/or incorporation into tRNA (Fig. 3.1D). Owing to the complexity of these chemical reactions and/or limited tRNA substrate specificity of many of these modification enzymes, this chapter will focus on the enzymology of more widespread tRNA modifications that are typically (but not always) found in multiple tRNA substrates.

Lessons from the tRNA methyltransferases – a representative enzyme family

The addition of a methyl group to RNA bases or ribose sugars is, on the surface, a relatively simple enzyme reaction. Yet, a closer look at the tRNA modifying enzymes that catalyse this seemingly simple reaction is highly informative, yielding a large amount of diversity and revealing many features of tRNA biology and biochemistry that are generalizable to other families of tRNA modifying enzymes. In yeast, 14 unique tRNA methyltransferases have been identified that catalyse all major examples of methylated nucleotides found in yeast tRNA (Table 3.1). A comparison of these methylations with those observed in other domains of life demonstrates that while many of the modifications themselves are found in all three domains of life, for example 2′-O methylation of ribose sugars, these same modifications are not always conserved at the same position in tRNA species from all three domains of life. Even for modifications that appear to be conserved, based on their

Table 3.1 tRNA methyltransferases in all three domains of life

Modification	*S. cerevisiae* (Eukarya)	*E. coli* (Bacteria)	Archaea	Methyltransferase family[h]
Nm_4	Trm13	No examples[c]	No examples	Unknown (predicted Class I)
m^2G_6	Trm14[d] (not in yeast, in other eukaryotes)	No examples	Trm14	Class I
m^1G_9, *(m^1A_9)*[a]	Trm10	No examples	SacI_1677(m^1G) TK0422 (m^1G/m^1A)	Unknown (predicted Class IV)
m^2G_{10}, *($m^{2,2}G10$)*[a]	Trm11/Trm112	No examples	TrmG10	Class I (THUMP domain)
Gm_{18}	Trm3	TrmH	C/D box RNP	Class IV (in Eukarya and Bacteria)
m^1A_{22}	No examples	(TrmK) not in *E. coli*, in other bacteria	No examples	Unknown
$m^{2,2}G_{26}$	Trm1	No examples	TrmG26	Class I
m^3C_{32}	Trm140	No examples	No examples	Unknown (predicted Class I)
Nm_{32}, Nm_{34}	Trm7	TrmJ	Trm7	Class I-Trm7 Class IV-TrmJ
$\underline{m}cm^5U_{34}$[b]	Trm9/Trm112	No examples	No examples	Unknown
$\underline{m}nm^5U_{34}$[b]	No examples	TrmC	No examples	Unknown
$c\underline{m}o^5U_{34}$[b]	No examples	CmoB (CmoA)[e]	No examples	Unknown (predicted Class I)
m^1G_{37} (m^1I, yW)	Trm5	TrmD	aTrm5	Class I-Trm5 Class IV-TrmD
m^2A_{37}	No examples	TrmG	No examples	Unknown
m^6A_{37}	No examples	TrmX	No examples	Unknown
m^5C	Trm4 (34, 40, 48, 49)[f]	No examples	PAB1947 (39, 40, 48–50)[f]	Class I
Um_{44}	Trm44	No examples	No examples	Unknown
m^7G_{46}	Trm8/Trm82	TrmB	No examples	Class I
m^5U_{54} (rT)	Trm2	TrmA, TrmFO	TrmU54[g]	Class I (except for TrmFO)
$m^1\Psi_{54}$	No examples	No examples	TrmY	
Cm_{56}, Um_{56}	No examples	No examples	aTrm56 (PAB1040)	Class IV
m^1A_{58}	Trm6/Trm61	(TrmI) not in *E. coli*, in other bacteria	TrmI	Class I

[a]Modifications indicated in parentheses and italics only occur in Archaea.
[b]Methyl group added by the methyltransferase is indicated by underlining.
[c]No examples found among sequenced tRNAs in tRNA database (Juhling *et al.*, 2009) consisting of 34 yeast tRNAs, 139 total bacterial tRNAs (46 from *E. coli*), and 76 total archaeal tRNAs.
[d]Eukaryotic Trm14 orthologues and the m^2G6 modification have been identified in other species (including humans), but m^2G6 has not been observed in *S. cerevisiae*.
[e]CmoA is a putative methyltransferase based on evidence from accumulated nucleotides in the *cmoA* deletion strain.
[f]Numbers in parentheses indicate positions at which modification is observed; all instances catalysed by the indicated multisite specific enzyme.
[g]The m^5U54 modification is so far only observed in a small number of archaeal species.
[h]Structural class of methyltransferase (Class I, Rossman or Class IV, SPOUT) for the enzymes from different domains is as indicated, where it has been determined by structural characterization. Enzymes that have not been characterized structurally are indicated as unknown; however, where a structural class has been predicted bioinformatically this is indicated.

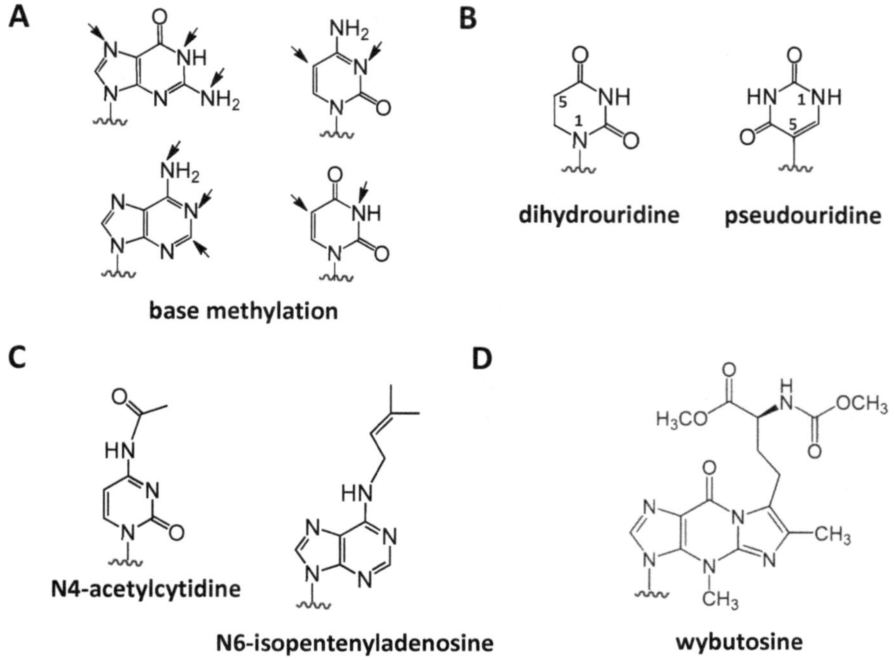

Figure 3.1 Representative examples of modified nucleotides found in tRNA. (A) Methylation of each of the four nucleotide bases can be observed in tRNA species; arrows indicate the positions of methylation for each base in tRNAs. Individual bases are only methylated at a single position, for example, no tRNA species contains a simultaneous methylation of both *N*1 and *N*2 of guanosine. (B) Representative examples of other abundant modified nucleotides, pseudouridine, a modified form of uridine, where the glycosidic linkage to *N*1 is broken and reformed to the *C*5 carbon, and which is the most abundant modified nucleotide in tRNA, and dihydrouridine, commonly found at multiple positions of tRNAs from Eukarya and Bacteria. (C) Representative examples of more rare tRNA modifications, typically found at only a single site in relatively few tRNA species. *N*4-acetylcytidine is found selectively at position 12 of leucine and serine tRNA of eukaryotes, and a handful of other tRNA species in Bacteria and Archaea, while *N*6-isopentenyladenosine occurs at position 37 of primarily selenocysteine and serine tRNA species in Eukarya and Bacteria. (D) An example of a complex tRNA modification (wybutosine) catalysed by a series of enzymes that participate in a multi-enzyme complex. Wybutosine is found exclusively at position 37 of phenylalanine tRNAs.

presence at the same position of tRNAs from all three domains of life, there can be significant differences in the extent to which the modification is observed among sequenced tRNAs. For example, an adenosine residue is nearly universally found at position 58 of tRNA, and more than half of tRNAs that have been sequenced in yeast contain the m^1A_{58} methylation of this adenosine, while only 2 of the 76 sequenced and 7 of the 139 sequenced bacterial tRNAs (including none from *E. coli*) contain this same modification (Juhling *et al.*, 2009). In general, the observed trend is for more modifications to be observed in eukaryotic systems than in the bacterial or archaeal ones, with the notable exceptions of hyperthermophilic prokaryotes, which have been observed to possess an unusually high number of modifications as judged by the relatively few tRNA sequences that are currently available. Similarly differences in the position and extent of modification are observed with other types of modified nucleotides in addition to methylations.

Differences in composition and cofactor usage

The tRNA methyltransferases are classified as members of EC group 2.1.1, reflecting their common ability to transfer one-carbon methyl groups from a donor to an acceptor. However, among this group, there are many examples of distinct enzyme chemistry and catalytic strategies. The majority of tRNA methyltransferases are the product of a single methyltransferase gene, although families differ in whether the functional form of the enzyme is monomeric, dimeric or even tetrameric. However, a few exceptions to this rule are found among the eukaryotic methyltransferases that form m^2G_{10}, m^7G_{46} and m^1A_{58}, which each contain one subunit (Trm11, Trm8 and Trm61, respectively) housing the actual catalytic activity and a second non-catalytic subunit (Trm112, Trm82 and Trm6, respectively) that is nonetheless required for activity *in vitro* and *in vivo* (Anderson *et al.*, 1998; Anderson *et al.*, 2000; Alexandrov *et al.*, 2002; Purushothaman *et al.*, 2005). For the Trm6/Trm61 complex, the Trm6 subunit is important for tRNA binding and the two subunits appear to have arisen by a gene duplication event. The precise functions of Trm112 and Trm82 are less well understood, and these proteins share no obvious sequence similarity to their catalytic partners. Trm112 appears to serve as a 'hub' protein for other tRNA-modifying enzymes, since it also stimulates catalytic activity of the Trm9 methyltransferase, although it does not necessarily bind simultaneously to both catalytic subunits (Purushothaman *et al.*, 2005).

The methyl donor for known tRNA methyltransferases is overwhelmingly observed to be *S*-adenosylmethionine (SAM), although again, exceptions to the rule are notable. The m^5U(rT) methylation at position 54 is one of the few highly conserved modifications found at the same position in tRNAs from all three domains of life (although the presence of rT is far from universal in Archaea, which generally contain pseudouridine at position 54 of tRNAs) (Juhling *et al.*, 2009). The m^5U modification is catalysed by homologous enzymes in all three domains (known as Trm2 in Eukarya, TrmA in Bacteria and TrmU54 in Archaea) (Table 3.1). Nonetheless, a number of bacterial species have been identified that lack TrmA, and instead use the methyltransferase TrmFO to catalyse methylation of U_{54}, which depends on a N_5,N_{10}-methylenetetrahydrofolate cofactor to serve as methyl donor for the reaction instead of SAM (Urbonavicius *et al.*, 2005). Interestingly, m^5U_{54} methyltransferase activity has been observed in extracts of the bacterium *G. stearothermophilus*, but neither trmA nor trmFO genes have been identified in this species, raising the possibility of yet additional mechanisms for methyltransfer to be discovered.

SAM-dependent methyltransferases are thought to have evolved along at least five different lineages, as evidenced by distinct structural folds that have been identified among structurally characterized enzymes (Schubert *et al.*, 2003). As with methyltransferases in general, tRNA methyltransferases are predominantly members of the so-called Class I enzyme family, which is characterized by the presence of a Rossman fold motif previously associated with nucleotide binding. However, a small, but growing number of tRNA methyltransferases are members of the Class IV family, characterized by the presence of a SPOUT motif, which contains a knotted structure that contributes residues to SAM binding (Anantharaman *et al.*, 2002). Indeed, this structural class is named according to its founding members SpoU(now TrmH) and TrmD, both tRNA methyltransferases. The molecular mechanisms employed by various tRNA methyltransferases have been investigated in great detail, as described elsewhere (Nakanishi and Nureki, 2005; Hou and Perona, 2010; Motorin *et al.*, 2010).

Diverse modes of tRNA recognition by tRNA-modifying enzymes

Substrate recognition is an important aspect of tRNA modification enzyme biochemistry, and efforts to delineate properties that allow tRNA modification enzymes to distinguish between highly similar tRNA substrates reveal some general principles. Aside from a handful of multisite specific tRNA modifying enzymes (including those that catalyse formation of m^5C and Ψ), distinct enzymes are generally responsible for catalysing chemically identical modifications when they are found at more than one position on the tRNA (for example, m^1G at position 9 vs. position 37, $m^{2,2}G$ at position 10 vs. position 26, m^1A at position 22 vs. position 58) (Table 3.1 and Fig. 3.2A). Thus, tRNA modification enzymes must frequently exhibit modification site specificity, and in some cases, this may even require discrimination between two identical nucleotides located adjacent to each other in a given tRNA substrate (Ochi *et al.*, 2010).

A second problem of tRNA recognition arises

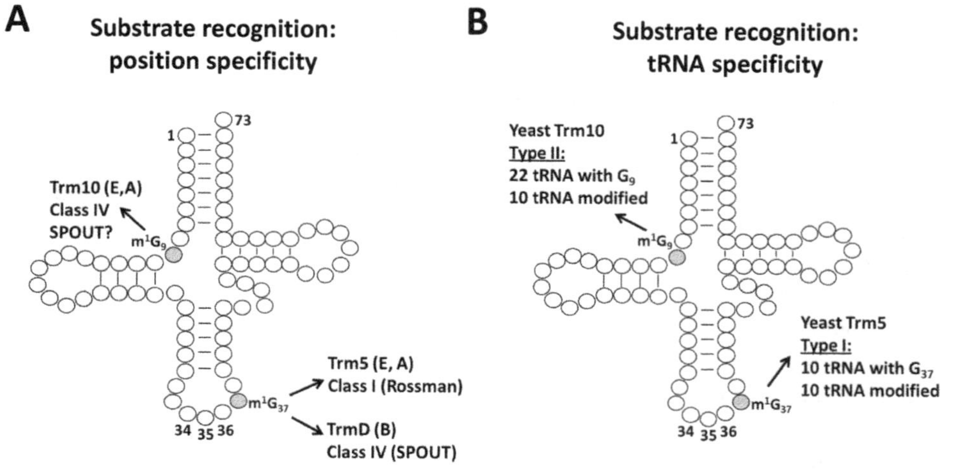

Figure 3.2 Modification enzyme substrate specificity. (A) Modifications that occur at different sites on tRNAs are catalysed by different enzymes. In the example, the m^1G modification is found at two different positions of yeast tRNAs, and two different tRNA methyltransferases (that do not share significant sequence similarity with each other) catalyse the formation of m^1G. Likewise, even when the same modification occurs at the same position on the tRNA (such as the m^1G_{37} in Eukarya and Bacteria), the methylation reaction may be catalysed by distinct methyltransferase enzymes that do not share any identifiable evolutionary relationship. (B) Modification enzymes can be further classified according to their tRNA substrate specificities into Type I (catalyse modification of any tRNA substrate if the correct target nucleotide is at the position to be modified, for example, see Trm5) or Type II (will catalyse modification of a subset of tRNA substrates from among the larger group of tRNAs that contain the correct target nucleotide, for example, see Trm10).

for some modification enzymes in terms of the tRNA isotype specificity for a given modification. tRNA modification enzymes can be divided into three groups along these lines. Type I enzymes are those for which the mere presence of the correct nucleotide at the position to be modified is sufficient to specify activity of the enzyme, and thus the modification enzyme could be viewed as relatively insensitive to the identity of the tRNA (Fig. 3.2B). Examples of these types of enzymes are Trm5, which modifies 10 out of 10 G_{37}-containing yeast tRNAs to m^1G, and Trm1, which modifies 22/23 G_{26}-containing yeast tRNAs to $m^{2,2}G$ (Juhling *et al.*, 2009). Type II enzymes are characterized by more complicated tRNA substrate recognition properties, since among tRNA species that contain the correct target nucleotide at the position to be modified, only a subset of tRNAs are actually substrates for the action of the enzyme (Fig. 3.2B) (Juhling *et al.*, 2009). This group contains enzymes such as the Trm6/Trm61 methyltransferase, which catalyses m^1A_{58} modification of 23 out of 34 sequenced yeast tRNAs, and the Trm10 methyltransferase, which catalyses m^1G_9 modification of only 10 out of 23 G_9-containing yeast tRNAs. Interestingly, enzymes of distinct tRNA recognition types can be found even among homologous groups, since the TrmH/Trm3 enzyme family contains both Type I Gm_{18}-methyltransferase (from *T. thermophilus*) and Type II members of the enzymes family (enzymes from *E. coli* and yeast) (Hori *et al.*, 2003). At the extreme end of the spectrum of tRNA substrate specificity are the type III enzymes, which catalyse a specific modification to a single tRNA isoacceptor. Examples of these enzymes are the Tyw enzymes in yeast that catalyse wybutosine formation specifically on $tRNA^{Phe}_{GAA}$ substrates, or the Thg1 enzyme in yeast that catalyses selective addition of a single G_{-1} residue to $tRNA^{His}_{GUG}$ substrates (Juhling *et al.*, 2009). Perhaps not surprisingly, the rules for substrate selection for these single-tRNA specific enzymes have been observed in some cases to mirror those of aminoacyl-tRNA synthetases, and specific identity elements, such as the tRNA anticodon, are used to specify which tRNA substrate is modified (Jackman and Phizicky, 2006a).

Evolution of tRNA-modifying enzymes

The tRNA methyltransferases are also an abundant source of examples of diverse patterns of evolution of tRNA modification enzyme activities. Some tRNA modifications have been considered 'primordial' due to their presence in tRNA species from all three domains of life, thus indicating that they may have been present very early in the evolution of tRNA (Grosjean, 2009). These include the m^1A_{58}, m^5U_{54}, m^1G_{37}, Nm_{32} and Gm_{18} modifications. Accordingly, the enzymes that catalyse m^1A_{58} (Trm2, TrmA and TrmU54) and m^5U_{54} (Trm6/61 and TrmI) modifications share sequence homology within each group that suggests common evolutionary origins (Table 3.1). Yet the m^1G_{37} and Gm_{18} modifications are catalysed by enzymes in different domains of life that appear to have evolved by convergence, rather than from a common ancestor. The m^1G_{37} methyltransferase Trm5 (found in Eukarya and Archaea) is a member of the common Rossman fold (Class I) methyltransferase enzyme family, while the analogous methyltransferase TrmD in Bacteria is a member of the SPOUT (Class IV) methyltransferase family (Bystrom and Bjork, 1982; Bjork *et al.*, 2001). The Cm/Um_{32} methyltransferases follow an identical pattern, with Rossman fold enzymes (Trm7) in Eukarya and Archaea and SPOUT enzymes (TrmJ) in Bacteria (Pintard *et al.*, 2002; Purta *et al.*, 2006). Finally, while the Trm3 and TrmH Gm_{18} 2'-O methyltransferases are related members of the SPOUT enzyme family, the only examples of Gm_{18} modification in Archaea identified to date are catalysed by ribonucleoprotein enzymes of the BoxC/D RNA-guided methyltransferase enzyme family (Persson *et al.*, 1997; Cavaille *et al.*, 1999; Ziesche *et al.*, 2004). The occurrence of modified nucleotides themselves are generally consistent with the relatively closer relationship that has been proposed between Eukarya and Archaea, since many modifications are common to Eukarya and Archaea, but not found in any sequenced tRNAs from Bacteria (such as m^1G_9, m^2G_{10}, $m^{2,2}G_{26}$, and m^5C at multiple positions). However, a few notable exceptions to this rule exist, particularly the widespread observation of dihydrouridine (D) in tRNAs from Eukarya and Bacteria, while no examples of D modification in Archaea are described among 76 total sequenced archaeal tRNA species (Juhling *et al.*, 2009).

Biological function of modified nucleotides in tRNA

The ubiquitous and often highly conserved nature of nucleotide modifications in tRNA could be interpreted to suggest that modifications are critical for the function of tRNA. However, as the gene products responsible for introducing each modification were identified, and subsequently tested for essentiality, the results indicated that the role of tRNA modifications was more complex. Several tRNA modification enzymes have been identified that are essential for viability; (for example, the m^1A_{58} methyltransferase in Eukarya, m^5U_{54} and m^1G_{37} methyltransferases in Bacteria and I_{34} deaminase in all domains of life) (Persson *et al.*, 1992; Garcia-Barrio *et al.*, 1995; Gerber and Keller, 1999; Bjork *et al.*, 2001). In the case of I_{34}, the essential nature of the modification is clearly due to the need for A-to-I alteration to expand the number of codons recognized by a single tRNA isoacceptor by wobble decoding. Likewise, loss of the m^1G_{37} modification leads to significant +1 frameshifting, and although the eukaryotic m^1G_{37} modification enzyme Trm5 is not strictly essential in yeast, the *trm5Δ* strain grows extremely poorly (Bjork *et al.*, 1989, 2001). The role of the m^1A_{58} methylation is more subtle, lack of this modification upon deletion of *trm6* or *trm61* causes death due to degradation of hypomodified $tRNA^{Ini}$, but other tRNAs that normally contain the modification are not affected, and thus viability can be restored by overexpression of the unmodified initiator tRNA in the deletion strain (Anderson *et al.*, 1998; Kadaba *et al.*, 2004). For TrmA-catalysed m^5U_{54} modification, point mutants in *trmA* that result in complete lack of the modification, but remain viable, have been identified, suggesting that the essential function of TrmA in *E. coli* is separate from the tRNA modification function of the gene (Persson *et al.*, 1992). For other modifications in the tRNA anticodon loop and stem, effects on growth are frequently observed upon deletion of the gene that catalyses the modification. The causes of these growth defects have been comprehensively reviewed by others, and largely stem from effects on translation and fidelity of decoding (Phizicky and Alfonzo, 2010; Phizicky and Hopper, 2010). In the remainder of the tRNA body, a few examples of modified nucleotides that exert substantial effects on tRNA structure and/or folding have also been identified (for example, the m^1A_9 in mitochondrial $tRNA^{Lys}$ that prevents formation of an alternative incorrect structure) (Helm and Attardi, 2004; Voigts-Hoffmann *et al.*, 2007; Motorin and Helm, 2010).

For the majority of tRNA modifications found throughout the rest of the tRNA body, however, deletion of the enzyme(s) responsible for formation of any single modified nucleotide does not cause a readily observable growth defect. Insight into the biological functions of these types of modified tRNA nucleotides came from the recent observation of synthetic lethal effects due to loss of more than one modification in yeast (Alexandrov *et al.*, 2006). Initially, an array of yeast strains was generated that contained pairwise combinations of deletions of nine different yeast tRNA modification genes. The tested genes all showed at least one synthetic growth defect in combination with another modification gene, suggesting that, while loss of a single modification could be compensated for, loss of multiple different combinations of modifications could affect tRNA function, specifically by causing the degradation of a single hypomodified tRNA substrate (Alexandrov *et al.*, 2006; Kotelawala *et al.*, 2008). In this respect, the effects of the loss of multiple modifications mirror the situation that occurs with loss of m^1A_{58}, since in both cases only one out of many possible substrate tRNAs appears to be targeted for degradation. However, the pathways involved with degradation of the hypomodified tRNA substrates in each case are distinct. In the case of the m^1A_{58}-lacking $tRNA^{Ini}$, the tRNA becomes polyadenylated and is degraded in a Trf4(or Trf5)/Air1/Air2/Mtr4 (TRAMP)-dependent process requiring the participation of the exosome, while tRNAs lacking combinations of core mutations are all degraded due to their lack of structural stability by a rapid tRNA decay (RTD) pathway involving the 5′–3′ exonucleases Rat1 and Xrn1 (Kadaba *et al.*, 2004; Chernyakov *et al.*, 2008; Whipple *et al.*, 2011). These unexpected insights into essential functions for tRNA modifications are exciting and have provoked many to take a new look at the impact and significance of modified nucleotides in tRNA for biology.

tRNA editing

The phenomenon of tRNA editing was discovered much more recently than tRNA modification; the first example of editing of precursor tRNA transcripts, by removal and replacement of nucleotides in the tRNA 5′-end, was reported in 1993, and was soon followed by the identification of several additional types of tRNA editing activities (Lonergan and Gray, 1993a). Several of these editing reactions occur throughout the body of the tRNA, including insertion of nucleotides to create canonical stem and loop structures and deamination of C to U at specific sites to create basepairs or change decoding properties of certain tRNAs (Janke and Paabo, 1993; Marechal-Drouard *et al.*, 1993, 1996; Antes *et al.*, 1998; Alfonzo *et al.*, 1999). However, since tRNAs are not the only RNA substrates for insertion-type editing in protozoan mitochondria, and C-to-U deamination was already considered above as a type of tRNA modification reaction, the following discussion is focused largely on editing at tRNA 5′- and 3′-ends, which in several cases appears to utilize novel tRNA-specific enzymatic activities. Because an editing reaction results in incorporation of one of the standard four RNA nucleotides (as opposed to a chemically modified nucleotide) into the tRNA, evidence that an editing reaction has occurred requires comparison of mature tRNA sequences with those originally encoded by the genome. Here we review the progress of recent efforts towards categorizing the types of editing reactions that occur and identifying enzymes that appear to be involved in catalysing them.

5′-End editing reactions – observation of 5′-editing in protozoan mitochondria

5′-tRNA editing is defined by nucleotide changes made in the 5′ half of the acceptor stem of a tRNA. These forms of editing have so far only been observed in mitochondrial tRNAs (mt-tRNA) in a variety of protozoan species. Correct Watson–Crick basepairing of nucleotides in the acceptor stems of tRNAs is required for proper folding of tRNAs and recognition by most aminoacyl-tRNA synthetases (Rich and RajBhandary, 1976; Rould *et al.*, 1989). Hence, the presence of mis-paired nucleotides in the mitochondrially encoded sequences of tRNAs in the amoeboid protist, *Acanthamoeba castellanii*, was the first clue towards identification of a replacement-type 5′-tRNA editing pathway in this organism (Lonergan and Gray, 1993a). Mitochondrial genome sequencing in *A. castellanii* initially revealed a cluster of five tRNA genes, four of which ($tRNA^{Met1}$, $tRNA^{Ala}$, $tRNA^{Asp}$ and $tRNA^{Met2}$), encoded non-Watson–Crick basepairs in any of the first three basepairs of the acceptor stem. Yet, primary RNA sequences of these tRNAs obtained by dideoxy sequencing revealed nucleotide changes, all of which appeared to involve reversions of the mismatched nucleotides to correctly basepairing nucleotides. Interestingly, these changes were restricted to the 5′ half with no changes observed in the 3′ half of the tRNA. Soon after this initial observation, 12 of the 15 mitochondrial encoded tRNAs in *A. castellanii* were demonstrated to undergo 5′-tRNA editing to replace mismatched nucleotides encoded by the genomic sequences, thus yielding functional tRNAs with completely basepaired acceptor stems (Lonergan and Gray, 1993a,b; Price and Gray, 1999a).

Biochemical characterization of the 5′-editing activity was initially carried out using fractionated mitochondrial extracts, which demonstrated that the 5′-editing reaction appeared to involve at least two steps. First, the mismatched bases at the 5′-end of the encoded tRNA were removed (Lonergan and Gray, 1993a). Then the correct Watson–Crick basepaired nucleotides were added to the 5′-truncated tRNA, apparently using a 3′–5′ templated polymerase activity that could synthesize nucleic acids in the opposite direction of all known polymerases (Lonergan and Gray, 1993a; Price and Gray, 1999b). The identity of such 3′–5′ polymerases was not revealed until the discovery of members of the Thg1 enzyme superfamily, as described below.

Surprisingly, the editing machinery in *A. castellanii* was observed to 'repair' a $U_3:U_{70}$ mismatch to an $A_3:U_{70}$ Watson–Crick basepair in $tRNA^{Ala}$. This observation is intriguing as a $G_3:U_{70}$ wobble is a conserved identity element for $tRNA^{Ala}$, yet sequencing data for $tRNA^{Ala}$ clones indicated the presence of a stable, mature and edited $tRNA^{Ala}$ with a CCA end suggesting the presence of the edited tRNA among a functional pool of translation-competent $tRNA^{Ala}$ in the mitochondria of *A. castellanii*. Although the editing activity also repaired the $U_3:G_{70}$ wobble basepair to a $C_3:G_{70}$ pair in $tRNA^{Leu2}$, the $U_4:G_{69}$ wobble in $tRNA^{Ala}$ is not replaced with a Watson–Crick basepair (Lonergan and Gray, 1993a; Price and Gray, 1999a). Whether this position restriction is due to the nuclease, the nucleotide addition enzyme, or both remains an open question requiring further biochemical characterization.

Since the initial discovery of the 5′-tRNA editing activity in *A. castellanii*, a number of other protozoan species have been identified in which a similar type of editing reaction is likely to occur. Inspection of the genomically encoded mt-tRNA sequences from newly available protozoan mitochondrial genomes has provided a number of new candidate species that appear likely to require 5′-editing to produce functional tRNAs, as inferred by the presence of varied mismatched nucleotide bases in any of the first three positions of the tRNA acceptor stem (Ogawa *et al.*, 2000; Takano *et al.*, 2001; Laforest *et al.*, 2004; Jackman *et al.*, 2012). However, the number of organisms in which 5′-editing has been verified biochemically by sequencing of mature tRNA remains relatively small, consisting of only *Spizellomyces punctatus*, *Harpotrichium* (2 species), *Monoblepharella* and *Physarum polycephalum* (Laforest *et al.*, 1997; Laforest *et al.*, 2004; Gott *et al.*, 2010). Even taking into account the larger number of species where editing appears likely to occur but has not yet been confirmed biochemically, 5′-tRNA editing does not appear to be a ubiquitous process.

Although most fungi do not encode predicted 5′-tRNA editing substrates, 5′-tRNA editing does occur in the mitochondria of *S. punctatus* (a member of the Chytridiomycota family), and this system has been useful for additional characterization of the biochemical activities associated with the process (Laforest *et al.*, 1997; Bullerwell and Gray, 2005). The mitochondria of *S. punctatus* encode only eight tRNAs ($tRNA^{Leu}$, $tRNA^{Gln}$, $tRNA^{Tyr}$, $tRNA^{Lys}$, $tRNA^{Met}$, $tRNA^{Asp}$, $tRNA^{Trp}$ and $tRNA^{Pro}$), all of which contain mismatches in the acceptor stem, most of which would be predicted to completely disrupt correct folding of the tRNA molecules. Sequencing of the mature tRNAs revealed that all six tested tRNAs were at least edited at one position, and the mechanism for this 5′-tRNA editing appeared to be very similar to that seen in *A. castellanii*. Some of the major biochemical features common to 5′-tRNA editing in *S. punctatus* and *A. castellanii* include (1) addition of correct Watson–Crick basepaired nucleotides in the 3′–5′ direction, (2) use of the 3′-arm of the tRNA acceptor stem as a template for addition of the correct nucleotides and (3) restriction of editing to the first three positions of the 5′-arm of the tRNAs (Laforest *et al.*, 1997; Price and Gray, 1999a,b; Bullerwell and Gray, 2005). These similar biochemical characteristics could imply a common evolutionary link in the development of the 5′-tRNA editing system. Yet *A. castellanii* and *S. punctatus,* which belong to phyla *Amoebozoa* and *Chytridiomycota* respectively, are quite distantly related phylogenetically (Laforest *et al.*, 2004). This observation raises the possibility of independent evolution of similar editing activities in these two lineages.

Physarum polycephalum is an extensively studied model system for the phenomenon of RNA editing (Grosjean and Benne, 1998; Gott, 2003; Byrne *et al.*, 2007). Various RNA editing events, including nucleotide insertions and deletions, U to C substitutions and C to U conversion have been demonstrated to occur in mitochondrial mRNA transcripts (Mahendran *et al.*, 1991; Gott *et al.*, 1993; Takano *et al.*, 2001). These RNA editing events generate open reading frames in primary transcripts that would otherwise lead to shorter polypeptides due to presence of aberrant premature stop codons. Therefore, it was perhaps not surprising to detect similar editing reactions occurring with structural mitochondrial RNAs, such as rRNA and tRNA, in *P. polycephalum*. Recently, the types of editing reactions in *P. polycephalum* were expanded to also include 5′-tRNA editing of the nucleotide replacement type.

Two methionyl-tRNAs, $tRNA^{Met1}$ and $tRNA^{Met2}$ in *P. polycephalum* contain genomically encoded mismatched basepairs at the 1:72 position. $tRNA^{Met1}$ has a U_1:C_{72} mismatch and $tRNA^{Met2}$ is C_1:C_{72} (Gott *et al.*, 2010). Although a mismatch at the 1:72 position is an identity element for initiator $tRNA^{Met}$ in bacteria and organelles, the hypothesis that these mismatches are repaired to restore correct basepairing in the acceptor stem of the methionyl-tRNAs was largely based on the observation that $tRNA^{Met2}$ is transcribed as a part of a longer transcript wherein the 3′-end of the upstream non-coding RNA, ppoRNA, overlaps with the 5′-end of $tRNA^{Met2}$ by a single nucleotide (Fig. 3.3). 3′-end sequencing of the mitochondrial ppoRNA suggested that RNA processing to generate the individual transcripts would generate a 5′-truncated (lacking G_1) $tRNA^{Met2}$ (Bullerwell *et al.*, 2010; Gott *et al.*, 2010). Using tRNA circularization and sequencing approaches, it was shown that the mitochondrial $tRNA^{Met2}$ indeed undergoes 5′-tRNA editing wherein the missing nucleotide is added to $tRNA^{Met2}$, restoring the G_1:C_{72} basepair, in a manner similar to that observed in *A. castellanii.* Subsequently, investigation of the other $tRNA^{Met}$ transcript revealed that $tRNA^{Met1}$ was also a substrate for 5′-tRNA editing. Circularization and cloning resulted in clones that either contained the non-encoded G_1, or that were truncated by one base at the 5′-end, with none of the clones containing the genomically encoded C. This occurrence can only be explained by the involvement of a 5′–3′ nuclease that removes the mismatched base for addition by a 3′–5′ nucleotidyltransferase, thus extending its similarity to *Acanthamoeba*-like 5′-tRNA editing, and suggesting that multiple types of nucleolytic events could lead to 5′-truncated tRNAs that could be substrates for the replacement type editing activity. Given the precedent for co-transcriptional editing of other transcripts in *P. polycephalum,* the possibility of an alternative mechanism that could account for the edited 5′-ends of $tRNA^{Met1}$ and $tRNA^{Met2}$ through insertion of a G in the longer RNA transcript prior to RNA processing and cleavage must be considered. But conclusive evidence for absence of co-transcriptional editing coupled with *in vitro* nucleotide incorporation based 'editing' assays prove that generation of mature $tRNA^{Met1}$ and $tRNA^{Met2}$, with fully

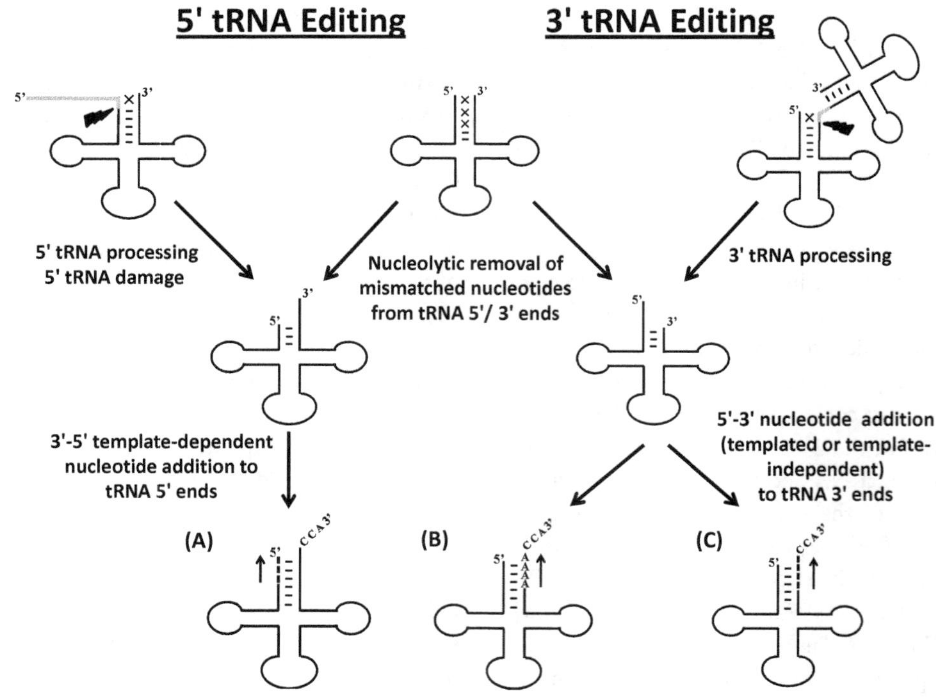

Figure 3.3 Model for nucleotide-replacement type tRNA 5′ and 3′ editing pathways. The left half of the figure shows 5′-tRNA editing, which involves nucleotide changes that occur in the 5′-half of the tRNA aminoacyl-acceptor stem. Removal of mismatched bases from the tRNA 5′-end occurs via 5′exo/endonuclease activity (in *A. castellanii*, *S. punctatus* and *P. polycephalum*) or via 5′ – tRNA processing (lightning bolt; in *P. polycephalum*) of upstream RNA in a primary transcript. (A) The 5′-truncated tRNAs are repaired using a template dependent 3′–5′ nucleotide addition activity (dashed line; *A. castellanii*, *S. punctatus* and *P. polycephalum*). This biochemical activity is consistent with 3′–5′ polymerase activities catalysed by TLPs. The right half of the figure shows 3′-tRNA editing, where the nucleotide changes occur in the 3′-half of the tRNA aminoacyl-acceptor stem. Removal of mismatched nucleotides and generation of 3′-truncated tRNAs is catalysed by nucleases and/or 3′-tRNA processing (indicated by the lightning bolt) of bicistronic tRNA transcripts with overlapping sequences (indicated by the grey bar; extreme right). Nucleotide addition(s) to the 3′-ends occur in the 5′–3′ direction and are predicted to be catalysed by (B) a template-independent RNA polymerase (Poly A polymerase or other non-templated polymerase activity; e.g. *E. herklotsi, L. bleekeri and H. sapiens*) or (C) a template-dependent RNA dependent RNA polymerase (dashed line; *L. forficatus* and *S. ecuadoriensis*).

basepaired acceptor stems are likely a product of nucleotide replacement type 5′-tRNA editing (Gott *et al.*, 2010).

Mechanisms of 5′-tRNA editing

Characteristics of the nuclease component

The non-Watson–Crick basepaired mismatches repaired as a part of 5′-tRNA editing have so far always been found within the first one to three positions of the aminoacyl-acceptor stem in substrate tRNAs. Thus, the first step for 5′-tRNA editing would require removal of the mismatched nucleotides prior to 3′–5′ nucleotide addition, although the enzyme(s) that catalyse the nuclease reaction remain unknown (Fig. 3.3). One candidate for removal of the mismatched nucleotides is the 5′-tRNA processing enzyme ribonuclease P (RNase P). Aberrant cleavage by RNase P has been previously observed, since the enzyme is known to catalyse cleavage of the leader sequence between the –2 and –1 position of bacterial $tRNA^{His}$ species (instead of between –1 and +1), in order to preserve the essential G_{-1} identity element for this tRNA (Orellana *et al.*, 1986). It is conceivable that a similar RNase P-mediated miscleavage in the acceptor stem of tRNAs could lead to removal of the mismatched nucleotides, although the case of *P. polycephalum* $tRNA^{Met2}$ may represent an alternative pathway in which the 5′-nucleotide(s) are removed as a consequence of maturation of an upstream RNA (Bullerwell and Gray, 2005; Gott *et al.*, 2010; Fig. 3.3).

In the case of 5′-tRNA editing observed in *A. castellanii* and *S. punctatus*, most tRNA sequencing clones showed either a low abundance or absence of unedited tRNAs and none were partially edited. However, a significant degree of partial editing was observed in tRNAs from *Monoblepharella* and *Harpochytrium*, where only some of the 5′-mismatched nucleotides had been repaired (Laforest *et al.*, 2004). These data suggest the use of exonucleolytic removal of the mismatched bases as opposed to an endonucleolytic process by arguing that in the case of an endonuclease activity either all or none of the first three bases would be removed to enable further nucleotide addition. This argument is further supported by the observation that, in most instances, the partial editing of tRNAs involves the third position of the tRNA (Laforest *et al.*, 2004). Yet even though these data seem to favour an exonucleolytic cleavage pathway for removal of the mismatches, the identification of the actual mechanism awaits identification of the necessary gene products.

Identification of 3′–5′ polymerases that could function in 5′-editing

The novelty of nucleotide addition in 5′-tRNA editing was due to the fact that template-dependent nucleotide addition to the 5′-end of tRNAs occurred in the 3′–5′ direction, opposite to all known DNA and RNA polymerases (Lonergan and Gray, 1993a; Price and Gray, 1999b). The only known enzymes that have been demonstrated to catalyse 3′–5′ nucleotide addition belong to the $tRNA^{His}$ guanylyltransferase (Thg1) enzyme family (Jackman and Phizicky, 2006b; Abad *et al.*, 2010). Although Thg1 was first identified in *S. cerevisiae* as the enzyme responsible for the post transcriptional addition of an essential guanosine residue at the −1 position of $tRNA^{His}$ (Gu *et al.*, 2003), Thg1 family members had been identified in some archaea and bacteria where the function of the enzyme was apparently unrelated to $tRNA^{His}$ maturation (Rao *et al.*, 2011). Biochemical characterization of the bacterial and archaeal family members suggested that these enzymes (known as Thg1-like proteins (TLPs)) instead participate in repair of 5′-truncated tRNAs, using their 3′–5′ polymerase activity to restore a fully basepaired aminoacyl-acceptor stem, as is seen in 5′-editing (Rao *et al.*, 2011). Importantly, unlike yeast Thg1, which is highly specific for addition to $tRNA^{His}$ over other tRNA substrates, the bacterial and archaeal TLPs exhibited several properties that closely paralleled those needed for the 3′–5′ polymerase component of the 5′-editing enzymes. First, the bacterial TLP from *B. thuringiensis* (BtTLP) exhibited a catalytic preference for addition of nucleotides to tRNA substrates that contain 5′-end truncations at positions 1–3, as opposed to addition to full-length tRNAs. Second, BtTLP catalysed Watson–Crick-dependent addition of nucleotides to the 5′-end of the acceptor stem using the nucleotides in the 3′ half of the stem as a template. Third, unlike yeast Thg1, which is highly specific for addition to $tRNA^{His}$, BtTLP catalysed nucleotide addition to other 5′-truncated tRNAs, including $tRNA^{Phe}$, with comparable steady state kinetic rates, suggesting a broader tRNA substrate specificity for these enzymes (Rao *et al.*, 2011).

Discovery of TLP genes in protozoa

Since 5′-tRNA editing substrates have not yet been identified in *B. thuringiensis* or other bacteria, bacterial TLPs have been hypothesized to play a role in tRNA 5′-end repair that is beyond the scope of this chapter. Yet these observations provided the foundation for investigating the functions of TLPs in organisms that are known to catalyse 5′-tRNA editing. For these studies, the eukaryotic slime mould *Dictyostelium discoideum* was used as a model organism, since 5′-tRNA editing was predicted to occur in this species based on the presence of non-Watson–Crick mismatches in 9 out of its 18 mitochondrially encoded tRNAs (Ogawa *et al.*, 2000; Abad *et al.*, 2010). These mt-tRNA mismatches are similar to the ones observed in *A. castellanii* and *S. punctatus* (mismatches in multiple tRNAs at positions 1–3), and thus are predicted to require similar enzymatic activities to restore functional mt-tRNA. *D. discoideum* also has a well-annotated genome from which one Thg1 (DdiThg1) and three different TLP genes (DdiTLP2, DdiTLP3 and DdiTLP4) could be readily identified (Abad *et al.*, 2010). Although all four Thg1 family enzymes in *D. discoideum* are nuclear encoded, TLP2 and TLP3 are predicted to contain mitochondrial targeting sequences that could indicate possible function in the mitochondria, where 5′-editing occurs. Although the expression and subcellular localization of DdiThg1 and the three DdiTLPs in *D. discoideum* is still under investigation, the *in vitro* biochemical activities of two of the TLPs (DdiTLP3 and DdiTLP4) suggest that either or both of these two enzymes are capable of participating in 5′-editing by virtue of their abilities to repair 5′-truncated mt-tRNA substrates using 3′–5′ polymerase activity (Abad *et al.*, 2010). Recently, two nuclear encoded TLPs (one of which contains a mitochondrial targeting sequence) have also been identified in *A. castellanii*; testing of the activities of these predicted gene products will reveal whether they possess similar catalytic activities. Surprisingly, the *A. castellanii* genome does not encode an identifiable homologue of the canonical eukaryotic Thg1 that plays the essential role in G_{-1} addition to $tRNA^{His}$ (Jackman *et al.*, 2012). Hence it will be interesting to determine whether there is an unusual division of labour between $tRNA^{His}$ maturation and 5′-tRNA editing catalysed by TLPs in this organism.

3′-End editing reactions

The addition of nucleotides to the 3′-end of the tRNA acceptor stem during 3′-editing is catalysed in the 5′–3′ direction, as is observed with all known DNA and RNA polymerases. Thus, while the presence of previously unknown enzymatic activities is less likely for 3′-editing, these systems are equally remarkable due to their ability to apparently take advantage of a variety of existing tRNA processing enzymes to ensure the production of functional tRNA species for mitochondrial translation.

Mitochondrial genomes commonly reveal a high degree of condensation in the genome and generally encode only a relatively small number of transcripts. These transcripts are often transcribed as larger, polycistronic molecules, and further processed to give rise to individual coding and/or structural RNAs. Owing to this high level of condensation, in multiple cases ranging from protozoa to humans, there are instances of mt-RNAs that have only a single nucleotide boundary between two functional RNAs or even of two adjacent RNA molecules whose sequences are overlapping by one or more nucleotides. Therefore, optimal post-transcriptional processing of these RNAs is essential for proper function and impaired processing events can cause deleterious effects on cell physiology.

As the name suggests, 3′-tRNA editing refers to editing events that lead to changes in the 3′-half of the acceptor stem of tRNAs, thus giving rise to a sequence that is different from that encoded by the genomic DNA sequence. The focus of this chapter will be on cases of 3′ – nucleotide replacement type tRNA editing in mitochondria, now known to occur in several species.

Template-independent 3′-tRNA editing reactions

The first observation of 3′-tRNA editing came from reviewing the mitochondrially encoded sequences of tRNAs in land snails. Similar to the protozoan examples described above, mitochondrial genome sequencing of land snails *Cepaea nemoralis, Euhadra herklotsi* and *Albinaria turrita* revealed the unusual presence of one to four mismatches in the acceptor stems of tRNAs (Yokobori and Paabo, 1995; Terrett *et al.*, 1996; Yamazaki *et al.*, 1997). After further investigation, a novel editing mechanism was identified that could repair the mismatches and restore the tRNAs to a functional pool of RNAs. In *E. herklotsi*, the 3′ end of the gene for $tRNA^{Gly}$ overlaps with the 5′ end of the downstream gene for $tRNA^{His}$ by four nucleotides while that of $tRNA^{Lys}$ overlaps with subunit I of cytochrome oxidase mRNA by six nucleotides. The inferred acceptor stem of $tRNA^{Gly}$ contains two mismatches while that of $tRNA^{Lys}$ contains three. Similarly, although no overlap of $tRNA^{Tyr}$ sequence was observed with a downstream gene, it too contained three mismatches in its acceptor stem. Analysis of the cDNA clones derived from circularized *in vivo* isolated tRNAs indicated that the mismatches in $tRNA^{Gly}$, $tRNA^{Lys}$ and $tRNA^{Tyr}$ were all altered to adenosine residues, which also restored basepairing to most positions in the aminoacyl acceptor stem (Yokobori and Paabo, 1995). The nucleotide changes observed with the tRNAs were not restricted to just the basepairs in the acceptor stem, but also involved a change of the unpaired discriminator nucleotide to an A_{73}. Interestingly, the genomic sequence of $tRNA^{His}$ suggests the presence of two additional mismatches at the base of the acceptor stem, but the cDNA sequencing indicated that these were not repaired. The consequences of this inability to repair these two mismatches in $tRNA^{His}$ are not known.

The exact mechanism for the 3′-tRNA editing of the substrates in land snails is currently unknown although the final products of the editing reaction provide clues towards the identity of the potential key players. In the case of $tRNA^{Gly}$ and $tRNA^{Lys}$, the removal of the mismatched bases from the 3′-end of the tRNA acceptor stem seems to occur via 5′-end processing of the downstream RNA (Fig. 3.3). Hence $tRNA^{His}$ and the mRNA for subunit I for cytochrome oxidase receive their 5′ regions from the upstream $tRNA^{Gly}$ and $tRNA^{Lys}$ respectively, thus 3′-truncating $tRNA^{Gly}$ and $tRNA^{Lys}$ (Yokobori and Paabo, 1995). Since $tRNA^{Tyr}$ is the last gene in the precursor transcript, the mechanism for removal of the 3′-end mismatched nucleotides from this tRNA is still unknown, although in *C. nemoralis* and *A. turrita*, the $tRNA^{Tyr}$ gene overlaps with $tRNA^{Trp}$ at its 3′ end (Terrett *et al.*, 1996).

Since all the nucleotide changes observed in the substrate tRNAs involve changes to A residues, the nucleotide addition step of this 3′-editing reaction is hypothesized to be catalysed by an enzyme of the poly-A polymerase (PAP) family (Fig. 3.3). The possibility of a template-dependent RNA polymerase reaction could be considered, since the overwhelming majority of the observed 3′-adenosine additions also restore basepairing in the acceptor stem. However, the observations that all of the unpaired discriminator nucleotides are also converted to A residues, and that an A–A mismatch is generated at the 1:72 position of $tRNA^{Tyr}$ from a pre-existing A–C mismatch both suggest that the nucleotide adding enzyme is a template-independent polymerase, such as a PAP-type enzyme (Yokobori and Paabo, 1995). The presence of the 3′-CCA on these edited tRNAs indicates that they are restored to a pool of translation-competent tRNAs that are aminoacylated. Yet, the mechanisms responsible for controlling the extent of A additions to tRNA 3′-ends and/or a role for additional processing enzymes that could remove excess nucleotides to generate an optimal 3′-end for CCA addition are questions that require further investigation.

Similar to the phenomenon observed in land snails, 3′-tRNA editing has also been documented in the mitochondrially derived tRNA in other species, including the squid *Logilo bleekeri* as well as higher eukaryotes such as humans and chickens (Tomita *et al.*, 1996; Yokobori and Paabo, 1997; Reichert *et al.*, 1998; Reichert and Morl, 2000; Levinger *et al.*, 2004). As with snails, most of these cases appear to involve overlapping mt-tRNA genes in which the RNA processing machinery favours cleavage of the primary transcript such that the downstream tRNA retains the encoded bases at its 5′-end, thus giving rise to a 3′-truncated tRNA that is repaired by an editing mechanism (Tomita *et al.*, 1996; Reichert *et al.*, 1998; Fig. 3.3). However, there are some differences between these various organisms related to the apparent nature of the nucleotide addition activity associated with each example of editing. In *L. bleekeri*, the nucleotide additions that replace the two nucleotides lost from the upstream tRNA (mt-$tRNA^{Tyr}$) are predominantly A residues, suggesting a PAP-type enzyme is involved, similar to that proposed for snails (Tomita *et al.*, 1996). The human mitochondrial $tRNA^{Tyr}$ 3′-end overlaps with the 5′-end of $tRNA^{Cys}$ by a single adenosine residue that becomes part of the downstream $tRNA^{Cys}$ molecule after processing, leaving the $tRNA^{Tyr}$ lacking a discriminator nucleotide at position 73 (Reichert *et al.*, 1998; Reichert and Morl, 2000). Repair of this 3′-truncated $tRNA^{Tyr}$ occurs by addition of an A_{73} discriminator nucleotide (followed by normal 3′-CCA addition), and since this is also an A-addition reaction, similar involvement of a PAP-type enzyme might be assumed. However, in the human case, *in vitro* investigations of the 3′-repair activity in mitochondrial extracts in the presence of all four NTPs demonstrated a significant amount of C_{73} addition (in 50% of sequenced clones) as opposed to A_{73} addition (29%) (Reichert *et al.*, 1998). From this observation, the possibility that a CCA adding enzyme-type polymerase is involved in this repair must also be considered, since in bacteria as well as eukaryotic systems, the CCA enzyme catalyses repair of 3′-CCA ends in case of their nucleolytic degradation (Zhu and Deutscher, 1987; Aebi *et al.*, 1990). Notably, in the case of the human mt-tRNA^{Tyr}, the last two nucleotides in the acceptor stem prior to N_{73} are C-residues, thus a mechanism involving recognition of the 3′-end of the stem to add A_{73}, followed by repositioning to add the correct 3′-CCA end could be envisaged. In this respect, the recent demonstration of the ability of CCA-adding enzyme to add multiple CCA residues to some RNAs is intriguing (Wilusz *et al.*, 2011). Finally, several small RNAs like U2, U6, 5S and 7SL are now known to carry a single non-encoded adenosine residue at their 3′ ends, which is hypothesized to increase stability and reduce the turnover of these RNAs (Sinha *et al.*, 1998). At least one enzyme with the ability to catalyse a single A addition to the 3′-end of these RNA substrates has been identified that could similarly catalyse addition of the discriminator A residue to these 3′-truncated tRNA^{Tyr} molecules (Sinha *et al.*, 1999; Perumal *et al.*, 2001). It is still not known yet if this enzyme can localize to the mitochondria, as it is currently known to localize to the nucleus and the cytoplasm.

Regardless of the identity of the polymerase involved in 3′-end repair during editing, these reactions appear likely to be catalysed by ubiquitous enzymes required for maturation and processing of multiple classes of RNA. Hence, a function for these enzymes in tRNA editing appears to be an example of the use of an existing activity for evolution of novel RNA processing events that ensure RNA quality control and perhaps to also aid genome condensation.

Template-dependent 3′-editing reactions

The editing of tRNAs observed in the centipede *Lithobius forficatus* is novel as it was the first case of 3′-tRNA editing that involved addition of nucleotides to the 3′-arm of the acceptor in an apparently template-dependent manner, using the 5′-half of the acceptor stem as a template (Lavrov *et al.*, 2000). Complete mitochondrial genome sequencing of *L. forficatus* identified 22 tRNA genes that encode tRNAs with unusual secondary structures and only one has a complete fully paired tRNA acceptor stem. The remaining 21 tRNAs encode up to five non-Watson–Crick mismatches in the tRNA acceptor stem while some also overlap in sequence with downstream genes. Hence editing of these tRNAs was predicted to be obligatory for attaining a proper functional three dimensional structure. cDNA sequencing of circularized tRNAs indicated that most of the tRNAs have completely basepairing Watson–Crick nucleotides at the 3′-ends, in contrast to the nucleotides originally specified by the gene sequence. Moreover, the presence of intact 3′-CCA ends on the sequenced tRNAs suggested that they were translationally competent *in vivo*. In this case, nucleotide changes observed in the 3′-end of the acceptor stems were not restricted to adenosines, and instead involved additions of all four NTPs. Hence, a role for an RNA-dependent RNA polymerase (RdRp) has been hypothesized in the repair of the tRNA 3′-ends (Fig. 3.3). Although an RdRp has not been identified yet, several instances of viral derived and/or nuclear encoded RdRp have been documented, which otherwise function in host viral-defence pathways (Koga *et al.*, 1998; Cogoni and Macino, 1999). Mitochondrially associated RdRp have also been identified in some fungi and *Arabidopsis* (Hong *et al.*, 1998; Cole *et al.*, 2000). Although the nature of the nucleotidyl transferase activity is relatively well defined (albeit the identity of the gene(s) that catalyse the reaction are still a mystery), less is known about the pathway for removal of the mismatched nucleotides. As for the template-independent processes described above, some of the tRNAs with encoded mismatches are predicted to overlap with the 5′-ends of downstream genes and the mechanism for nucleotide removal could include activities of the tRNA processing enzymes. For non-overlapping tRNAs, the mechanism(s) for nucleotide removal is unknown.

A similar example of template-dependent 3′-tRNA editing has also been reported to occur in the mitochondria of the jakobid flagellate *Seculamonas ecuadoriensis* (Leigh and Lang, 2004). Jakobid flagellates have the most ancestral mitochondrial genomes which appear to closely resemble genomes of α-proteobacteria (Lang *et al.*, 1997). Although never observed in α-proteobacteria or its closest jakobid relative, the mitochondrial genome sequencing of *S. ecuadoriensis* indicated the presence of two tRNAs that encode non-Watson–Crick mismatches in the tRNA acceptor stems, and sequencing of mature tRNAs revealed that the mismatched nucleotides are removed and replaced with Watson–Crick basepaired nucleotides in a reaction presumably catalysed by an RdRp-type enzyme (Leigh and Lang, 2004). However, given the extent of partial editing observed at some positions, and a strong bias towards addition of Cs and As at the completely edited sites, the involvement of a more C-A specific nucleotide transferase (perhaps related to the ubiquitous CCA adding enzyme) cannot be excluded (Li and Deutscher, 1996). This prediction could in part also explain the observed conversion of all edited discriminator nucleotides exclusively to A residues, although the possibility of an alternative activity that is involved exclusively in addition of the unpaired discriminator nucleotide cannot be completely disregarded.

In the case of *S. ecuadoriensis*, a significant amount of partial editing was observed at sites lower down on the 3′-half of the acceptor stem, suggesting that the mismatched bases are removed in an exonucleolytic manner (as was discussed above for the nuclease involved in template-independent editing). It has been argued that there could be a competition between the exonuclease and the 3′-repair activity which might explain the partial nature of the editing in some clones. On the basis of this observation, the exonuclease involved in the editing reaction has been hypothesized to be of bacterial origin since the partial editing also mirrors the interplay between the tRNA 3′-processing exonuclease(s) and the CCA adding enzyme which help maintain the integrity of the CCA end in bacterial tRNAs (Li and Deutscher, 1996).

Of six *S. ecuadoriensis* mt-tRNAs predicted to contain G:U wobble basepairing in the aminoacyl-acceptor stem, $tRNA^{Cys}$ and $tRNA^{Ala}$ were chosen as representatives of the class and tested for editing. Unlike some other cases of tRNA editing reviewed in this chapter, the G:U wobble basepairs in $tRNA^{Cys}$ and $tRNA^{Ala}$ are not repaired (Leigh and Lang, 2004). The G:U basepair in $tRNA^{Cys}$ is at the 1:72 position while that in $tRNA^{Ala}$ is at the 3:70 position and is a known identity element for alanyl-tRNA. Hence the lack of editing observed at these sites in the corresponding tRNAs could reflect the inability of these bases to perturb tRNA secondary structures sufficiently to warrant nucleolytic removal and/or the co-evolution of the 3′-tRNA editing machinery to account for existing biochemically distinct features of tRNAs.

Future trends

The fields of tRNA modification and tRNA editing are entering distinctly different phases that are likely to direct current and future research. Rapid progress towards identification of missing tRNA modification enzymes over the past several decades has resulted in a situation where enzyme functions have been assigned to nearly all known examples of modified nucleotides in tRNAs. Thus, the question with respect to tRNA modification has largely changed from 'who?' (identification of the enzymes catalysing a given modification) to 'how?' (understanding substrate recognition, catalytic mechanism and regulation of these activities). The ever-increasing number of structures of modification enzymes and enzyme–substrate complexes, combined with further biochemical and mechanistic analysis, will help researchers to realize the goal of understanding how these enzymes are able to catalyse unique chemical reactions with high selectivity. For tRNA editing, on the other hand, the current questions are still primarily focused on identification of key enzyme players that carry out the unusual editing reactions. The relatively more recent discovery of tRNA editing, along with the fact that tRNA editing reactions so far appear to be a unique property of eukaryotic organelles where genetic manipulation is not as easily carried out as for nuclear genes, are both likely factors in the slower pace of progress towards understanding the enzymology and biological functions of these reactions.

Identification of new modifications and editing events

Despite the fact that a large number of enzymes associated with these processes (particularly tRNA modification) have already been identified, other modification or editing events are surely yet to be discovered. A significant challenge in the study of tRNA modification is the lack of large scale approaches to determine sequences of fully modified tRNAs, particularly for the Archaea, for which there are still only a few sequences of mature tRNA species available. The lack of sequence information that addresses the modification status of large numbers of tRNAs may in fact prompt erroneous conclusions about the lack of a modification in a given organism or domain, and the lack of examples of a given modification (see Table 3.1) may reflect simply a paucity of sequences in some cases. The use of improved mass spectrometry technology, combined with innovative ways to carry out high-throughput screening of tRNAs for modified nucleotides will be important to try to address this issue. In fact, recent observations suggest that the presence of RNA modifications even on mRNA may be more widespread than previously thought, indicating that new technologies to address this question could have an impact beyond the field of tRNA biology (Li *et al.*, 2011).

Unusual examples of tRNA processing in mitochondria

Recent years have seen an increased interest in tRNA biology in eukaryotic organelles. Indeed, it seems that many 'exceptions' to the rule with respect to tRNAs have been identified in mitochondria, including the identification of the only known example of 5′-tRNA maturation by ribonuclease P that is catalysed by a protein-only enzyme as opposed to the ubiquitous ribonucleoprotein system (Holzmann *et al.*, 2008). Likewise, tRNA editing of the nucleotide replacement types discussed here is a process that has so far only been observed to occur in mitochondrial systems. This concentration suggests that some aspect of the environment in the mitochondria may facilitate use of unusual means to produce and maintain tRNAs. The emergence of methods for mitochondrial DNA transformation combined with the development of *in vitro* systems for studying mitochondrial RNA processing events should enhance the ability to study these intriguing reactions in greater detail.

Evolution of tRNA editing and modification

The appearance of tRNA editing systems themselves is puzzling; if an organism requires the presence of specific structural feature to produce a functional tRNA (such as a fully basepaired aminoacyl acceptor stem), then why have sequences that specify the required nucleotides to achieve this structure not been encoded in the genome? For tRNA editing that occurs as a result of overlapping genes, perhaps arguments based on economization of mitochondrial genome 'real estate' could be used to support the existence of these activities. However, particularly for the 5′-editing events that involve transcription of apparently 'junk' sequences that are removed and replaced later, those arguments are more problematic. An attractive possibility is that the presence of these editing enzymes is advantageous to cells for other reasons, perhaps because of their ability to repair cellular RNAs, and therefore, the evolutionary pressure to retain the enzymes for these other purposes is the driving force for their continued use in 5′-tRNA editing.

The evolution of tRNA modification systems also raises questions of apparent redundancy of function. If modification of only a single tRNA species is required for optimal biological function, as has recently been demonstrated with numerous modifications in yeast, why have many of the enzymes involved retained their abilities to modify multiple tRNAs? The limitation seems unlikely to be due to the need to evolve single tRNA-specific modification reactions since several examples of highly tRNA-specific modification enzymes are known. An interesting possibility is that different modifications, or combinations of modifications, are required for growth under different physiological conditions. More detailed investigation of the enzymes that catalyse tRNA editing and modification and determination of their full range of biological activities will help to address these critical questions of tRNA function and evolution.

References

Abad, M.G., Rao, B.S., and Jackman, J.E. (2010). Template-dependent 3′–5′ nucleotide addition is a shared feature of tRNAHis guanylyltransferase enzymes from multiple domains of life. Proc. Natl. Acad. Sci. U.S.A. *107*, 674–679.

Abad, M.G., Long, Y., Willcox, A., Gott, J.M., Gray, M.W., and Jackman, J.E. (2011). A role for tRNA(His) guanylyltransferase (Thg1)-like proteins from *Dictyostelium discoideum* in mitochondrial 5′-tRNA editing. RNA *17*, 613–623.

Aebi, M., Kirchner, G., Chen, J.Y., Vijayraghavan, U., Jacobson, A., Martin, N.C., and Abelson, J. (1990). Isolation of a temperature-sensitive mutant with an altered tRNA nucleotidyltransferase and cloning of the gene encoding tRNA nucleotidyltransferase in the yeast Saccharomyces cerevisiae. J. Biol. Chem. *265*, 16216–16220.

Alexandrov, A., Martzen, M.R., and Phizicky, E.M. (2002). Two proteins that form a complex are required for 7-methylguanosine modification of yeast tRNA. RNA *8*, 1253–1266.

Alexandrov, A., Chernyakov, I., Gu, W., Hiley, S.L., Hughes, T.R., Grayhack, E.J., and Phizicky, E.M. (2006). Rapid tRNA decay can result from lack of nonessential modifications. Mol. Cell *21*, 87–96.

Alfonzo, J.D., Blanc, V., Estevez, A.M., Rubio, M.A., and Simpson, L. (1999). C to U editing of the anticodon of imported mitochondrial tRNA(Trp) allows decoding of the UGA stop codon in *Leishmania tarentolae.* EMBO J. *18*, 7056–7062.

Anantharaman, V., Koonin, E.V., and Aravind, L. (2002). SPOUT: a class of methyltransferases that includes spoU and trmD RNA methylase superfamilies, and novel superfamilies of predicted prokaryotic RNA methylases. J. Mol. Microbiol. Biotechnol. *4*, 71–75.

Anderson, J., Phan, L., Cuesta, R., Carlson, B.A., Pak, M., Asano, K., Bjork, G.R., Tamame, M., and Hinnebusch, A.G. (1998). The essential Gcd10p-Gcd14p nuclear complex is required for 1 - methyladenosine modification and maturation of initiator methionyl-tRNA. Genes Dev. *12*, 3650–3662.

Anderson, J., Phan, L., and Hinnebusch, A.G. (2000). The Gcd10p/Gcd14p complex is the essential two-subunit tRNA(1 - methyladenosine) methyltransferase of *Saccharomyces cerevisiae.* Proc. Natl. Acad. Sci. U.S.A. *97*, 5173–5178.

Antes, T., Costandy, H., Mahendran, R., Spottswood, M., and Miller, D. (1998). Insertional editing of mitochondrial tRNAs of *Physarum polycephalum* and *Didymium nigripes.* Mol. Cell. Biol. *18*, 7521–7527.

Armengaud, J., Urbonavicius, J., Fernandez, B., Chaussinand, G., Bujnicki, J.M., and Grosjean, H. (2004). N2-methylation of guanosine at position 10 in tRNA is catalyzed by a THUMP domain-containing, S-adenosylmethionine-dependent methyltransferase, conserved in Archaea and Eukaryota. J. Biol. Chem. *279*, 37142–37152.

Bjork, G.R., Wikstrom, P.M., and Bystrom, A.S. (1989). Prevention of translational frameshifting by the modified nucleoside 1-methylguanosine. Science *244*, 986–989.

Bjork, G.R., Jacobsson, K., Nilsson, K., Johansson, M.J., Bystrom, A.S., and Persson, O.P. (2001). A primordial tRNA modification required for the evolution of life? EMBO J. *20*, 231–239.

Bullerwell, C., and Gray, M. (2005). *In vitro* characterization of a tRNA editing activity in the mitochondria of Spizellomyces punctatus, a Chytridiomycete fungus. J. Biol. Chem. *280*, 2463–2470.

Bullerwell, C.E., Burger, G., Gott, J.M., Kourennaia, O., Schnare, M.N., and Gray, M.W. (2010). Abundant 5S rRNA-like transcripts encoded by the mitochondrial genome in amoebozoa. Eukaryot. Cell *9*, 762–773.

Byrne, E.M., Visomirski-Robic, L., Cheng, Y.W., Rhee, A.C., and Gott, J.M. (2007). RNA editing in *Physarum* mitochondria: assays and biochemical approaches. Methods Enzymol. *424*, 143–172.

Byström, A.S., and Björk, G.R. (1982). Chromosomal location and cloning of the gene (trmD) responsible for the synthesis of tRNA (m1G) methyltransferase in *Escherichia coli* K-12. Mol. Gen. Genet. *188*, 440–446.

Cavaille, J., Chetouani, F., and Bachellerie, J.P. (1999). The yeast Saccharomyces cerevisiae YDL112w ORF encodes the putative 2′ - O-ribose methyltransferase catalyzing the formation of Gm18 in tRNAs. RNA *5*, 66–81.

Chatterjee, K., Blaby, I.K., Thiaville, P.C., Majumder, M., Grosjean, H., Yuan, Y.A., Gupta, R., and de Crecy-Lagard, V. (2012). The archaeal COG1901/DUF358 SPOUT-methyltransferase members, together with pseudouridine synthase Pus10, catalyze the formation of 1-methylpseudouridine at position 54 of tRNA. RNA *18*, 421–433.

Chernyakov, I., Whipple, J.M., Kotelawala, L., Grayhack, E.J., and Phizicky, E.M. (2008). Degradation of several hypomodified mature tRNA species in Saccharomyces cerevisiae is mediated by Met22 and the 5′–3′ exonucleases Rat1 and Xrn1. Genes Dev. *22*, 1369–1380.

Cogoni, C., and Macino, G. (1999). Gene silencing in *Neurospora crassa* requires a protein homologous to RNA-dependent RNA polymerase. Nature *399*, 166–169.

Cole, T.E., Hong, Y., Brasier, C.M., and Buck, K.W. (2000). Detection of an RNA-dependent RNA polymerase in mitochondria from a mitovirus-infected isolate of the Dutch Elm disease fungus, Ophiostoma novo-ulmi. Virology *268*, 239–243.

Czerwoniec, A., Dunin-Horkawicz, S., Purta, E., Kaminska, K.H., Kasprzak, J.M., Bujnicki, J.M., Grosjean, H., and Rother, K. (2009). MODOMICS: a database of RNA modification pathways. 2008 update. Nucleic Acids Res. *37*, D118–D121.

de Crecy-Lagard, V. (2007). Identification of genes encoding tRNA modification enzymes by comparative genomics. Methods Enzymol. *425*, 153–183.

de Crecy-Lagard, V., Marck, C., Brochier-Armanet, C., and Grosjean, H. (2007). Comparative RNomics and modomics in Mollicutes: prediction of gene function and evolutionary implications. IUBMB life *59*, 634–658.

de Crecy-Lagard, V., Brochier-Armanet, C., Urbonavicius, J., Fernandez, B., Phillips, G., Lyons, B., Noma, A., Alvarez, S., Droogmans, L., Armengaud, J., *et al.* (2010). Biosynthesis of wyosine derivatives in tRNA: an ancient and highly diverse pathway in Archaea. Mol. Biol. Evol. *27*, 2062–2077.

El Yacoubi, B., Lyons, B., Cruz, Y., Reddy, R., Nordin, B., Agnelli, F., Williamson, J.R., Schimmel, P., Swairjo, M.A., and de Crecy-Lagard, V. (2009). The universal YrdC/Sua5 family is required for the formation of threonylcarbamoyladenosine in tRNA. Nucleic Acids Res. *37*, 2894–2909.

Esberg, A., Huang, B., Johansson, M.J., and Bystrom, A.S. (2006). Elevated levels of two tRNA species bypass the requirement for elongator complex in transcription and exocytosis. Mol. Cell *24*, 139–148.

Garcia-Barrio, M.T., Naranda, T., Vazquez de Aldana, C.R., Cuesta, R., Hinnebusch, A.G., Hershey, J.W., and Tamame, M. (1995). GCD10, a translational repressor of GCN4, is the RNA-binding subunit of eukaryotic translation initiation factor-3. Genes Dev. *9*, 1781–1796.

Gerber, A.P., and Keller, W. (1999). An adenosine deaminase that generates inosine at the wobble position of tRNAs. Science *286*, 1146–1149.

Goto-Ito, S., Ishii, R., Ito, T., Shibata, R., Fusatomi, E., Sekine, S.I., Bessho, Y., and Yokoyama, S. (2007). Structure of an archaeal TYW1, the enzyme catalyzing the second step of wye-base biosynthesis. Acta Crystallogr. D Biol. Crystallogr. *63*, 1059–1068.

Gott, J.M. (2003). Expanding genome capacity via RNA editing. CR Biol. *326*, 901–908.

Gott, J.M., Visomirski, L.M., and Hunter, J.L. (1993). Substitutional and insertional RNA editing of the cytochrome c oxidase subunit 1 mRNA of *Physarum polycephalum*. J. Biol. Chem. *268*, 25483–25486.

Gott, J.M., Somerlot, B.H., and Gray, M.W. (2010). Two forms of RNA editing are required for tRNA maturation in *Physarum* mitochondria. RNA *16*, 482–488.

Grosjean, H. (2009). DNA and RNA Modification Enzymes: Structure, Mechanism, Function and Evolution (Landes Bioscience, Austin, TX).

Grosjean, H., and Benne, R. (1998). Modification and Editing of RNA (ASM Press, Washington, DC).

Gu, W., Jackman, J.E., Lohan, A.J., Gray, M.W., and Phizicky, E.M. (2003). tRNAHis maturation: an essential yeast protein catalyzes addition of a guanine nucleotide to the 5′ end of tRNAHis. Genes Dev. *17*, 2889–2901.

Harris, K.A., Jones, V., Bilbille, Y., Swairjo, M.A., and Agris, P.F. (2011). YrdC exhibits properties expected of a subunit for a tRNA threonylcarbamoyl transferase. RNA *17*, 1678–1687.

Helm, M., and Attardi, G. (2004). Nuclear control of cloverleaf structure of human mitochondrial tRNA(Lys). J. Mol. Biol. 337, 545–560.

Hoagland, M.B., Stephenson, M.L., Scott, J.F., Hecht, L.I., and Zamecnik, P.C. (1958). A soluble ribonucleic acid intermediate in protein synthesis. J. Biol. Chem. *231*, 241–257.

Holley, R.W., Apgar, J., Everett, G.A., Madison, J.T., Marquisee, M., Merrill, S.H., Penswick, J.R., and Zamir, A. (1965). Structure of a ribonucleic acid. Science *147*, 1462–1465.

Holzmann, J., Frank, P., Loffler, E., Bennett, K.L., Gerner, C., and Rossmanith, W. (2008). RNase P without RNA: identification and functional reconstitution of the human mitochondrial tRNA processing enzyme. Cell *135*, 462–474.

Hong, Y., Cole, T.E., Brasier, C.M., and Buck, K.W. (1998). Evolutionary relationships among putative RNA-dependent RNA polymerases encoded by a mitochondrial virus-like RNA in the Dutch elm disease fungus, Ophiostoma novo-ulmi, by other viruses and virus-like RNAs and by the Arabidopsis mitochondrial genome. Virology *246*, 158–169.

Hori, H., Kubota, S., Watanabe, K., Kim, J.M., Ogasawara, T., Sawasaki, T., and Endo, Y. (2003). Aquifex aeolicus tRNA (Gm18) methyltransferase has unique substrate specificity. TRNA recognition mechanism of the enzyme. J. Biol. Chem. *278*, 25081–25090.

Hou, Y.M., and Perona, J.J. (2010). Stereochemical mechanisms of tRNA methyltransferases. FEBS Lett. *584*, 278–286.

Huang, B., Johansson, M.J., and Bystrom, A.S. (2005). An early step in wobble uridine tRNA modification requires the Elongator complex. RNA *11*, 424–436.

Iyer, L.M., Abhiman, S., de Souza, R.F., and Aravind, L. (2010). Origin and evolution of peptide-modifying dioxygenases and identification of the wybutosine hydroxylase/hydroperoxidase. Nucleic Acids Res. *38*, 5261–5279.

Jackman, J.E., and Phizicky, E.M. (2006a). tRNAHis guanylyltransferase adds G-1 to the 5′ end of tRNAHis by recognition of the anticodon, one of several features unexpectedly shared with tRNA synthetases. RNA *12*, 1007–1014.

Jackman, J.E., and Phizicky, E.M. (2006b). tRNAHis guanylyltransferase catalyzes a 3′–5′ polymerization reaction that is distinct from G-1 addition. Proc. Natl. Acad. Sci. U.S.A. *103*, 8640–8645.

Jackman, J.E., Montange, R.K., Malik, H.S., and Phizicky, E.M. (2003). Identification of the yeast gene encoding the tRNA m1G methyltransferase responsible for modification at position 9. RNA *9*, 574–585.

Jackman, J.E., Gott, J.M., and Gray, M.W. (2012). Doing it in Reverse: 3′-to-5′ Polymerization by the Thg1 Superfamily. RNA *18*, 886–899.

Janke, A., and Paabo, S. (1993). Editing of a tRNA anticodon in marsupial mitochondria changes its codon recognition. Nucleic Acids Res. *21*, 1523–1525.

Juhling, F., Morl, M., Hartmann, R.K., Sprinzl, M., Stadler, P.F., and Putz, J. (2009). tRNAdb 2009: compilation of tRNA sequences and tRNA genes. Nucleic Acids Res. *37*, D159–D162.

Kadaba, S., Krueger, A., Trice, T., Krecic, A.M., Hinnebusch, A.G., and Anderson, J. (2004). Nuclear surveillance and degradation of hypomodified initiator tRNAMet in S. cerevisiae. Genes Dev. *18*, 1227–1240.

Kalhor, H.R., and Clarke, S. (2003). Novel methyltransferase for modified uridine residues at the wobble position of tRNA. Mol. Cell. Biol. *23*, 9283–9292.

Kalhor, H.R., Penjwini, M., and Clarke, S. (2005). A novel methyltransferase required for the formation of the hypermodified nucleoside wybutosine in eucaryotic tRNA. Biochem. Biophys. Res. Commun. *334*, 433–440.

Kato, M., Araiso, Y., Noma, A., Nagao, A., Suzuki, T., Ishitani, R., and Nureki, O. (2011). Crystal structure of a novel JmjC-domain-containing protein, TYW5, involved in tRNA modification. Nucleic Acids Res. *39*, 1576–1585.

Kempenaers, M., Roovers, M., Oudjama, Y., Tkaczuk, K.L., Bujnicki, J.M., and Droogmans, L. (2010). New archaeal methyltransferases forming 1-methyladenosine or 1-methyladenosine and 1-methylguanosine at position 9 of tRNA. Nucleic Acids Res. *38*, 6533–6543.

Koga, R., Fukuhara, T., and Nitta, T. (1998). Molecular characterization of a single mitochondria-associated double-stranded RNA in the green alga Bryopsis. Plant Mol. Biol. *36*, 717–724.

Kotelawala, L., Grayhack, E.J., and Phizicky, E.M. (2008). Identification of yeast tRNA Um(44) 2′-O-methyltransferase (Trm44) and demonstration of a Trm44 role in sustaining levels of specific tRNA(Ser) species. RNA *14*, 158–169.

Laforest, M.J., Roewer, I., and Lang, B.F. (1997). Mitochondrial tRNAs in the lower fungus Spizellomyces punctatus: tRNA editing and UAG 'stop' codons recognized as leucine. Nucleic Acids Res. *25*, 626–632.

Laforest, M.J., Bullerwell, C.E., Forget, L., and Lang, B.F. (2004). Origin, evolution, and mechanism of 5′ tRNA editing in chytridiomycete fungi. RNA *10*, 1191–1199.

Lang, B.F., Burger, G., O'Kelly, C.J., Cedergren, R., Golding, G.B., Lemieux, C., Sankoff, D., Turmel, M., and Gray, M.W. (1997). An ancestral mitochondrial DNA resembling a eubacterial genome in miniature. Nature *387*, 493–497.

Lavrov, D.V., Brown, W.M., and Boore, J.L. (2000). A novel type of RNA editing occurs in the mitochondrial tRNAs of the centipede Lithobius forficatus. Proc. Natl. Acad. Sci. U.S.A. *97*, 13738–13742.

Leigh, J., and Lang, B.F. (2004). Mitochondrial 3′ tRNA editing in the jakobid Seculamonas ecuadoriensis: a novel mechanism and implications for tRNA processing. RNA *10*, 615–621.

Levinger, L., Morl, M., and Florentz, C. (2004). Mitochondrial tRNA 3′ end metabolism and human disease. Nucleic Acids Res. *32*, 5430–5441.

Li, M., Wang, I.X., Li, Y., Bruzel, A., Richards, A.L., Toung, J.M., and Cheung, V.G. (2011). Widespread RNA and DNA sequence differences in the human transcriptome. Science *333*, 53–58.

Li, Z., and Deutscher, M.P. (1996). Maturation pathways for E. coli tRNA precursors: a random multienzyme process *in vivo*. Cell *86*, 503–512.

Lonergan, K.M., and Gray, M.W. (1993a). Editing of transfer RNAs in Acanthamoeba castellanii mitochondria. Science *259*, 812–816.

Lonergan, K.M., and Gray, M.W. (1993b). Predicted editing of additional transfer RNAs in *Acanthamoeba castellanii* mitochondria. Nucleic Acids Res. *21*, 4402.

Lu, J., Huang, B., Esberg, A., Johansson, M.J., and Bystrom, A.S. (2005). The Kluyveromyces lactis gamma-toxin targets tRNA anticodons. RNA *11*, 1648–1654.

Mahendran, R., Spottswood, M.R., and Miller, D.L. (1991). RNA editing by cytidine insertion in mitochondria of *Physarum polycephalum*. Nature *349*, 434–438.

Mandel, L.R., and Borek, E. (1963). The biosynthesis of methylated bases in ribonucleic acid. Biochemistry 2, 555–560.

Marechal-Drouard, L., Cosset, A., Remacle, C., Ramamonjisoa, D., and Dietrich, A. (1996). A single editing event is a prerequisite for efficient processing of potato mitochondrial phenylalanine tRNA. Mol. Cell. Biol. *16*, 3504–3510.

Marechal-Drouard, L., Ramamonjisoa, D., Cosset, A., Weil, J.H., and Dietrich, A. (1993). Editing corrects mispairing in the acceptor stem of bean and potato mitochondrial phenylalanine transfer RNAs. Nucleic Acids Res. *21*, 4909–4914.

Martzen, M.R., McCraith, S.M., Spinelli, S.L., Torres, F.M., Fields, S., Grayhack, E.J., and Phizicky, E.M. (1999). A biochemical genomics approach for identifying genes by the activity of their products. Science *286*, 1153–1155.

Menezes, S., Gaston, K.W., Krivos, K.L., Apolinario, E.E., Reich, N.O., Sowers, K.R., Limbach, P.A., and Perona, J.J. (2011). Formation of m2G6 in *Methanocaldococcus jannaschii* tRNA catalyzed by the novel methyltransferase Trm14. Nucleic Acids Res. *39*, 7641–7655.

Motorin, Y., and Grosjean, H. (1999). Multisite-specific tRNA:m5C-methyltransferase (Trm4) in yeast *Saccharomyces cerevisiae*: identification of the gene and substrate specificity of the enzyme. RNA *5*, 1105–1118.

Motorin, Y., and Helm, M. (2010). tRNA stabilization by modified nucleotides. Biochemistry *49*, 4934–4944.

Motorin, Y., Lyko, F., and Helm, M. (2010). 5-Methylcytosine in RNA: detection, enzymatic formation and biological functions. Nucleic Acids Res. *38*, 1415–1430.

Nakanishi, K., and Nureki, O. (2005). Recent progress of structural biology of tRNA processing and modification. Mol. Cells *19*, 157–166.

Noma, A., Kirino, Y., Ikeuchi, Y., and Suzuki, T. (2006). Biosynthesis of wybutosine, a hyper-modified nucleoside in eukaryotic phenylalanine tRNA. EMBO J. 25, 2142–2154.

Ny, T., and Bjork, G.R. (1980). Cloning and restriction mapping of the trmA gene coding for transfer ribonucleic acid (5-methyluridine)-methyltransferase in *Escherichia coli* K-12. J. Bacteriol. *142*, 371–379.

Ny, T., Lindstrom, H.R., Hagervall, T.G., and Bjork, G.R. (1988). Purification of transfer RNA (m5U54)-methyltransferase from *Escherichia coli*. Association with RNA. Eur. J. Biochem. *177*, 467–475.

Ochi, A., Makabe, K., Kuwajima, K., and Hori, H. (2010). Flexible recognition of the tRNA G18 methylation target site by TrmH methyltransferase through first binding and induced fit processes. J. Biol. Chem. *285*, 9018–9029.

Ogawa, S., Yoshino, R., Angata, K., Iwamoto, M., Pi, M., Kuroe, K., Matsuo, K., Morio, T., Urushihara, H., Yanagisawa, K., *et al.* (2000). The mitochondrial DNA of *Dictyostelium discoideum*: complete sequence, gene content and genome organization. Mol. Gen. Genet. *263*, 514–519.

Orellana, O., Cooley, L., and Soll, D. (1986). The additional guanylate at the 5′ terminus of *Escherichia coli* tRNAHis is the result of unusual processing by RNase P. Mol. Cell. Biol. *6*, 525–529.

Persson, B.C., Gustafsson, C., Berg, D.E., and Bjork, G.R. (1992). The gene for a tRNA modifying enzyme, m5U54-methyltransferase, is essential for viability in *Escherichia coli*. Proc. Natl. Acad. Sci. U.S.A. *89*, 3995–3998.

Persson, B.C., Jager, G., and Gustafsson, C. (1997). The spoU gene of *Escherichia coli*, the fourth gene of the spoT operon, is essential for tRNA (Gm18) 2′-O-methyltransferase activity. Nucleic Acids Res. *25*, 4093–4097.

Perumal, K., Sinha, K., Henning, D., and Reddy, R. (2001). Purification, characterization, and cloning of the cDNA of human signal recognition particle RNA 3′-adenylating enzyme. J. Biol. Chem. *276*, 21791–21796.

Phizicky, E.M., and Alfonzo, J.D. (2010). Do all modifications benefit all tRNAs? FEBS Lett. *584*, 265–271.

Phizicky, E.M., and Hopper, A.K. (2010). tRNA biology charges to the front. Genes Dev. *24*, 1832–1860.

Pintard, L., Lecointe, F., Bujnicki, J.M., Bonnerot, C., Grosjean, H., and Lapeyre, B. (2002). Trm7p catalyses the formation of two 2′-O-methylriboses in yeast tRNA anticodon loop. EMBO J. *21*, 1811–1820.

Price, D.H., and Gray, M.W. (1999a). Confirmation of predicted edits and demonstration of unpredicted edits in *Acanthamoeba castellanii* mitochondrial tRNAs. Curr. Genet. 35, 23–29.

Price, D.H., and Gray, M.W. (1999b). A novel nucleotide incorporation activity implicated in the editing of mitochondrial transfer RNAs in *Acanthamoeba castellanii*. RNA 5, 302–317.

Purta, E., van Vliet, F., Tkaczuk, K.L., Dunin-Horkawicz, S., Mori, H., Droogmans, L., and Bujnicki, J.M. (2006). The yfhQ gene of *Escherichia coli* encodes a tRNA:Cm32/Um32 methyltransferase. BMC Mol. Biol. 7, 23.

Purushothaman, S.K., Bujnicki, J.M., Grosjean, H., and Lapeyre, B. (2005). Trm11p and Trm112p are both required for the formation of 2-methylguanosine at position 10 in yeast tRNA. Mol. Cell. Biol. *25*, 4359–4370.

Rao, B.S., Maris, E.L., and Jackman, J.E. (2011). tRNA 5′-end repair activities of tRNAHis guanylyltransferase (Thg1)-like proteins from Bacteria and Archaea. Nucleic Acids Res. *39*, 1833–1842.

Reichert, A.S., and Morl, M. (2000). Repair of tRNAs in metazoan mitochondria. Nucleic Acids Res. *28*, 2043–2048.

Reichert, A., Rothbauer, U., and Morl, M. (1998). Processing and editing of overlapping tRNAs in human mitochondria. J. Biol. Chem. *273*, 31977–31984.

Renalier, M.H., Joseph, N., Gaspin, C., Thebault, P., and Mougin, A. (2005). The Cm56 tRNA modification in archaea is catalyzed either by a specific 2′-O-methylase, or a C/D sRNP. RNA *11*, 1051–1063.

Rich, A., and RajBhandary, U.L. (1976). Transfer RNA: molecular structure, sequence, and properties. Annu. Rev. Biochem. *45*, 805–860.

Roovers, M., Wouters, J., Bujnicki, J.M., Tricot, C., Stalon, V., Grosjean, H., and Droogmans, L. (2004). A primordial RNA modification enzyme: the case of tRNA (m1A) methyltransferase. Nucleic Acids Res. *32*, 465–476.

Rould, M.A., Perona, J.J., Soll, D., and Steitz, T.A. (1989). Structure of *E. coli* glutaminyl-tRNA synthetase complexed with tRNA(Gln) and ATP at 2.8 A resolution. Science *246*, 1135–1142.

Schubert, H.L., Blumenthal, R.M., and Cheng, X. (2003). Many paths to methyltransfer: a chronicle of convergence. Trends Biochem. Sci. *28*, 329–335.

Sinha, K., Perumal, K., Chen, Y., and Reddy, R. (1999). Post-transcriptional adenylation of signal recognition particle RNA is carried out by an enzyme different from mRNA Poly(A) polymerase. J. Biol. Chem. *274*, 30826–30831.

Sinha, K.M., Gu, J., Chen, Y., and Reddy, R. (1998). Adenylation of small RNAs in human cells. Development of a cell-free system for accurate adenylation on the 3′-end of human signal recognition particle RNA. J. Biol. Chem. *273*, 6853–6859.

Soll, D. (1971). Enzymatic modification of transfer RNA. Science *173*, 293–299.

Srinivasan, M., Mehta, P., Yu, Y., Prugar, E., Koonin, E.V., Karzai, A.W., and Sternglanz, R. (2011). The highly conserved KEOPS/EKC complex is essential for a universal tRNA modification, t6A. EMBO J. *30*, 873–881.

Suzuki, Y., Noma, A., Suzuki, T., Senda, M., Senda, T., Ishitani, R., and Nureki, O. (2007). Crystal structure of the radical SAM enzyme catalyzing tricyclic modified base formation in tRNA. J. Mol. Biol. *372*, 1204–1214.

Suzuki, Y., Noma, A., Suzuki, T., Ishitani, R., and Nureki, O. (2009). Structural basis of tRNA modification with CO_2 fixation and methylation by wybutosine synthesizing enzyme TYW4. Nucleic Acids Res. *37*, 2910–2925.

Takano, H., Abe, T., Sakurai, R., Moriyama, Y., Miyazawa, Y., Nozaki, H., Kawano, S., Sasaki, N., and Kuroiwa, T. (2001). The complete DNA sequence of the mitochondrial genome of *Physarum polycephalum*. Mol. Gen. Genet. *264*, 539–545.

Terrett, J.A., Miles, S., and Thomas, R.H. (1996). Complete DNA sequence of the mitochondrial genome of *Cepaea nemoralis* (Gastropoda: Pulmonata). J. Mol. Evol. *42*, 160–168.

Tkaczuk, K.L., Dunin-Horkawicz, S., Purta, E., and Bujnicki, J.M. (2007). Structural and evolutionary bioinformatics of the SPOUT superfamily of methyltransferases. BMC Bioinformatics *8*, 73.

Tomita, K., Ueda, T., and Watanabe, K. (1996). RNA editing in the acceptor stem of squid mitochondrial tRNA(Tyr). Nucleic Acids Res. *24*, 4987–4991.

Umitsu, M., Nishimasu, H., Noma, A., Suzuki, T., Ishitani, R., and Nureki, O. (2009). Structural basis of AdoMet-dependent aminocarboxypropyl transfer reaction catalyzed by tRNA-wybutosine synthesizing enzyme, TYW2. Proc. Natl. Acad. Sci. U.S.A. *106*, 15616–15621.

Urbonavicius, J., Skouloubris, S., Myllykallio, H., and Grosjean, H. (2005). Identification of a novel gene encoding a flavin-dependent tRNA:m5U methyltransferase in bacteria – evolutionary implications. Nucleic Acids Res. *33*, 3955–3964.

Voigts-Hoffmann, F., Hengesbach, M., Kobitski, A.Y., van Aerschot, A., Herdewijn, P., Nienhaus, G.U., and Helm, M. (2007). A methyl group controls conformational equilibrium in human mitochondrial tRNA(Lys). J. Am. Chem. Soc. *129*, 13382–13383.

Whipple, J.M., Lane, E.A., Chernyakov, I., D'Silva, S., and Phizicky, E.M. (2011). The yeast rapid tRNA decay pathway primarily monitors the structural integrity of the acceptor and T-stems of mature tRNA. Genes Dev. *25*, 1173–1184.

Wilkinson, M.L., Crary, S.M., Jackman, J.E., Grayhack, E.J., and Phizicky, E.M. (2007). The 2′-O-methyltransferase responsible for modification of yeast tRNA at position 4. RNA *13*, 404–413.

Wilusz, J.E., Whipple, J.M., Phizicky, E.M., and Sharp, P.A. (2011). tRNAs marked with CCACCA are targeted for degradation. Science *334*, 817–821.

Xing, F., Martzen, M.R., and Phizicky, E.M. (2002). A conserved family of *Saccharomyces cerevisiae* synthases effects dihydrouridine modification of tRNA. RNA *8*, 370–381.

Yamazaki, N., Ueshima, R., Terrett, J.A., Yokobori, S., Kaifu, M., Segawa, R., Kobayashi, T., Numachi, K., Ueda, T., Nishikawa, K., *et al.* (1997). Evolution of pulmonate gastropod mitochondrial genomes: comparisons of gene organizations of Euhadra, Cepaea and Albinaria and implications of unusual tRNA secondary structures. Genetics *145*, 749–758.

Yokobori, S., and Paabo, S. (1995). Transfer RNA editing in land snail mitochondria. Proc. Natl. Acad. Sci. U.S.A. *92*, 10432–10435.

Yokobori, S., and Paabo, S. (1997). Polyadenylation creates the discriminator nucleotide of chicken mitochondrial tRNA(Tyr). J. Mol. Biol. *265*, 95–99.

Young, A.P., and Bandarian, V. (2011). Pyruvate is the source of the two carbons that are required for formation of the imidazoline ring of 4-demethylwyosine. Biochemistry *50*, 10573–10575.

Zhu, L., and Deutscher, M.P. (1987). tRna nucleotidyltransferase is not essential for *Escherichia coli* viability. EMBO J. *6*, 2473–2477.

Ziesche, S.M., Omer, A.D., and Dennis, P.P. (2004). RNA-guided nucleotide modification of ribosomal and non-ribosomal RNAs in Archaea. Mol. Microbiol. *54*, 980–993.

Coordination of RNA Editing with Other RNA Processes in Kinetoplastid Mitochondria

4

Jorge Cruz-Reyes and Laurie K. Read

Abstract

The extraordinary RNA editing by U insertion and U deletion in mitochondrial mRNAs is arguably the best characterized process in kinetoplastids. However, much less is known about ancillary factors of the editing multiprotein enzyme core. This enzyme architecture and basic catalysis guided by small non-coding gRNAs have enjoyed central stage, compared to other aspects in the biology of editing substrates, from biogenesis to translation. Many mRNAs and thousands of gRNAs are undoubtedly targeted by numerous factors that regulate unwinding, annealing, stability, assembly into editing enzymes, and translation. Recent years have seen a virtual explosion in the discovery of editing accessory factors. This chapter discusses the progress in this area, and frames a working model whereby the editing machinery is functionally and physically linked to pre and post editing events through a dynamic higher-order network of protein and RNA interactions.

Introduction

Kinetoplastid protozoa

These medically relevant flagellates, including species of *Trypanosoma* and *Leishmania,* contain a distinctive brightly stained DNA structure known as kinetoplast or kDNA in their single mitochondrion. Genome mapping has revealed orthologues for over 6000, out of about 8000 genes (Berriman *et al.*, 2005; El-Sayed *et al.*, 2005; Ivens *et al.*, 2005; Choi and El-Sayed, 2012) Nonetheless, kinetoplastids cause quite different devastating diseases such as human African trypanosomiasis (HAT) or sleeping sickness, Chagas disease and various forms of leishmaniasis (Stuart *et al.*, 2008). Their complex life cycles involve bloodstream and insect stages with dramatic adaptations to host and in energetic metabolism (Vickerman, 1985). For example, *T. brucei* uses the abundant glucose supply in blood to generate energy exclusively by glycolysis, however successful infection of the vector (tsetse fly) requires activation of its mitochondrion. Kinetoplastids are also known for their ancient origin (Sogin, 1991; Simpson and Maslov, 1999) and unique molecular mechanisms of antigenic variation, trans-splicing of nuclear polycistronic units, and remarkable mitochondrial DNA replication and U insertion/deletion RNA editing directed by small guide RNAs (gRNAs) (Benne *et al.*, 1986; Blum and Simpson, 1990; Lee and Van der Ploeg, 1997; Sturm *et al.*, 1999; Campbell *et al.*, 2003; Liu *et al.*, 2005; Cruz-Reyes, 2007). Notably, nuclear and mitochondrial transcription is largely polycistronic; hence kinetoplastid gene expression is primarily regulated after transcription making it fundamentally distinct from most organisms (Koslowsky and Yahampath, 1997; Grams *et al.*, 2000; Campbell *et al.*, 2003; Pays, 2005; Palenchar and Bellofatto, 2006; Clayton and Shapira, 2007). This chapter examines expression of the kDNA (mitochondrial genome) focusing on RNA editing, arguably the most studied molecular process in kinetoplastids (Simpson *et al.*, 2004; Stuart *et al.*, 2005; Cruz-Reyes and Hernandez, 2008). Following a brief account of kDNA organization and RNA editing, we discuss a rapidly expanding network of accessory factors and mechanisms that physically or functionally link editing with other aspects of mitochondrial RNA biology. Most data and nomenclature used here derive from *T. brucei* studies but *Leishmania* systems are considered to be very similar.

Mitochondrial genome

The kDNA is typically a giant catenated network of circular molecules. In *T. brucei,* it consists of a few dozen copies of a maxicircle (~23–40 kb) and a heterologous population of ~10,000 minicircles (~1 kb) representing over a hundred different classes (Simpson *et al.*, 2000; Hong and Simpson, 2003; Ochsenreiter *et al.*, 2007). Maxicircle genes are tightly packed in both strands, including two rRNAs, 18 mRNAs, and two gRNAs, all of which are transcribed in polycistronic units except for one of the gRNAs. rRNAs and mRNAs are similar to those in other mitochondria, but the former are the smallest known examples in eukaryotes, and most mRNAs require editing to create a translatable sequence (Fig. 4.1). Significant overlapping of several genes evidently calls for controlled mechanisms of mutual exclusion in RNA termini formation, since production of a functional transcript could truncate and effectively destroy another (Clement *et al.*, 2004). Minicircles contain the vast majority of gRNA genes identified to date. They occur on the major strand only, either as clusters of three to five gRNAs transcribed in multicistronic precursors in *T. brucei* or as a single gene in *Leishmania*. Interestingly, besides the above-recognized kDNA organization, recent studies have revealed novel small gRNA-sized molecules of unknown function in the two strands of both maxicircles and minicircles, including areas not previously known to be coding (Madej *et al.*, 2007, 2008; Aphasizheva and Aphasizhev, 2010). Since kDNA lacks tRNA genes, all tRNAs are imported into mitochondria after maturation in the cytosol (Alfonzo and Soll, 2009).

RNA editing

Most mRNAs in the single mitochondrion of trypanosomes are edited, with U insertion alone producing most of the final sequence in some transcripts (Feagin *et al.*, 1988; Decker

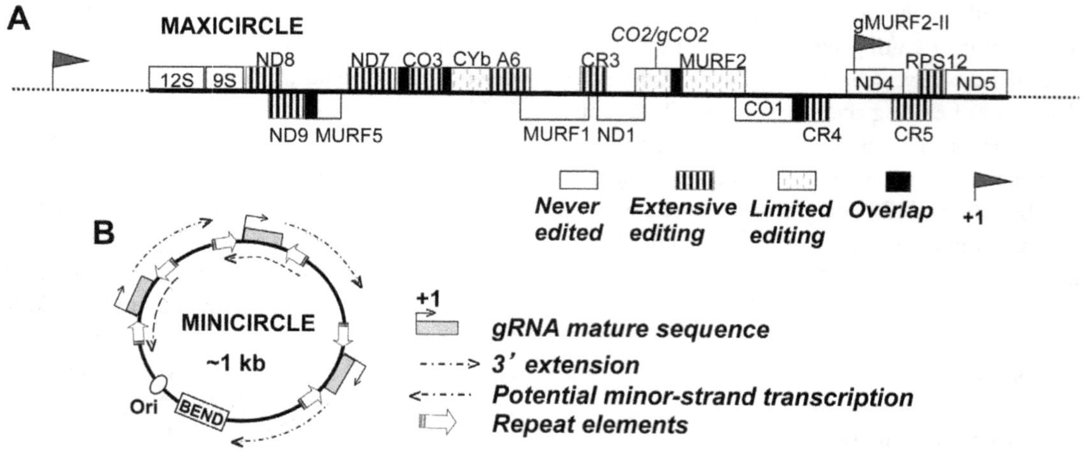

Figure 4.1 kDNA in *T. brucei* mitochondria. (A) Maxicircle gene cluster (~15 kb) on the major and minor strands. The extent of editing and gene overlaps are indicated (see key). Transcription initiation ~12 kb upstream from the 12S rRNA gene is thought to generate polycistronic units across the gene cluster. Mature rRNA and mRNAs arise via 5′ and 3′ processing. The cis-acting gRNA, gCO2, is appended to the 3′ end of its mRNA target. The only known monocistron in kDNA is the trans-acting gRNA gMURF2-II made from its own promoter. MURF, maxicircle unidentified reading frame; CR, C-rich; CO, cytochrome oxidase; A6, ATPase 6; ND, NADH dehydrogenase; Cyb, apocytochrome *b*; RPS12, ribosomal protein subunit 12. (B) Minicircle gRNA genes on the major strand, each encoding a primary transcript. The nascent +1 position marks the mature 5′ end, thereby precursor extensions (clockwise dotted arrows) require 3′ processing. Typical gRNAs, with some exceptions, occur in cassettes with imperfect ~18 bp inverted repeats (open arrows). Minor-strand transcription of unknown function possibly at multiple locations (anticlockwise arrows) remains to be annotated. The *ori* and a region of bent helical DNA are indicated. The kDNA lacks tRNA genes, and therefore the nuclear-encoded transcripts are imported into mitochondria

and Sollner-Webb, 1990; Sturm *et al.*, 1992). Complete editing of an mRNA entails a vast and intricate series of RNA-RNA and RNA–protein interactions, which must be executed in an accurate and efficient fashion. For example, precise gRNA/mRNA pairing is a hallmark of the editing process, without which creation of functional open reading frames would be impossible. gRNAs interact with cognate mRNAs to form an anchor duplex that initiates a gRNA-defined editing block. It is estimated that there are nearly 1000 gRNAs in *T. brucei,* including both redundant and extensively overlapping gRNAs (Ochsenreiter *et al.*, 2007), and over three thousand editing sites in the transcriptome (Burgess RNA 2000). A single mRNA is remodelled from 3′ to 5′ in blocks of 10 to 15 sites, each block been defined by continuous basepairing, including G-to-U, between edited mRNA and encoded gRNA sequence. A downstream edited block often forms a specific anchor duplex with the 5′ end of the incoming upstream gRNA (Koslowsky *et al.*, 1991; Maslov and Simpson, 1992). The editing enzyme, represented by a stable multiprotein core structure termed RECC or 20S editosome (Rusche *et al.*, 1997; Stuart *et al.*, 2005; Li *et al.*, 2009; Simpson *et al.*, 2010), remodels individual mRNA bonds in three basic steps directed by the annealed gRNA: first, endonucleolytic cleavage; then U-addition or U-removal; and finally resealing by ligation (Cruz-Reyes and Sollner-Webb, 1996; Kable *et al.*, 1996; Seiwert *et al.*, 1996; Kang *et al.*, 2006). Different sets of enzymes catalyse the U insertions and U deletions (Huang *et al.*, 2001; Cruz-Reyes *et al.*, 2002; Ernst *et al.*, 2003; Schnaufer *et al.*, 2003; Carnes *et al.*, 2005; Trotter *et al.*, 2005; Rogers *et al.*, 2007).

How a functional cognate substrate reaches RECC is currently unknown. In the mitochondrial milieu, the diverse gRNA pool faces an mRNA population of exceeding complexity, consisting of 'never edited', 'unedited', 'edited', and a hugely heterogeneous population of 'partially edited' mRNAs, besides rRNAs and tRNAs. Predicted anchor duplexes range from 4 to 12 nucleotides. Given these complexities, RECC presumably requires numerous editing accessory factors with RNA binding, annealing, and chaperone activities to facilitate accurate and effective gRNA/mRNA anchor duplex formation. Furthermore, RNA editing reactions *in vitro*, originally reconstituted using synthetic cognate substrates and mitochondrial extract (Seiwert and Stuart, 1994), only represent a small fraction of the natural process. In line with this notion, purified RECC: (1) is most effective with small synthetic substrates, intended to minimize the entropy penalty of cognate RNA structures (Cruz-Reyes *et al.*, 2001; Cifuentes-Rojas *et al.*, 2006); (2) minimally requires single helical turns near the editing site for cleavage and full-round editing (Cifuentes-Rojas *et al.*, 2006; Hernandez *et al.*, 2008); (3) recognizes interconverted deletion and insertion sites upon manipulation of simple features (Cifuentes-Rojas *et al.*, 2005); and (4) edits up to two contiguous sites efficiently (Alatortsev *et al.*, 2008). However, purified RECC is unable to discern cognate edited from never-edited substrates, and lacks the processivity required *in vivo*, indicating that additional factors confer substrate specificity and the natural regulation of this process.

Besides the multitude of gRNA/mRNA duplexes with differing structures that are recognized by RECC *in vivo*, accurate interactions between mRNA, gRNA, and RECC-associated endonucleases, TUTase, exonuclease, and RNA ligases must ensue, such that mRNA cleavage takes place at the appropriate site, and the precise number of Us get inserted or deleted. Accessory factors may facilitate the dynamics of the RNA-RNA, RNA-protein, and protein–protein interaction networks of RECC, and ensure that the reactive sites of the pre-mRNA are properly positioned for catalysis. This process is further complicated by the existence of three different types of RECCs for insertion or deletion, each containing one of three specialized endonucleases and associated proteins (RENs) (Carnes *et al.*, 2007; Guo *et al.*, 2012). It is unknown whether the different types of RECCs exist as stable complexes or if RENs and their partner proteins shuttle on and off a stable core. Either way, transiently associating proteins may facilitate transitions of different classes of RECCs or act as chaperones for the different RENs.

Since RNAs must be extensively rearranged during editing, intermolecular gRNA/mRNA interactions are in a continual state of rearrangement within a gRNA-defined editing block as the gRNA interacts with an increasingly edited mRNA (Leung and Koslowsky, 2001). The latter necessitates that intramolecular structures are also in flux as editing proceeds. Both gRNAs and mRNAs possess secondary structural elements that may compete with or impede productive gRNA/mRNA interactions. gRNAs typically form a two stem–loop structure (Schmid *et al.*, 1995), while mRNA anchor regions can be engaged in quite stable intramolecular stems (Reifur *et al.*, 2010). Numerous proteins that modulate RNA/RNA interactions presumably assist in RNA rearrangements as editing progresses from one site to the next within a gRNA-defined block. Once editing of a block is completed, gRNAs must be exchanged in a process that entails unwinding of a ~40-nt stretch of gRNA/mRNA complementarity so that a gRNA that has completed its guiding function is removed. Displacement of one gRNA must occur prior to annealing of the next gRNA, which is complementary to the edited sequence that was directed by (and, thus, paired with) the previous gRNA. Because an extensive gRNA/mRNA duplex must be disrupted, gRNA exchange almost certain involves the actions of RNA helicases. Finally, editing accessory factors are likely required to optimize editing fidelity and perhaps mediate the discard of improperly edited RNAs, analogous to the role of several helicases in pre-mRNA splicing (Will and Luhrmann, 2011).

Reverse genetic approaches, typically tetracycline-regulatable RNAi in *T. brucei*, have resulted in the identification of numerous non-RECC proteins whose repression impacts the editing process. These editing accessory factors include RNA helicases, RNA chaperones, proteins that facilitate RNA annealing, and RNA binding proteins, which likely modulate both intra – and intermolecular RNA–RNA interactions. In addition to unwinding double-stranded RNA, helicases may act as RNPases, utilizing ATP hydrolysis to facilitate RNA–protein rearrangements involving both RECC and non-RECC editing factors. RNAi-mediated repression of some factors leads to defects in the editing of specific RNAs, while others affect a broad range of RNAs. In some cases, effects on editing initiation at the 3′ end versus editing 3′ to 5′ progression have been established. Some of these proteins reportedly associate transiently with RECC. All identified factors in these categories are discussed later in this chapter. Notably, a subset of editing accessory factors function within the context of the dynamic macromolecular MRB1 complex (mitochondrial RNA-binding protein complex 1), which is also involved in the coordination of editing with other RNA processing and stability events (Fig. 4.2). The expanding collection of editing accessory factors presumably provides the editing machinery with flexibility and the ability to respond quickly to environmental conditions and life cycle progression. An added level of modulation may also be afforded by post-translational modifications, although this subject is in its infancy.

Besides the mechanistic challenges of the basic editing reaction, the process is further complicated by the need for a tight developmental regulation. Editing is vital as it helps generate a functional respiratory system in kinetoplastids, and in fact contributes to a dramatic switch in energetics in *T. brucei*. In particular, edited mRNAs for components of NADH dehydrogenase peak in the bloodstream form (BF) allowing utilization of reducing power even though this cell stage lacks cytochromes and active phosphorylation. Moreover, edited CYb and COII mRNAs are unique to the insect-stage procyclic form (PF) consistent with the activation of its mitochondrion. Although RNA editing was discovered over two decades ago (Benne *et al.*, 1986), not until recently have we seen a major focus on the molecular network that ties editing with other aspects of mitochondrial RNA biology. Central outstanding questions currently under intense study include: How are editing substrates transcribed and polycistronic units processed? How is editing catalysis and substrate stability/activation controlled? Is the 3′ to 5′ progression of RECC regulated? How is translation of unedited and partially edited mRNAs prevented? What levels of regulation control the dramatic energetic switch during development? The current understanding of these topics is discussed next.

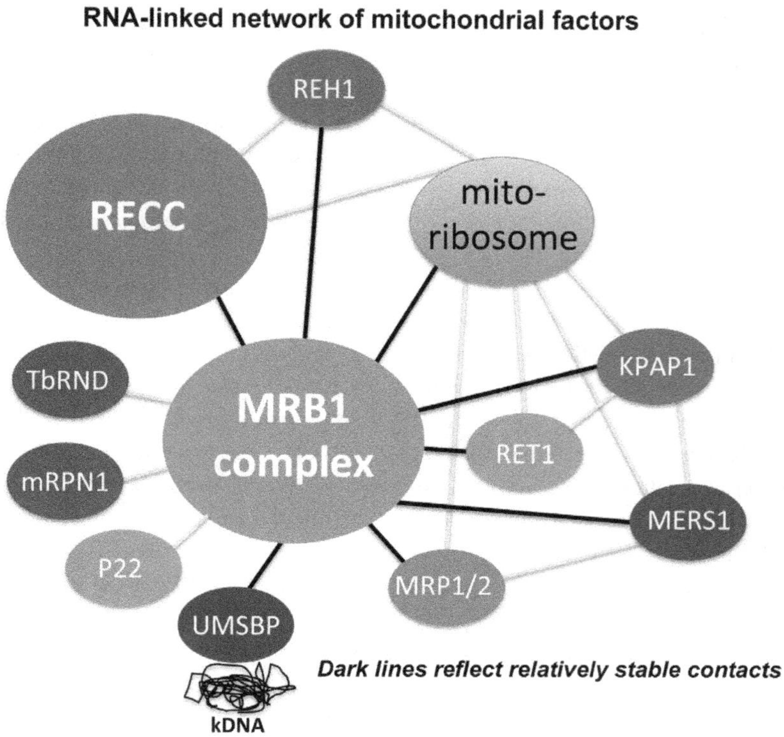

Figure 4.2 RNA-linked network of mitochondrial factors. Dark lines reflect relatively stable contacts. A colour version of this figure is located in the plate section at the back of the book.

Biogenesis

kDNA transcription

Mitochondrial RNA polymerase

The mitochondrial genome of many organisms including plants, animals and protozoa is transcribed by a nuclear-encoded mitochondrial RNA polymerase (mtRNAP), which belongs to a protein family represented by the single-subunit bacteriophage T7 RNAP. The Englund laboratory first reported a candidate mtRNAP that is essential for growth of the procyclic (insect-borne) stage of *T. brucei* (Wang *et al.*, 2000). This protein should also be indispensable in the bloodstream stage since maturation of mitochondrial mRNAs by RNA editing is needed for survival (Schnaufer *et al.*, 2001). The RNA level of mtRNAP appears to be differentially controlled during development (Clement and Koslowsky, 2001) and its sequence conservation is confined to the putative catalytic carboxyl terminus (Clement and Koslowsky, 2001; Grams *et al.*, 2002). Consistent with the expectation that a single mtRNAP is responsible for kDNA expression, RNAi-mediated repression of this protein resulted in the specific loss of mitochondrial mRNAs, rRNAs and gRNAs (Grams *et al.*, 2002; Hashimi *et al.*, 2009). Clement and Koslowsky made the point that all mtRNAPs known to date require one or more factors for specific transcription initiation, and further mused that the ancient origin of kinetoplastids may have brought about a unique transcription machinery in eukaryotes, even distinct transcription factors for minicircle and maxicircle expression. Moreover, mtRNAPs in other organisms normally synthesize the needed primers for replication of the mitochondrial genome. Interestingly, maxicircle but not minicircle replication was compromised upon RNAi down-regulation of the mtRNAP implying a distinct copy mechanism for the two DNA types. Since loss of mitochondrial transcripts was much faster than maxicircle reduction, the effect on RNA does not simply reflect a loss in kDNA

(Grams *et al.*, 2002). It is possible that minicircle replication uses a reported mtDNA primase (Li and Englund, 1997) rather than a specialized mtRNAP, since a second catalytic domain sequence was not found and two mtRNAPs in mitochondria would be unprecedented.

Mitochondrial transcriptional units and potential *cis*-elements

Although a single mtRNAP generates all rRNA, mRNA and gRNA the transcription mechanisms for each RNA class remain unknown. Most kDNA transcription in *T. brucei* is polycistronic but promoter elements and termination sites need to be precisely defined. Transcription of the major strand in maxicircles begins at least 1200 nucleotides upstream of the 12S rRNA, at the 5′ end of the coding region. Transcription initiation in the minor strand has not been identified, but both strands are presumed to create a single large polycistronic precursor that is rapidly processed into mature metabolically stable transcripts (Michelotti *et al.*, 1992). The only known monocistronic unit in the kDNA of *T. brucei* is the intragenic gRNA, gMURF2-II, which is exclusively expressed from its own locus within the ND4 gene (Clement *et al.*, 2004) (Fig. 4.1). The entire mature gMURF2-II sequence resides in the 5′ UTR of the mRNA, and the 5′ termini of these two transcripts are distanced by only six nucleotides. Thus, the gMURF2-II primary transcript derives from its promoter element, rather than through mutually exclusive processing of polycistronic precursors in maxicircles. Since no conserved sequence elements are evident near initiation sites, it is unclear whether the same or different transcription complexes produce the gMURF2-II monocistron and major/minor-strand polycistrons in maxicircles. Interestingly, maxicircles in *L. tarentolae* contain many non-coding RNAs (ncRNAs) of unknown function that resemble canonical gRNAs in size, stability and abundance but do not evidently target unedited mRNAs. Some ncRNAs may be independently transcribed at various identified locations, for example intragenically within unedited mRNA, in the strand complementary to never-edited mRNAs, or in the sense strand but outside the canonical coding region. Two ncRNAs are complementary to cognate gRNAs, including their anchor region, suggesting a regulatory role in editing. Also, a ncRNA in the divergent region near the replication origin could function in replication. Similar ncRNAs may also occur in *T. brucei* but the possibly role of these transcripts in kinetoplastids needs to be determined.

On *T. brucei* minicircles, the major strand contains a cluster of three to five canonical gRNAs genes with the same polarity orientation (Fig. 4.1). A typical gRNA gene is found in a cassette with imperfect inverted 18-bp repeats, and gRNAs were proposed to start with a conserved 5′-RYAYA-3′ motif, where the purine is the 5′ terminal residue. This residue representing +1 in the gene is located 31–32 bp away from the upstream repeat (Pollard *et al.*, 1990), implying a role of these highly conserved elements in transcription. However, some gRNAs characterized in a more recent study lacked this conservation, but it was unclear if this may be due to truncations of the 5′ terminus (Madej *et al.*, 2008). Also, a few gRNAs are not located in cassettes with inverted repeats, for example gCYB(558) and gCYB(560A) which direct the editing of the first block of sites in CYb mRNA. These gRNAs are also primary transcripts, but their 5′ ends are surprisingly heterogeneous (Riley *et al.*, 1994; Clement *et al.*, 2004). In line with the notion that the inverted repeats may function in transcription, the absence of these elements may cause imprecise initiation and thus heterogeneity in gCYB 5′ ends. Apart from canonical gRNAs, longer 200–220 nt transcripts were reported near the minicircle replication origin (Pollard *et al.*, 1990). These molecules may not represent potential replication primers, as minicircle synthesis most likely uses a primase (Grams *et al.*, 2002). In contrast to *T. brucei*, each minicircle in *L. tarentolae* is thought to encode a single canonical gRNA. However, like maxicircles in this kinetoplastid, both strands of minicircles encode stable gRNA-sized ncRNAs of uncertain function, in a few instances complementary to canonical gRNAs (Madej *et al.*, 2007). In *T. brucei,* a small ncRNA species in the minor strand of a minicircle was detected in hybridizations with a gRNA-like probe (Aphasizheva and Aphasizhev,

2010). Thus, it appears that the full coding and functional capacity of the kDNA, particularly novel gRNA-sized ncRNAs, has been underestimated.

Although, canonical trans-acting gRNAs in minicircles and maxicircles are only 50 to 60 nt long on average, these metabolically mature sequences derive from unstable 600–800 nt long precursors, including the maxicircle-encoded gMURF2-II (Aphasizheva and Aphasizhev, 2010). The *cis*-acting gCOII in maxicircles, which is appended to the COII mRNA is probably created from the same hypothetical precursor containing mRNAs and rRNAs. Thus, transcription of each gRNA precursor in a DNA minicircle most likely terminates at different locations, and most of the minicircle sequence (~1 kb) may be transcribed consistent with early PCR amplifications of cDNA fragments across minicircles, possibly with the exception of a short region containing the DNA replication origin (Grams *et al.*, 2000). Further understanding of kDNA transcription requires the development of functional assays with mitochondrial extract or purified components. These assays would allow testing putative promoter sequences and identification of functional cofactors of the native transcriptional machinery. Purified recombinant and native mtRNAP are needed to validate the expected enzymatic activity *in vitro*. Also, characterization of mtRNAP function and its accessory factors during development may reveal important insights in transcription regulation, and perhaps replication initiation in mitochondrial biogenesis.

Processing of primary transcripts

As discussed above, the current model of kDNA expression posits that metabolically stable rRNAs and mRNAs 5′-monophosphate and 3′-hydroxyl termini arise from the processing of long polycistronic precursors in maxicircles. Similarly, minicircle gRNAs in *T. brucei* derive from extended polycistrons. However, because mitochondrial RNAs are not capped *in vivo*, and gRNAs are primary transcripts, the resulting mature gRNAs possess 5′-tri(di)phosphate and 3′ hydroxyl termini, indicative of 3′ but not 5′ processing. Also, poorly characterized gRNA-sized ncRNAs of unknown function in *T. brucei* and *L. tarentolae* appear to bear these ends. In fact, 5′-tri(di)phosphate ends in the pool of gRNAs and ncRNAs are selectively labelled with vaccinia enzyme guanylyltransferase and [α-^{32}P]GTP in preparations of total mitochondrial RNA (Blum *et al.*, 1990). Since these small RNAs are similar in size they are readily detected by guanylyltransferase labelling as a discrete 50–70 nt region in denaturing acrylamide gels.

The molecular basis of RNA processing remain a challenging problem in kinetoplastid mitochondria, particularly the mutually exclusive events that govern rRNA and mRNA expression in maxicircles. An additional complication is that unprocessed or stalled intermediates seem very unstable *in vivo* (Michelotti *et al.*, 1992) often resulting in steady-state levels that are substantially lower than their mature counterparts. Not surprisingly, guanylyltransferase labelling and Northern blots of mitochondrial RNA missed these molecules in several early studies, and PCR amplification of precursor fragments is preferred (Koslowsky and Yahampath, 1997; Grams *et al.*, 2000; Acestor *et al.*, 2009; Ammerman *et al.*, 2011; Madina *et al.*, 2011).

As to production of mature gRNAs, an early *in vitro* study by Grams *et al.* (2000) in the Hajduk laboratory proposed that *T. brucei* mitochondrial extracts contain an activity that sediments at ~19S and specifically processes the 5′-most gRNA in synthetic dicistronic substrates through 3′ end nucleolytic cleavage. In this model, the excised sequence containing downstream gRNAs should be degraded. The authors reported *in vitro* processing of a synthetic 284 nt dicistron containing gCO3 and gND7 at its 5′ and 3′ ends, respectively. This substrate released a ~50 bp fragment containing gCO3, and the activity was only detected upon 5′ labelling suggesting that the 3′ terminus is degraded. The entire gND7 at the 3′ terminus was dispensable for activity but its presence was somewhat stimulatory, and deletion of a few nucleotides in the intergenic region just upstream of gND7 obliterated the activity. A major ~178 nt fragment, that also accumulated in the assay, generated gCO3 when used as the starting substrate. Although, the *in vivo* occurrence of this fragment was not confirmed, this observation suggests a stepwise processing involving endonucleolytic and/or exonucleotyic events.

Furthermore, the 3′ ends of gCO3 generated by *in vitro* cleavage and those mapped *in vivo* appeared to be the same. Other tested dicistronic combinations from the same or different minicircles were also reported to support accurate processing of the 5′ gRNA. Finally, Grams *et al.* (2000) found the gRNA processing activity in biochemically purified editing complexes that sediment at 20S. The authors concluded that the reported gRNA processing is either a moonlight activity of RECC or is catalysed by accessory factors.

Recently, Madina *et al.* (2011) in the Cruz-Reyes laboratory identified an RNase III family endonuclease, termed mRPN1 (mitochondrial RNA processing nuclease 1), which is required for efficient gRNA processing *in vivo*. Accordingly, RNAi-mediated knockdown of mRPN1 in *T. brucei* resulted in reduced gRNA levels and accumulation of pre-gRNA precursors, with a concomitant inhibition of RNA editing. The mRPN1 repression was not lethal presumably because of partial RNAi induction. mRPN1 is not a subunit of RECC but, surprisingly, it shares the general domain architecture of REN editing endonucleases, indicating that mRPN1 and RENs are duplications of a common ancestral gene. Recombinant mRPN1 is dimeric based on structure modelling and gel filtration analyses, and its activity is fully dependent on dsRNA, Mg^{2+}, and an invariable catalytic carboxylate in its nuclease domain. It cleaves within model duplex substrates generating 2-nt 3′ overhangs. This specificity is typical of bacterial RNase III and eukaryotic Dicer in the same family of enzymes (Zhang *et al.*, 2004), but fundamentally distinct from the peculiar cleavage by purified RECC just 5′ of short duplexes (Cruz-Reyes and Sollner-Webb, 1996; Seiwert *et al.*, 1996). The only purified editing endonuclease, REN1, exhibited this specificity (Kang *et al.*, 2006), which is almost certainly conserved for the other REN proteins. mRPN1 may be related to the activity originally described by the Hajduk laboratory, but further studies are needed to delineate the mechanistic details of processing *in vivo*. If helical structures are targeted, these could be provided either *in cis* via fold-back conformations of gRNA precursors, consistent with the observations from the Hajduk laboratory, or *in trans* as duplexes of gRNA precursor sequence with complementary RNAs that may originate in the minor strand of minicircles, as recently proposed (Aphasizheva and Aphasizhev, 2010). Importantly, mRPN1 co-purified with editing proteins through RNA-dependent associations, which may involve gRNA or its precursors (Madina and Cruz-Reyes, personal communication). Furthermore, native mRPN1 in *T. brucei* is associated with three proteins TbRGG2, MRB8170 and MRB4160 in immunopurifications treated with RNases, and reciprocal purifications also suggest a stable RNA-independent interaction with mRPN1. Interestingly, TbRGG2, MRB8170 and MRB4160 are present in compositionally related complexes collectively known as MRB1 complexes, which impact RNA editing at multiple levels (described in 'The MRB1 complex and coordination of editing with pre-editing and post-editing functions', below). While down-regulation of TbRGG2 by RNAi had no visible effect on the gRNA pool detected by guanylyltransferase labelling, this factor was proposed to control editing via regulation of gRNA function (Ammerman *et al.*, 2010). The function of MRB8170 and MRB4160 remains to be defined. In line with these observations, the Cruz-Reyes laboratory proposed the 'hand-over' model, whereby mature gRNAs may shuttle from mRPN1 complexes to downstream MRB1 complexes via mechanisms involving transient interactions. It is conceivable that the shared components TbRGG2, MRB8170 and/or MRB4160 help bridge these interactions.

Addition of 3′ tails and its impact on RNA processing and stability

Mature transcripts in *T. brucei* mitochondria contain diverse non-encoded 3′ tails, either short extensions which are either oligo(A), oligo(U), or A-rich with a few U residues, or long A/U heteropolymers (Simpson and Shaw, 1989; Blum and Simpson, 1990; Decker and Sollner-Webb, 1990; Adler *et al.*, 1991; Bhat *et al.*, 1992; Kao and Read, 2007). This section discusses contrasting functional effects of the short tails on stability of mature RNAs, and the impact of 3′ uridylylation on the processing of precursor primary transcripts. A subsequent conversion of short tails into long (~ 120–250 nt) 3′ A/U extensions promotes translation, but this will be discussed in 'Translation', below.

The 3′ terminus of *T. brucei* rRNAs, gRNAs and some mRNAs contains a 10–15 nt U-tail, whereas most mRNAs receive a short 20–25 nt A-tail. A transcriptome study in *L. tarentolae* by Madej *et al.* in the Huttenhofer laboratory proposed that all mitochondrial RNAs in this organism, including ncRNAs of unknown function in both strands of minicircles and maxicircles undergo 3′ U addition, but 3′ adenylylation was not reported (Madej *et al.*, 2007). In minicircles of *T. brucei*, minor-strand transcripts detected in a Northern blot also appear to be modified by 3′ U addition (Aphasizheva and Aphasizhev, 2010). Notably, terminal 3′ uridylylation in mitochondria is largely, if not solely, carried out by the terminal U-specific transferase RET1 (RNA editing TUTase 1), whereas terminal 3′ adenylylation requires the terminal A-specific transferase KPAP1 (kinetoplastid poly-A polymerase 1) (Aphasizhev *et al.*, 2002; Etheridge *et al.*, 2008). It is conceivable that terminal 3′ uridylylation and adenylylation are competing activities during mRNA 3′ tail biogenesis. Likewise, homeostasis between synthesis and trimming of 3′ tails may occur, for example between the opposite activities of RET1 and the exonuclease TbRND on gRNAs. These possibilities will be further examined in 'Effects on RNA stability', below. In any case, *in vivo* or *in vitro* manipulations that may alter a fine balance of factors involved in 3′ terminal maturation or their substrate specificity are likely to have a dramatic impact on transcript function, biogenesis or stability. Adding further complexity, the steady-state level of some mRNAs differs substantially in procyclic and bloodstream stages. These developmental differences correlate with the structure of 3′ termini, particularly the extent of polyadenylation (Bhat *et al.*, 1992; Read *et al.*, 1994) but the regulatory mechanisms involved need to be determined.

Effects on RNA processing

Aphasizheva *et al.* reported that RET1's TUTase activity in *T. brucei* is required for the nucleolytic processing of maxicircle and minicircle RNA precursors (Aphasizheva and Aphasizhev, 2010). RNAi-induced down-regulation of RET1 led to the accumulation of unprocessed precursors for gRNAs and rRNAs, and loss of mature transcripts. In the few mRNAs examined, induced accumulation of precursors was also accompanied by a decrease in mature transcripts. Northern blots of maxicircle gMurf2-II and a few minicircle gRNAs showed a predominant ~ 600 nt species that accumulates under RET1 repression. Conditions of transcription inhibition confirmed that the concomitant decline in mature RNAs is not due to accelerated decay. Furthermore, guanylyltransferase labelling of total mitochondrial RNA showed an increase of a heterogeneous 600–800-nt population, and decrease in the gRNA-containing pool, which is shorter than usual due to the loss of 3′ uridylylation in the RET1-depleted cells. Also, overexpression of a catalytically inactive RET1 induced a modest processing inhibition and KPAP1-repression had no visible effects. Together, this suggests that RET1 TUTase activity promotes processing, albeit most likely indirectly. Northern blots of 9S rRNA revealed > 2000 nt species that accumulated, but these species were not detected with a 12S rRNA probe. The observed large transcripts may represent stalled intermediates from hypothetical precursors across the encoded maxicircle region. As expected, a loss of both mature rRNAs was observed upon RET1 repression whether or not conditions of transcription-inhibition were included.

The encoded 3′ end of rRNAs is usually quite uniform, however mapping analyses of 3′ termini showed an association between RET1-repression and an increased incidence of abnormal extensions up to 18 nt long, some of which were uridylylated. This observation has the important implication that 3′ uridylylation either *in cis* or *in trans* may influence the selection of rRNA 3′ ends. The Aphasizhev laboratory speculated that processing involves complementary transcripts in a process indirectly affected by 3′ uridylylation. Following the small ncRNA transcriptome study by Madej *et al.* (2007) in *Leishmania*, the same authors also detected ncRNA species with complementarity to a minicircle gRNA in *T. brucei*, including an abundant ~ 600 s nt species and a ~ 60 nt species at very low level (Aphasizheva and Aphasizhev, 2010). The small ncRNA declined under RET1 repression, and the larger species may have also decreased slightly (either because it is not a precursor or its 3′ Us are stabilizing). This *trans*-model contrasts with a *cis*-model that would be more consistent with the gRNA processing data from the Hajduk laboratory (Grams *et al.*, 2000). Additional studies are needed to understand the indirect role of 3′ U addition in RNA precursor processing in maxicircle and minicircles, and whether this activity may be required *in cis* or *in trans*.

As to mRNAs, precursors of never-edited MURF5 accumulated under RET1-repression and transcription inhibition, including a 600–800 nt heterogeneous species, as well as a much larger species. In turn, the mature mRNA clearly decreased. Incidentally, mature MURF5 is one of the few transcripts with a (destabilizing) short 3′ U tail rather than the (stabilizing) short A – or A-rich tail in most mRNAs, as is discussed below. Thus, even without a U-tail, mature MURF5 mRNA is lost due to inhibited precursor processing in RET1-deplected cells. Also in these cells, the editing substrate RPS12 mRNA, accumulated putative intermediates of ~500, 600, 900 nt long and much larger species detected with an unedited oligonucleotide sequence. Thus, examples of all three major classes of mitochondrial RNAs, and possibly ncRNAs of unknown function in minicircles, are involved in a processing mechanism controlled by 3′ uridylylation.

Effects on RNA stability

mRNA

The regulation of mature mitochondrial mRNA levels by the addition of 3′ tails is complex as it involves seemingly disparate effects and underlying molecular mechanisms that remain to be identified. With the identification of RET1 and KPAP1 as 3′ terminal transferases and their genetic repression by RNAi, significant progress has been achieved in understanding the normal roles of 3′ tails *in vivo* (Aphasizhev *et al.*, 2003c; Etheridge *et al.*, 2008). Namely, specific effects on steady-state levels or stability of mRNAs where tied with the type of 3′ modification found in these transcripts as well as their requirement for editing *in vivo*. A short 3′ A-tail or A-rich sequence seems to be required and sufficient to maintain the steady state levels of partially edited, fully edited, and some never-edited mRNAs, as RNAi-induced depletion of KPAP1 reduced their abundance. In contrast, unedited mRNAs were not evidently destabilized by KPAP1 depletion, suggesting that short 3′ A-tails are dispensable in unedited, but protective in edited molecules. Notably, editing directed by the first gRNA appears sufficient to activate the protective properties of A-tails. This is consistent with earlier *in vitro* data using purified mitochondrial fractions demonstrating that an RNA edited at only six editing sites at its 3′ end was stabilized by an oligo(A) tail as effectively as was a fully edited RNA (Kao and Read, 2005). Still enigmatic are the mechanisms that couple editing with the conversion of 3′ A-tails into protective structures, and the mechanisms by which oligo(A) tails protect edited and never edited RNAs from decay.

On the other hand, at least half of the annotated never-edited mRNAs (ND1, MURF1, and MURF5) lack A-tails. Instead the mature transcripts contain a short continuous 3′ U-tail which turned out to be destabilizing, as these molecules are significantly up-regulated by RNAi repression of RET1. An early study of the extensively edited mRNA COIII, showed that unedited molecules have short continuous 3′ U extensions (instead of A tails) but their role in stability was not examined (Decker and Sollner-Webb, 1990). Whether COIII mRNA provides the only example of unedited sequences with 3′ oligo(U) termini is unclear. Overall, it appears that the nature and functional impact of 3′ termini in mature mRNAs *in vivo* is transcript-specific even though they originate from nucleolytic processing of the same hypothetical precursor. The *in vivo* studies described above, were preceded by studies *in vitro* that collectively provided the first indications of a critical interplay between 3′ adenylylation and 3′ uridylylation (Ryan and Read, 2005). For example, it was found that exogenous addition of UTP *in organello*, induced significant degradation of the polyadenylylated mRNA pool. This *in organello* activity required active polymerization of UTP, as the supplement of uridine analogues that should block U-addition also inhibited mRNA degradation.

Furthermore, in the same system, but derived from RET1-repressed cells, the UTP-dependent mRNA degradation activity was diminished confirming that TUTase activity is required. In the light of the recent *in vivo* studies, it appears that flooding the organelle with UTP disrupted a normal balance between RET1 and KPAP1, forcing terminal U addition in mRNAs that normally have a short continuous 3′ A-tail. The artificially incorporated Us *in vitro* may act as a decay signal directing mRNAs to the normal pathway that degrades 3′ U-tailed ND1, MURF1 and MURF5 mRNAs. In the studies *in organelle* using RET1-repressed cells, the observed loss of mRNA could also reflect inhibition of precursor processing, but the original study did not resolve whether the molecules being degraded in a U-dependent fashion included precursors or only mature sequences. Thus, the early *in vitro* studies underscore the importance of a fine regulation of RET1 and KPAP1 activity in the creation of 3′ tails and thereby the metabolic fate of mitochondrial mRNAs.

gRNA

Mature gRNAs normally possess a precise encoded 3′ end with heterogeneous 3′ extensions of 10–15 Us, whose *in vivo* function has been debated since the discovery of these molecules (Blum and Simpson, 1990). The 3′ U extensions could tether purine-rich stretches in unedited sequence in functional mRNA/gRNA duplexes, but could also represent stabilizing structures. The following studies support seemingly contrasting views of the role of 3′ U tails on gRNA stability. In one case, the impediment of U-tail synthesis has no effect in stability, whereas in another the active degradation of U-tails is destabilizing.

On one hand, the Aphasizhev laboratory addressed the turnover of trans-acting maxicircle gMurf2-II and minicircle gRNAs under conditions of RET1 repression and transcription blockage (Aphasizheva and Aphasizhev, 2010). Surprisingly, the half-life of accumulated gRNAs with and without U-tails was very similar in time courses without transcription. This implied that U-tails are dispensable for stability, and that undefined gRNA features control turnover. Association of gRNAs with MRB1 (GRBC) complexes is thought to be stabilizing, as repression of GAP1(GRBC2) or GAP2 (GRBC1) leads to gRNA loss. Because recombinant GAP2 has RNA-binding activity, it was speculated that this factor could directly bind endogenous gRNA. Also, a synthetic gRNA with or without U-tail was found to bind GAP2 with similar affinities, although the specificity for other RNAs was not evaluated.

On the other hand, Zimmer *et al.* (2011) in the Read laboratory characterized a novel mitochondrial 3′ to 5′ exoribonuclease, TbRND, which specifically degrades poly(U) oligonucleotides *in vitro*. Guanylyltransferase labelling assays showed that RNAi repression of TbRND results in a 2–3 nt extension of the gRNA pool on average, but this had no visible effects on cell growth or RNA editing. However, overexpression of wild-type, but not catalytically inactive TbRND, induces gRNA loss and consequent inhibition of editing. Interestingly, COII mRNA editing, which uses a cis-acting gRNA was unaffected. The effect of TbRND appears gRNA-specific as the level of rRNAs, which are also 3′-uridylylated, was normal upon TbRND depletion or overexpression. Furthermore, tagged TbRND co-purified with a subset of proteins typically found in MRB1 complexes, GAP1, GAP2, TbRGG2 and MRB4160, which are thought to regulate gRNA stability and usage (Weng *et al.*, 2008; Ammerman *et al.*, 2010), consistent with a TbRND–gRNA interaction *in vivo*. Overall, these data suggest that active degradation of 3′ U tails can destabilize trans-acting gRNAs.

While RET1 and TbRND may have competing roles in gRNA U-tail formation explaining its heterogeneous size *in vivo*, the impact of this 3′ terminal structure in turnover needs further clarification. In this regard, gRNAs may be signalled to degrade through a U-tail removal event, rather than through an opposite synthetic mechanism. This would explain why RNAi repression of either TbRND or RET1 has no effect on gRNA stability. It is also possible that the 3′ exoribonuclease activity of TbRND is normally U specific, but overexpressed TbRND *in vivo* may partially invade stabilizing features in the encoded gRNA sequence. Furthermore, loss of gRNA could partly result from an indirect effect on gRNA processing, which requires 3′ uridylylation

(Aphasizheva and Aphasizhev, 2010), but this was not examined. However, the level of rRNAs, whose processing also requires terminal 3′ U addition, is not affected by TbRND expression consistent with rRNA processing being largely normal in these cells. It should also be kept in mind that induced RNAi-repression is generally incomplete, so that some 3′ uridylylation may remain in the gRNA pool after RET1 down-regulation. Some observations indicate that other exonucleases may participate in gRNA metabolism. For example, TbRND was not detected in bloodstream trypanosomes, but the steady-state level of some gRNAs can differ substantially during development (Koslowsky *et al.*, 1992). In any case, additional studies should clarify a possible interplay between synthetic and degradative events in gRNA metabolism.

rRNA

Mitochondrial rRNA precursors were first detected by the Hajduk laboratory using metabolic labelling analyses that revealed a fast turnover of these molecules (Michelotti *et al.*, 1992). This appears to be true for most primary transcripts in mitochondria, which may be up to hundreds of times less abundant than mature RNAs. Recent decay assays under conditions of transcription inhibition showed that rRNA stability is unaffected by RET1 expression (Aphasizheva and Aphasizhev, 2010). Surprisingly, the same study found that KPAP1 repression led to a moderate stabilization of mature 9S and 12S RNAs, but the reason for this is unclear since there is no evidence that these molecules undergo 3′ adenylylation. Since the steady-state abundance of mitochondrial rRNAs increases approximately 30-fold in bloodstream forms compared with the procyclic developmental stage (Michelotti *et al.*, 1992), features other than 3′ termini are more likely to control turnover of these molecules, or their precursor processing may be differentially regulated.

A major missing link in our understanding of mitochondrial RNA stability in trypanosomes is the identity(ies) of the degradative nuclease(s). Identification and characterization of these enzymes will then permit analysis of their modes of regulation and their *cis*-acting target sequences.

Translation

The uridine insertion/deletion editing of 12 of 18 maxicircle-encoded mRNAs generates an exceedingly complex steady state mRNA population. This population consists of (1) unedited mRNAs that have not yet begun editing, (2) fully edited RNAs that have completed editing, and (3) a highly heterogeneous mixture of partially edited RNAs that are edited to varying degrees at their 3′ ends and unedited at their 5′ ends. In addition, the six never-edited RNAs also add complexity to the mitochondrial RNA population. Thus, an intriguing question regarding mitochondrial translation is raised: faced with this multifarious collection of mRNAs, how does the ribosome manage to translate only those with complete open reading frames (ORFs)? It is critical that there be a means of discrimination because translation of unedited or partially edited mRNAs, which lack complete ORFs, would lead to the production of truncated and non-functional proteins.

Recently, a mechanism involving production of a long A/U 3′ extension was described that is apparently used to specify translation of fully edited RPS12 mRNA in *T. brucei* mitochondria (Aphasizheva *et al.*, 2011). As discussed above, it has long been known that edited and never edited mRNAs exist in two populations with short (~20 nt) and long (~200 nt) 3′ extensions, and that the proportion of a given mRNA with long or short tails is developmentally regulated such that long tails are often correlated with the life cycle stage in which their translation product is likely functional (Bhat *et al.*, 1992). Thus, 3′ tails were postulated to influence mitochondrial gene expression. Long 3′ tails were later revealed to comprise approximately 70% A and 30% U, and it was shown that their synthesis involves KPAP1 (Etheridge *et al.*, 2008). cRT-PCR and Northern blot evidence suggested that long A/U tail synthesis is restricted to completely edited mRNAs. In addition, the Schneider and Koslowsky laboratories showed that repression of the essential pentatricopeptide repeat (PPR) protein, PPR1/KPAF1, leads to disappearance of long tails on mitochondrial mRNAs *in vivo* (Mingler *et al.*, 2006; Pusnik *et al.*, 2007). These observations were later unified by Aphasizheva *et al.* (2011), who showed that long tails are synthesized by the combined action of KPAP1 and RET1. Moreover, *in vitro* PPR1/KPAF1 (potentially in combination with a second PPR protein KPAF2) modulates the enzymatic activities of KPAP1 and RET1 such that long tails with the lengths and composition of those observed *in vivo* are generated. To test whether long 3′ tails are correlated with translation, these authors isolated translating ribosomes and demonstrated that they were 4-fold enriched for long tailed mRNAs and depleted for short-tailed mRNAs and RNA editing factors. Together, these data are consistent with a model in which mRNAs acquire long A/U extensions following the completion of RNA editing, and long tailed mRNAs are subsequently recruited to ribosomes. Several interesting and important questions remain. For example, is this a general mechanism applicable to all mitochondrial RNAs? Do short A-tails and short U-tails, which precede the genesis of long A/U homopolymers, have different effect on the formation of long A/U tails? What is the signal by which PPR1/KPAF1 specifically recognizes fully edited RNAs as competent for long tail synthesis? Regarding the latter, it is conceptually satisfying that completion of the editing process physically signals long tail synthesis, but such a mechanism cannot be invoked for never edited RNAs, which are also present in short and long tailed populations. It is also unknown how long tail synthesis is restricted to only a portion of a given fully edited and never edited mRNA population (Etheridge *et al.*, 2008). That is, why does not 100% of the population of each fully edited or never edited RNA acquire a long tail? This may be related to the regulatory mechanism by which tail length is controlled in a life cycle specific manner. It is interesting in this regard that trypanosome mitochondria contain a large number of PPR proteins compared to other non-plant eukaryotes (Mingler *et al.*, 2006; Pusnik *et al.*, 2007). Thus, PPR proteins, which typically regulate RNA processing in a sequence specific manner, may play a role in regulating 3′ tail synthesis in an RNA and life cycle-specific manner. Additional experiments reported by Aphasizheva *et al.* (2011) entailing mass spectrometry of ribosomal large subunit (LSU) pulldowns also revealed that RECC, the MRB1 complex, RET1, unedited mRNA, and gRNA preferentially associate with the LSU compared with the small ribosomal subunit (SSU). Consequently, these authors speculated that RNA editing and 3′ tail addition occurs in association with the LSU, and that SSU joining occurs following long 3′ tail synthesis. It will be of interest in the future to determine the temporal relationships between these factors (mRNAs, gRNAs, editing factors, and RET1) and the LSU, and to determine what proportion are LSU associated, towards validation of this model. Consistent with a proposed coordination of translation with downstream RNA processes in mitochondria, an earlier study showed that RNA helicase REH2, a subunit of MRB1 complexes pulls down virtually entire mitoribosomes, as well as components of RECC, RET1, MERS1 and MRP complexes (Fig. 4.2). It is thus possible that at least MRB1 components remain associated with the full ribosome during translation (Hernandez *et al.*, 2010).

The MRB1 complex and coordination of editing with pre-editing and post-editing functions

Numerous RNA editing accessory factors associate in the mitochondrial RNA-binding complex 1 (MRB1, also called GRBC), which was independently discovered by three groups using different proteins as bait, and defined as components that co-purify with GAP1/2 in *T. brucei* or *L. tarentolae* mitochondria (Hashimi *et al.*, 2008; Panigrahi *et al.*, 2008; Weng *et al.*, 2008). The reported complexes in these studies included both common and distinct proteins. Subsequent tag purifications or antibody pulldowns, including extensive RNase treatments, confirmed several common components and revealed multiple differential RNA-based associations that appeared bait specific (Weng *et al.*, 2008; Hernandez *et al.*, 2010; Ammerman *et al.*, 2011, 2012). It is currently thought that MRB1 comprises at least 20 proteins that interact through a dynamic network of protein–protein and protein–RNA contacts (Ammerman *et al.*, 2012). Based on studies of RNase-resistant interactions, the Cruz-Reyes and Read laboratories proposed a very similar protein MRB1 core of 5–7 subunits (Hernandez *et al.*, 2010; Ammerman *et al.*, 2011). However, the possibility that some RNA linkers were protected from nuclease action was not eliminated in those studies. Recently, Ammerman *et al.* (2012) more precisely defined the MRB1 architecture through a comprehensive yeast two-hybrid screen combined with co-immunoprecipitations in the presence and absence of nucleases. These studies revealed a six-protein MRB1 core, mediated by protein–protein interactions, that contains GAP1/2, MRB5390, MRB3010, MRB8620 and MRB11870. Given that MRB5390 and MRB3010 appear to function at very early stages of editing and GAP1/2 bind gRNA (see below), the data are consistent with a model in which the MRB1 core is required for functional association of gRNA with RECC. Pulldown and gradient sedimentation data also provide evidence that GAP1/2 exists as a subcomplex apart from the core (Weng *et al.*, 2008; Hashimi *et al.*, 2009; Ammerman *et al.*, 2012). One possibility is that the GAP1/2 heterotetramer initially binds gRNA and subsequently associates with the remainder of the MRB1 core.

TbRGG2 interacts with the MRB1 core both directly and in an RNA-enhanced manner. One possible function of the MRB1 core is coordination of gRNA/mRNA annealing since recombinant core GAP1/2 subunits, and TbRGG2, which exhibits RNA annealing activity, UV cross-linked synthetic transcripts representing a gRNA and an mRNA fragment, respectively (Weng *et al.*, 2008; Ammerman *et al.*, 2010). Current evidence also suggests that TbRGG2 forms mutually exclusive subcomplexes within the MRB1 complex with MRB8170/MRB4160, and MRB8180 (Ammerman *et al.*, 2012). The significance of the different TbRGG2 subcomplexes is currently unknown. Future experiments in which TbRGG2 is uncoupled from its partner proteins by point mutations will be needed to define the roles of distinct TbRGG2 subcomplexes in RNA editing. The genetic analysis of TbRGG2 is particularly challenging as this factor is also present in several non-MRB1 particles, including those described in previous sections. Namely, with mRPN1, MRB8170 and MRB4160 (Madina *et al.*, 2011), with p22 (Sprehe *et al.*, 2010), and with MRB4160, GAP1/2 and TbRND (Fig. 4.2). In the latter case, TbRND binds stably via RNA, apparently to only a fraction of particles (Zimmer *et al.*, 2011). Thus, TbRGG2 is most likely a moonlight factor with multiple mitochondrial functions.

Several RNA binding proteins may interact with the MRB1 core exclusively through RNA linkers. Whether these proteins bind mRNA, gRNA, or both is unknown. For example, MRB6070 contains five RanBP type Zn fingers, suggesting an RNA-binding function. This protein has been identified in a subset of MRB1 complex pulldowns (Panigrahi *et al.*, 2008; Hernandez *et al.*, 2010; Ammerman *et al.*, 2012). Co-immunoprecipitations demonstrated that MRB6070 interacts with GAP1/2 in an RNA-dependent manner and with TbRGG2 and MRB6070 in an RNA-enhanced manner (Ammerman *et al.*, 2012). It will be of interest to determine the RNA targets of MRB6070 as well as the phenotypes of cells repressed for this protein. MRB1680 is another Zn finger-containing protein that may bind RNA. Like MRB6070, MRB1680 has been identified in a subset of MRB1 pulldowns. The association of MRB1680 with MRB1 was nuclease-resistant, in complexes purified via either GAP1, GAP2, or RNA helicase REH2 (Weng *et al.*, 2008; Hernandez *et al.*, 2010), but MRB1680 was undetected in two independent MRB1 pulldowns via MRB3010 (Panigrahi *et al.*, 2008; Ammerman *et al.*, 2011). This is consistent with either: MRB1680 absence, substoichiometric level or tenuous RNA linkages in complexes purified through MRB3010. As described ahead in section 6.4, cells repressed for MRB1680 exhibit a phenotype consistent with a defect in editing 3′ to 5′ progression. Like MRB6070 and MRB1680, numerous other proteins fail to demonstrate direct protein–protein interactions with MRB1 components by yeast two-hybrid screen (Ammerman *et al.*, 2012). Also, many of these proteins appear to be specific to the chosen bait in MRB1 purifications, as previously proposed (Hernandez *et al.*, 2010; Ammerman *et al.*, 2011). Together, this suggests that interactions of these proteins with the MRB1 core are mediated by either RNA, weak protein contacts in mitochondria, or are perturbed in the protein fusions in the yeast system.

In addition to the RNA-mediated and RNA-enhanced interactions within the MRB1 complex, numerous pulldown/mass spectrometry experiments suggest the involvement of a large, dynamic, RNA-mediated supercomplex containing proteins and complexes in numerous aspects of mitochondrial gene regulation (Fig. 4.2). For example, the MRB1 complex reportedly maintains physical contacts with both RECC (Panigrahi *et al.*, 2003a; Fisk *et al.*, 2008; Weng *et al.*, 2008; Hernandez *et al.*, 2010) and the mitochondrial ribosomes (mitoribosomes) (Hernandez *et al.*, 2010; Aphasizheva *et al.*, 2011), indicative of coordination between editing and translation of edited RNA. Mitoribosomes have been reported to co-purify with the RET1 terminal uridylyl transferase that adds Us to gRNA and mRNA 3′ ends, the KPAP1 polyadenylation complex, the MERS1 RNA decay complex, MRP1/2 annealing proteins (Aphasizheva *et al.*, 2011). Conversely, REH2-purified MRB1 complexes contained mitoribosomes, RET1, MERS1, KPAP1, MRP1/2 and RECC complexes (Hernandez *et al.*, 2010). Because RECC, RET1, short tailed mRNAs, and gRNAs co-purify primarily with the LSU, Aphasizheva and Aphasizhev (2011) proposed that RNA editing occurs in association with the LSU rather than with full ribosomes. The observed associations between the KPAP1 complex and MERS1 and RET1 suggest further coordination between RNA 3′ end formation and decay, consistent with earlier functional data on the role of polyadenylation in RNA decay in trypanosome mitochondria (Ryan *et al.*, 2003; Kao and Read, 2005; Etheridge *et al.*, 2008). The MERS1 RNA decay complex also co-purified with MRP1/2 (Weng *et al.*, 2008), which may be related to the role of MRP1/2 in blocking decay of some RNAs (Vanhamme *et al.*, 1998; Muller *et al.*, 2001; Fisk *et al.*, 2009) (see 'RNA annealing', below).

Most likely, the MRB1 complex is a central player in the network of dynamic protein and RNA interactions that control RNA stability, editing and translation in mitochondria. The extent of RNA involvement in select protein interactions within MRB1 may vary. For example, Weng *et al.* (2008) purified MRB1 through tagged GAP1/2 proteins and found the MRB1–MRP1/2 interaction to be largely RNA mediated, while Hernandez, *et al.* (2010) reported RNA-independent association of REH2-purified MRB1 with both MRP1 and the accessory RNA editing helicase REH1 (see 'Additional studies of editing factors cataloged by associated biochemical activities'). These data, and the fact that MRP1 and REH1 are rarely found associated with MRB1 components, suggest that the association of these editing accessory factors with MRB1 may be mediated through the REH2 helicase.

The evident plasticity of MRB1 complexes and appearance of some subunits in non-MRB1 complexes suggest a greater coordination of mitochondrial metabolism extending to transcription, processing of polycistronic precursors, and even replication (Fig. 4.2). For example, Ammerman, *et al.* (2012) identified the universal minicircle sequence-binding protein (UMSBP) (Milman *et al.*, 2007) in pulldowns of the MRB6070, an MRB1 subunit. UMSBP and other novel zinc-finger proteins identified in this study may specifically bind MRB1 complexes purified via MRB6070, as none of these proteins were detected in other MRB1 purifications. MRB5390, a core MRB1 subunit, and MRB1680 and the RNA-bound TbRGG1, which are found in some MRB1 purifications. MRB1680 appears to be required for efficient processing of maxicircle pre-RNAs. Furthermore, the mRPN1 complex with TbRGG2, MRB8170, and MRB4160 (Madina *et al.*, 2011) is involved in gRNA biogenesis. Since the MRB1 complex associates with RET1, gRNA processing and 3′ end formation may be coordinated temporally or spatially. For instance, RET1 could add U tails to gRNAs, directly following the action of the mRPN1 endonuclease although additional endo – or exoribonucleases may be also involved. Also, the possible role of MRB1 at the interface between RNA turnover, processing and translation needs further investigation. The interplay of RET1 and KPAP1 activity in mRNA metabolism, and whether their association with MRB1 is important in the addition of short tails to unedited or partially edited mRNAs, addition of long tails to fully edited RNAs for translation, or both is a subject for further investigation. Clearly, the myriad interactions described above must be subject to dynamic temporal and spatial coordination. A possible central factor involved in this coordination is MRB10130, which is a major node of protein–protein interactions within the MRB1 complex as well as mediating interactions with the KPAP1 and MERS complexes (Ammerman *et al.*, 2012). Structural predications indicate that MRB10130 is a member of the ARM/HEAT repeat family of proteins, whose members often function as organizers of protein complexes (Xu and Kimelman, 2007; Zhao *et al.*, 2009). Unravelling the mechanisms by which the complex protein–protein and protein–RNA interactions described above are regulated will be a fascinating subject for future study and will reveal how RNA editing is coordinated with pre-editing and post-editing gene regulatory events.

Additional studies of editing factors catalogued by associated biochemical activities

Below, we describe in more detail the current knowledge regarding kinetoplastid editing accessory factors in terms of their reported biochemical activities, which likely reflect their functions during the editing process.

RNA annealing

It has long been presumed that proteins that facilitate RNA annealing will be required to mediate the myriad gRNA/mRNA annealing reactions that must occur for productive editing. Several mitochondrial RNA annealing proteins have been described, with varying impacts on the RNA editing process.

MRP1/2

The first mitochondrial protein reported to display gRNA/mRNA annealing activity was MRP1 (Muller *et al.*, 2001), which was later shown to exist in association with the related MRP2 protein in an a_2/b_2 heterotetrameric complex (Aphasizhev *et al.*, 2003b; Schumacher *et al.*, 2006), hereafter referred to as MRP1/2. The MRP1/2 crystal structure revealed that both subunits adopt the same 'Whirly' transcription factor fold, despite lacking any recognizable nucleic acid binding motifs. *In vitro*, MRP1 and MRP1/2 bind to stem–loop II of gRNAs, and facilitate gRNA/mRNA annealing using a matchmaker mechanism (Muller *et al.*, 2001; Schumacher *et al.*, 2006). Upon MRP1/2 binding, the gRNA is stabilized in a conformation in which stem–loop I is unfolded and exposed in a manner permitting RNA–RNA hybridization. More extensive RNA binding studies demonstrated, however, that MRP1/2 shows no sequence specificity and little preference for ssRNA and dsRNA. Repression of MRP1 and MRP2 in PF *T. brucei*, either singly or together, was used to examine their functions in mitochondrial RNA metabolism *in vivo* (Vondruskova *et al.*, 2005; Fisk *et al.*, 2009). Poisoned primer extension and qRT-PCR assays revealed a rather limited role for MRP1/2 in RNA editing. The major effect was on editing of CYb mRNA; the edited version of the transcript was dramatically decreased and the corresponding unedited RNA increased upon MRP1/2 repression. Edited MURF2 and RPS12 RNAs were also decreased, but reductions in the corresponding unedited RNAs suggested the effect could be at the level of RNA stabilization. Consistent with a role for MRP1/2 in RNA stabilization, the never-edited COI, ND4, and ND5 transcripts were significantly decreased in PF MRP1/2 knockdowns, and a widespread effect on RNA stability was observed in BF MRP1/2 knockdowns. Although knockdown studies do not support a widespread role for MRP1/2 annealing activity in RNA editing, MRP1/2 complexes immunoprecipitated or TAP purified from mitochondria contain detectable *in vitro* RNA editing activity (Allen *et al.*, 1998; Aphasizhev *et al.*, 2003b), suggesting that MRP1/2 transiently associates with RECC. Collectively, the available data suggest that MRP1/2 is a multifunctional RNA binding complex that may function in both RNA editing and stability. Given the limited phenotype of cells depleted of MRP1/2 in PF *T. brucei*, this complex must either act redundantly with other annealing factors or possess specificity for a small subset of RNAs, possibly limited to CYb. How this specificity might be achieved is a subject for further study.

RPB16

RBP16 is a mitochondrial cold shock domain protein that was originally suggested to play a role in RNA editing based on its ability to bind gRNA *in vitro* through the oligo(U) tail, and its association with gRNA *in vivo* (Hayman and Read, 1999; Militello *et al.*, 2000). Subsequent studies revealed the capacity of RBP16 to modulate RNA–RNA interactions (Ammerman *et al.*, 2008). RBP16 possesses gRNA/mRNA annealing activity *in vitro*, acts a an RNA chaperone as evidenced by its ability to complement low-temperature growth in an *E. coli* cold shock protein mutant, and it is capable of melting an RNA hairpin loop in an *E. coli* model system. Interestingly, PF *T. brucei* cells repressed for RBP16 have a phenotype reminiscent of MRP1/2 knockdowns (Pelletier and Read, 2003; Fisk *et al.*, 2009). The most dramatic effect of RBP16 knockdown on editing involves CYb mRNA, and numerous other mitochondrial RNAs are destabilized. In BF, RBP16 appears to have a greater effect on RNA editing than does MRP1/2, since edited A6, COIII, and MURF2 RNAs are significantly decreased while their pre-edited counterparts are not upon RBP16 knockdown. *In vitro* RNA editing studies support a role for RBP16 in editing, as titration of recombinant protein into editing assays leads to a concentration-dependent stimulation of U insertion editing activity (Miller *et al.*, 2006). This stimulation is manifest as an increase in editing at one site, but not as an increase in processivity. On the other hand, RBP16 has never been detected in association with RECC, indicating that any such interactions must be highly transient. As with MRP1/2, analysis of editing defects in knockdown cells indicate that if RBP16 annealing activity functions during editing, it is either redundant or highly specific for a subset of RNAs. Fisk *et al.* (2009) addressed potential redundancy between RBP16 and MRP1/2 by simultaneously knocking down MRP1, MRP2, and RBP16 in PF *T. brucei* and monitoring the effect on RNA editing by qRT-PCR. While these studies revealed redundant functions for RBP16 and MRP1/2 in stabilization of edited A6 and COIII RNAs, there was no evidence that these two factors perform redundant functions in RNA editing.

TbRGG2

The strongest candidate for a wide-ranging annealing factor acting in trypanosome RNA editing is TbRGG2. This RNA binding protein comprises an N-terminal G-rich region with GWG and RG repeats and a C-terminal RRM domain. It was first identified by mass spectrometry of immunopurified editosomes (Panigrahi *et al.*, 2003a), and subsequent co-immunoprecipitations confirmed a transient TbRGG2–RECC interaction (Fisk *et al.*, 2008). TbRGG2 appears to function in the context of the MRB1 and other complexes, which include proteins involved in RNA editing, stabilization and gRNA processing (Hashimi *et al.*, 2008; Panigrahi *et al.*, 2008; Weng *et al.*, 2008; Hernandez *et al.*, 2010; Sprehe *et al.*, 2010; Madina *et al.*, 2011; Zimmer *et al.*, 2011; Ammerman *et al.*, 2012).

In vitro, TbRGG2 binds gRNA, unedited mRNA, and edited mRNA, and it exhibits robust gRNA/mRNA annealing activity (Fisk *et al.*, 2008; Ammerman *et al.*, 2010). TbRGG2 can also melt RNA secondary structure in an *E. coli* model system (Ammerman *et al.*, 2010). RNAi studies demonstrated that TbRGG2 is essential for growth and RNA editing of both PF and BF *T. brucei* (Fisk *et al.*, 2008; Acestor *et al.*, 2009). Upon TbRGG2 repression, editing of pan-edited RNAs is dramatically decreased, while that of minimally edited RNAs appears unaffected. Whether TbRGG2 only interacts with pan-edited RNAs or the lack of impact on minimally edited RNAs results from incomplete knockdown is unclear. Abundance of total gRNAs, as measured by 5′ guanylyltransferase labelling, is unaffected by TbRGG2 repression (Ammerman *et al.*, 2010), suggesting a role during the editing process itself. To begin to define which steps in the editing process are facilitated by TbRGG2, Ammerman, *et al.* (2010) examined RNAs from uninduced and induced TbRGG2 knockdown cells using poisoned primer extension, full gene PCR, and cDNA sequencing to distinguish effects at editing initiation (i.e. the 3′ most editing site) from those that occur during 3′ to 5′ progression of editing. Analysis of two different pan-edited RNAs clearly demonstrated that TbRGG2 impacts both initiation and 3′ to 5′ progression of editing. Interestingly, editing pauses at discrete sites upon TbRGG2 repression, and at least in one case, the site of pausing coincides with the 3′ end of the information domain of a specific gRNA. TbRGG2 repression also leads to a very striking decrease in the lengths and occurrence of 'junction regions' thought to be regions of active editing distinguished by their partially edited character. Together, these data point to a function for TbRGG2 in promoting productive gRNA/mRNA interactions during editing, an effect which likely requires the protein's RNA annealing and/or RNA melting activities. Current data strongly suggest that TbRGG2 is a key RNA annealing factor in RNA editing, and that its role in this regard is not redundant with other mitochondrial factors. In depth RNA sequencing studies will provide insight into whether TbRGG2 annealing activity is utilized within an editing block defined by one gRNA, during gRNA exchange, or both.

KREPA4

Although KREPA4 is an integral component of RECC, it is worth mentioning in this context, that this protein also displays gRNA/mRNA annealing activity (Kala and Salavati, 2010). KREPA4 annealing activity is associated with its OB-fold domain and is dependent on gRNA oligo(U) tails. Thus, KREPA4 is well positioned to accelerate gRNA/mRNA annealing during editing, and may contribute to this activity. However, the exceeding weak *in vitro* editing activity of isolated RECC and the apparently essential role of TbRGG2 annealing suggest that KREPA4-mediated annealing is not sufficient for RNA editing *in vivo*.

RNA melting and unwinding

Innumerable events involving destabilization and unwinding of RNA–RNA interactions, both intra – and intermolecular, must occur during RNA editing. gRNAs form a common two stem–loop secondary structure that, at least in some cases, requires partial unwinding to expose the anchor sequence. As stated above, MRP1/2 can perform this role *in vitro*, but appears to have a limited impact *in vivo*. mRNA secondary structure also plays an important role, since the mRNA portion of anchor binding sequences are sometimes sequestered in secondary structure. Reifur *et al.* (2010) examined several unedited mRNAs with regard to their structures, abilities to undergo gRNA pairing, and capacity for specific RECC-mediated endonucleolytic cleavage. These authors demonstrated that mRNA structure can be a critical factor in the efficiency of gRNA targeting and subsequent editing. These data suggest that, given the myriad secondary and tertiary RNA structures that need to be disrupted to allow gRNA binding, different RNAs may require different sets of editing accessory factors to permit effective gRNA/mRNA interactions. Differential effects of mRNA structure on editing efficiency is interesting with regard to RBP16 and MRP1/2 and their apparent specificity towards CYb RNA *in vivo*. The initiating anchor binding sequence of CYb mRNA is sequestered in a stable stem (Reifur *et al.*, 2010); thus, the RNA melting activity of RBP16 may be especially critical for this RNA to allow disruption of the stem and gRNA binding. The ability of MRP1/2 to present the gRNA anchor may be required in this case in which the availability of the anchor binding sequence is somewhat compromised. gRNA exchange almost certainly requires RNA unwinding. This process likely involves ATP-dependent RNA helicases, and two such proteins with differing roles in the editing process have been described (Missel *et al.*, 1997; Hashimi *et al.*, 2009; Hernandez *et al.*, 2010; Li *et al.*, 2011). These are designated RNA editing helicase (REH)1 and 2.

REH1

REH1 (originally mHel61) was identified bioinformatically as a mitochondrial DEAD-box protein, and both alleles were knocked out in PF *T. brucei* to assess a potential role in RNA editing (Missel *et al.*, 1997). Editing of the two RNAs monitored, CYb and COII, was decreased in the REH1 knockout, while never-edited and nuclear RNAs were unaffected, consistent with a role for this enzyme in editing. More recently, Li *et al.* (2011) revisited the role of REH1, confirming that it possesses the predicted ATP-dependent RNA unwinding activity. In cells repressed for REH1 by RNAi, editing of two out of seven RNAs tested was significantly decreased, while gRNA abundance remained constant. If REH1 depletion of these edited RNAs reflects a critical function in gRNA exchange, the prediction is that editing of the first gRNA-directed block should be unaffected by REH1 repression, while editing of subsequent blocks should be diminished. To test this hypothesis Li *et al.* (2011) used both cDNA amplification with PCR primer pairs assessing progressive 5′ editing and a combined RNase protection/primer extension assay. They found that, for A6 mRNA, REH1 was necessary for optimal progression from editing of block one to more 5′ editing blocks. Hence, REH1 appears to function in gRNA exchange. However, it is required for only a small subset of edited mRNAs, and the basis for this specificity is a mystery. Consistent with a role for REH1 in RNA editing, the protein has been detected by mass spectrometry in immunoprecipitated and biochemically purified RECC preparations, although not in more highly purified TAP isolated preparations from *T. brucei* (Panigrahi *et al.*, 2003b, 2008). Studies with SBP-tagged REH1 in *L. tarentolae* revealed an RNA-dependent association with RECC (Li *et al.*, 2011).

REH2

REH2 was originally identified through its association with the GAP1/2 gRNA binding proteins in the MRB1 complex (Panigrahi *et al.*, 2008; Weng *et al.*, 2008), and independently via purifications of native editing complexes (Hernandez *et al.*, 2010). Hernandez, *et al.* subsequently demonstrated largely RNA-dependent interactions between REH2 and RECC. The absence of REH2 from RECC preparations reported by other groups, typically via affinity tags, likely reflects this transient RNA-based association (Aphasizhev *et al.*, 2003a; Panigrahi *et al.*, 2006). Tagged REH2 precipitated from mitochondrial lysates possesses ATP-dependent gRNA/mRNA unwinding activity that is dependent on both its double-stranded RNA binding domain (dsRBD) and motif I, the latter being typically associated with ATP hydrolysis in DExH-box RNA helicases (Hernandez *et al.*, 2010). Unlike REH1, repression of REH2 leads to a substantial decrease in the levels of mature gRNAs and a consequent repression of editing of the majority of RNAs with the exception of COII, whose editing does not rely on *cis*-acting gRNAs (Hashimi *et al.*, 2009; Hernandez *et al.*, 2010; Madina *et al.*, 2011). Moreover, wild-type REH2 is associated with mature gRNAs in trypanosome mitochondria. Interestingly, pulldowns of dsRBD or motif I mutant REH2 contain significantly less mature gRNA than do wild-type pulldowns, and overexpression of either of these mutants results in a dominant negative effect on trypanosome growth (Hernandez *et al.*, 2010). However, gRNA precursors do not accumulate in REH2-repressed cells, suggesting that the protein does not impact gRNA biogenesis (Madina *et al.*, 2011). Upon gradient sedimentation, REH2 is detected in high density particles that are stabilized by RNA and its double-stranded RNA binding domain (dsRBD), as well as in relatively low density RNase-resistant particles. Mass spectrometry revealed RNase-resistant association with numerous components of the MRB1 complex in addition to GAP1/2. REH2 also displays an RNA-dependent association with mitoribosomes. Thus, REH2 appears to function in stabilization of mature gRNAs, while its physical association with proteins involved in numerous mitochondrial gene regulatory processes suggest it could be multifunctional. Indeed, gRNA destabilization may simply reflect an impediment to proper gRNA utilization, such that gRNAs that are unable to engage in productive editing enter a decay pathway.

RNA binding

In addition to the proteins described above, to which some RNA-based activities have been ascribed, a few additional RNA-binding proteins have been implicated in RNA editing. It is likely that this group of proteins will expand as additional proteins impacting editing are discovered. RNA binding proteins may contribute to editing at steps such as gRNA and/or mRNA transport and association with RECC or RNA stabilization.

TbRGG1

As its name implies, TbRGG1 contains an extended RGG box indicative of RNA binding. Early studies demonstrated its ability to bind poly(U), suggesting a possible role in binding the gRNA tail (Vanhamme *et al.*, 1998). Recent studies showed that TbRGG1 repression does not affect gRNA levels; however, edited RNAs are substantially decreased and some pre-edited RNAs are increased (Hashimi *et al.*, 2008). Because edited COII RNA levels are decreased in TbRGG1 knockdowns, an effect on gRNA utilization is unlikely. A role in either edited RNA stabilization or in editing efficiency has been postulated for this RNA-binding protein. TbRGG1 has never been reported to be associated with RECC. However, it does interact through an RNA linker with the MRB1 complex, consistent with a role in RNA editing or stabilization (Hashimi *et al.*, 2008).

GAP1 and GAP2

GAP1 and GAP2 (also known as GRBC1/2) were originally identified as proteins associated with MRP1/2 in *L. tarentolae,* although this interaction does not appear to be as stable in *T. brucei* (Aphasizhev *et al.*, 2003b; Weng *et al.*, 2008). They were later shown to associate in an a2/b2 heterotetramer (hereafter called GAP1/2), and this association was validated *in vivo* by the loss of both proteins upon repression of one or the other RNA (Weng *et al.*, 2008; Hashimi *et al.*, 2009; Aphasizheva and Aphasizhev, 2010). Despite the absence of any recognizable RNA-binding domain, UV cross-linking experiments demonstrated that both GAP1 and GAP2 bind the encoded portion of gRNAs *in vitro* (Weng *et al.*, 2008; Aphasizheva and Aphasizhev, 2010). *In vivo*, GAP1/2 exhibits a low level of RNA-dependent interaction with RECC (Weng *et al.*, 2008). Moreover, isolation of tagged GAP1 or GAP2 from *L. tarentolae* and *T. brucei* defined the novel MRB1 (also known as GRBC) complex described above. GAP1/2 appear to interact with a range of large macromolecular particles, as indicated by its heterodisperse distribution on glycerol gradients. Both RNase treatment of mitochondrial extracts prior to gradient sedimentation or isolation of RNA from dyskinetoplastic trypanosomes that lack mitochondrial RNA demonstrate that the association of GAP1/2 with very large mitochondrial particles is RNA-dependent (Acestor *et al.*, 2009; Hashimi *et al.*, 2009). Repression of GAP1/2 in PF *T. brucei* leads to a very striking phenotype. In the absence of these proteins, the gRNA population becomes significantly decreased, as measured by both bulk labelling of the gRNA population by guanylyltransferase activity and Northern blotting of individual gRNAs (Weng *et al.*, 2008; Hashimi *et al.*, 2009). Weng *et al.* (2008) showed that decay of specific gRNAs is accelerated in the absence of GAP1/2. Consequently, either GAP1 or GAP2 repression leads to a dramatic decrease in editing of all RNAs, with one important exception. Editing of COII RNA, whose single gRNA acts *in cis*, is unaffected. Thus, GAP1/2 are critical in the maintenance of trans acting gRNAs. A major question in the field is the mechanistic role of these proteins in RNA editing. Is their primary role to prevent decay of gRNAs, or do they function intimately in the utilization of gRNAs during editing? If so, at which step(s) do they act and is it important that they act in the context of the MRB1 complex?

Other proteins that affect RNA editing

p22

p22 is the homologue of human p32 protein, and a member of the mam33p protein family that is found in all cellular compartments and implicated in pre-mRNA splicing and ribosome biogenesis (Petersen-Mahrt *et al.*, 1999; Sprehe *et al.*, 2010; Yoshikawa *et al.*, 2011). Repression of p22 in PF *T. brucei* results in a very specific and dramatic inhibition of COII RNA editing (Sprehe *et al.*, 2010). COII RNA editing is unique in two respects. First, it is effected by a dedicated RECC subclass containing REN3 and KREPB6 (Carnes *et al.*, 2007). Second, COII RNA is not guided by a *trans* acting gRNA. Rather, the single COII gRNA is contained in the COII mRNA 3′ untranslated region and acts in *cis*, and thus COII RNA editing requires recruitment of only one RNA to the RECC. While the function of p22 in COII RNA editing is unclear, it does not bind RNA directly. One possible role for p22 in RNA editing is as a chaperone that recruits REN3 and/or KREPB6 to RECC. It may also recruit COII RNA through an unidentified RNA binding protein or impact more subtle rearrangements within the COII RNA-loaded RECC. Co-immunoprecipitation experiments demonstrated RNA-independent association of p22 with both TbRGG2, also found in MRB1 complexes, and the KREL1 component of RECC, consistent with a direct role in COII RNA editing (Sprehe *et al.*, 2010). Surprisingly, reported converse purifications did not identify p22, suggesting that p22 is substoichiometric, or binds stably only to a subset of MRB1 and RECC complexes. Other editing factors used as control, MRP1/2 and RBP16, were not detected, indicating that the above p22 associations are specific.

MRB1680, MRB5390 and MRB3010

MRB1680, MRB5390 and MRB3010 (names based on Ammerman *et al.*, 2012) were first identified based on their physical interactions with GAP1/2 in the MRB1 complex, although MRB1680 was identified in only one of three original studies (Hashimi *et al.*, 2008; Panigrahi *et al.*, 2008; Weng *et al.*, 2008). Interactions of MRB1680, MRB5390, and MRB3010 with GAP1/2 suggested that these proteins might function in RNA editing, and their impacts on editing were tested in RNAi studies. Repression of each of these proteins caused a dramatic growth defect and abnormalities in RNA editing, although the latter differed in some respects that may provide clues as to the proteins' functions.

MRB1680 repression had a dramatic effect on the editing of three out of four pan-edited RNAs tested, but was similar to TbRGG2 in that no effect on minimally edited RNAs was observed (Acestor *et al.*, 2009). The small corresponding increase in pre-edited RNAs observed in MRB1680 knockdown cells suggests that in the absence of wild-type MRB1680 levels, RNAs can enter the editing pathway but do not progress efficiently to fully edited status. Thus, MRB1680 may, like TbRGG2, impact the 3′ to 5′ progression of editing, although this has not been directly tested. MRB1680 comprises five C2H2 zinc fingers; thus, analysis of its RNA binding capacity will be an important direction towards understanding its mechanism of action.

MRB5390 repression led to a modest effect on editing, with two of four pan-edited RNAs and one of three minimally edited RNAs tested being impacted. Interestingly, pre-edited RNAs accumulated to a much greater extent in MRB5390 knockdowns than in MRB1680 knockdown. This suggests that MRB5390 may have a greater impact than MRB1680 on the initial stages of editing, such that RNAs never enter the editing pathway in cells depleted of MRB5390 and pre-edited RNAs accumulate. Effects on gRNA levels were not tested, so it is unclear if the impact could be at the level of gRNA abundance, similar to GAP1/2. Increased stability of one never-edited RNA and several dicistronic precursor RNAs was also observed in MRB5390 cells. Thus, MRB5390 may have a function in RNA processing that secondarily impacts editing. However, the identification of MRB5390 in all MRB1 purifications to date, and its placement in the MRB1 core (Ammerman *et al.*, 2012) argues for a more central role in the editing process. Like MRB1680, MRB5390 may interact with RNA as it contains a YbaK/ProRS domain, which is thought to be involved in oligonucleotide binding.

MRB3010 repression impacted editing of all RNAs tested in both PF and BF *T. brucei,* while gRNAs remained at wild-type levels (Ammerman *et al.*, 2011). Editing of pan-edited RNAs was dramatically compromised, while minimally edited RNAs were more modestly affected, in MRB3010 knockdowns. Like MRB5390, repression of MRB3010 caused a significant increase in pre-edited RNAs coincident with depletion of edited RNAs, suggesting a role at the earliest stages of editing. To directly assess this, Ammerman, *et al.* (2011) used full-gene RT-PCR, which permits visualization of the entire population of a given RNA including pre-edited, partially edited, and fully edited, upon subsequent electrophoresis. Analysis of two pan-edited RNAs, A6 and COII, demonstrated that cells repressed for MRB3010 accumulate unedited RNAs and partially edited RNAs. Comparison to RNAs isolated from TbRGG2 repressed cells clearly showed that repression of MRB3010 leads to defects at an earlier stage in editing than does TbRGG2 repression. MRB3010 comprises a component of the MRB1 core. Together with the dramatic effect on editing upon MRB3010 knockdown, these findings suggest that MRB3010 plays a central role in RNA editing. The effect of MRB3010 on RNA editing does not appear to involve direct RNA binding, however, as UV cross-linking experiments failed to detect any gRNA – or mRNA-binding activity.

References

Acestor, N., Panigrahi, A.K., Carnes, J., Zikova, A., and Stuart, K.D. (2009). The MRB1 complex functions in kinetoplastid RNA processing. RNA *15*, 277–286.

Adler, B.K., Harris, M.E., Bertrand, K.I., and Hajduk, S.L. (1991). Modification of *Trypanosoma brucei* mitochondrial rRNA by posttranscriptional 3′ polyuridine tail formation. Mol. Cell. Biol. *11*, 5878–5884.

Alatortsev, V.S., Cruz-Reyes, J., Zhelonkina, A.G., and Sollner-Webb, B. (2008). *Trypanosoma brucei* RNA editing: coupled cycles of U deletion reveal processive activity of the editing complex. Mol. Cell. Biol. *28*, 2437–2445.

Alfonzo, J.D., and Soll, D. (2009). Mitochondrial tRNA import – the challenge to understand has just begun. Biol. Chem. *390*, 717–722.

Allen, T.E., Heidmann, S., Reed, R., Myler, P.J., Goringer, H.U., and Stuart, K.D. (1998). Association of guide RNA binding protein gBP21 with active RNA editing complexes in *Trypanosoma brucei*. Mol. Cell. Biol. *18*, 6014–6022.

Ammerman, M.L., Fisk, J.C., and Read, L.K. (2008). gRNA/pre-mRNA annealing and RNA chaperone activities of RBP16. RNA.

Ammerman, M.L., Presnyak, V., Fisk, J.C., Foda, B.M., and Read, L.K. (2010). TbRGG2 facilitates kinetoplastid RNA editing initiation and progression past intrinsic pause sites. RNA *16*, 2239–2251.

Ammerman, M.L., Hashimi, H., Novotna, L., Cicova, Z., McEvoy, S.M., Lukes, J., and Read, L.K. (2011). MRB3010 is a core component of the MRB1 complex that facilitates an early step of the kinetoplastid RNA editing process. RNA *17*, 865–877.

Ammerman, M.L., Downey, K.M., Hashimi, H., Fisk, J.C., Tomasello, D.L., Faktorova, D., Kafkova, L., King, T., Lukes, J., and Read, L.K. (2012). Architecture of the trypanosome RNA editing accessory complex, MRB1. Nucleic Acids Res.

Aphasizhev, R., Aphasizheva, I., Nelson, R.E., Gao, G., Simpson, A.M., Kang, X., Falick, A.M., Sbicego, S., and Simpson, L. (2003a). Isolation of a U-insertion/deletion editing complex from Leishmania tarentolae mitochondria. EMBO J. *22*, 913–924.

Aphasizhev, R., Aphasizheva, I., Nelson, R.E., and Simpson, L. (2003b). A 100-kD complex of two RNA-binding proteins from mitochondria of Leishmania tarentolae catalyzes RNA annealing and interacts with several RNA editing components. RNA *9*, 62–76.

Aphasizhev, R., Aphasizheva, I., and Simpson, L. (2003c). A tale of two TUTases. Proc. Natl. Acad. Sci. U.S.A. *100*, 10617–10622.

Aphasizhev, R., Sbicego, S., Peris, M., Jang, S.H., Aphasizheva, I., Simpson, A.M., Rivlin, A., and Simpson, L. (2002). Trypanosome mitochondrial 3′ terminal uridylyl transferase (TUTase): the key enzyme in U-insertion/deletion RNA editing. Cell *108*, 637–648.

Aphasizheva, I., and Aphasizhev, R. (2010). RET1-catalyzed uridylylation shapes the mitochondrial transcriptome in *Trypanosoma brucei*. Mol. Cell. Biol. *30*, 1555–1567.

Aphasizheva, I., Maslov, D., Wang, X., Huang, L., and Aphasizhev, R. (2011). Pentatricopeptide repeat proteins stimulate mRNA adenylation/uridylation to activate mitochondrial translation in trypanosomes. Mol. Cell *42*, 106–117.

Benne, R., Van den Burg, J., Brakenhoff, J.P., Sloof, P., Van Boom, J.H., and Tromp, M.C. (1986). Major transcript of the frameshifted coxII gene from trypanosome mitochondria contains four nucleotides that are not encoded in the DNA. Cell *46*, 819–826.

Berriman, M., Ghedin, E., Hertz-Fowler, C., Blandin, G., Renauld, H., Bartholomeu, D.C., Lennard, N.J., Caler, E., Hamlin, N.E., Haas, B., *et al.* (2005). The genome of the African trypanosome *Trypanosoma brucei*. Science *309*, 416–422.

Bhat, G.J., Souza, A.E., Feagin, J.E., and Stuart, K. (1992). Transcript-specific developmental regulation of polyadenylation in *Trypanosoma brucei* mitochondria. Mol. Biochem. Parasitol. *52*, 231–240.

Blum, B., and Simpson, L. (1990). Guide RNAs in kinetoplastid mitochondria have a nonencoded 3′ oligo(U) tail involved in recognition of the preedited region. Cell *62*, 391–397.

Blum, B., Bakalara, N., and Simpson, L. (1990). A model for RNA editing in kinetoplastid mitochondria: 'guide' RNA molecules transcribed from maxicircle DNA provide the edited information. Cell *60*, 189–198.

Campbell, D.A., Thomas, S., and Sturm, N.R. (2003). Transcription in kinetoplastid protozoa: why be normal? Microbes Infect. *5*, 1231–1240.

Carnes, J., Trotter, J.R., Ernst, N.L., Steinberg, A., and Stuart, K. (2005). An essential RNase III insertion editing endonuclease in *Trypanosoma brucei*. Proc. Natl. Acad. Sci. U.S.A. *102*, 16614–16619.

Carnes, J., Trotter, J.R., Peltan, A., Fleck, M., and Stuart, K. (2007). RNA Editing in *Trypanosoma brucei* requires three different editosomes. Mol. Cell Biol.

Choi, J., and El-Sayed, N.M. (2012). Functional genomics of trypanosomatids. Parasite Immunol *34*, 72–79.

Cifuentes-Rojas, C., Halbig, K., Sacharidou, A., De Nova-Ocampo, M., and Cruz-Reyes, J. (2005). Minimal pre-mRNA substrates with natural and converted sites for full-round U insertion and U deletion RNA editing in trypanosomes. Nucleic Acids Res. *33*, 6610–6620.

Cifuentes-Rojas, C., Pavia, P., Hernandez, A., Osterwisch, D., Puerta, C., and Cruz-Reyes, J. (2006). Substrate determinants for RNA editing and editing complex interactions at a site for full-round U insertion. J. Biol. Chem. *282*, 4265–4276.

Clayton, C., and Shapira, M. (2007). Post-transcriptional regulation of gene expression in trypanosomes and leishmanias. Mol. Biochem. Parasitol. *156*, 93–101.

Clement, S.L., and Koslowsky, D.J. (2001). Unusual organization of a developmentally regulated mitochondrial RNA polymerase (TBMTRNAP) gene in *Trypanosoma brucei*. Gene *272*, 209–218.

Clement, S.L., Mingler, M.K., and Koslowsky, D.J. (2004). An intragenic guide RNA location suggests a complex mechanism for mitochondrial gene expression in *Trypanosoma brucei*. Eukaryot. Cell 3, 862–869.

Cruz-Reyes, J. (2007). RNA–protein interactions in assembled editing complexes in trypanosomes. Methods Enzymol. *424*, 107–125.

Cruz-Reyes, J., and Hernandez, A. (2008). Protein–protein and RNA–protein interactions in U-insertion/deletion RNA editing complexes. In RNA and DNA Editing, Smith, H.C., ed. (John Wiley & Sons, Inc., New Jersey), pp. 71–98.

Cruz-Reyes, J., and Sollner-Webb, B. (1996). Trypanosome U-deletional RNA editing involves guide RNA-directed endonuclease cleavage, terminal

U exonuclease, and RNA ligase activities. Proc. Natl. Acad. Sci. U.S.A. *93*, 8901–8906.

Cruz-Reyes, J., Zhelonkina, A., Rusche, L., and Sollner-Webb, B. (2001). Trypanosome RNA editing: simple guide RNA features enhance U deletion 100-fold. Mol. Cell. Biol. *21*, 884–892.

Cruz-Reyes, J., Zhelonkina, A.G., Huang, C.E., and Sollner-Webb, B. (2002). Distinct functions of two RNA ligases in active *Trypanosoma brucei* RNA editing complexes. Mol. Cell. Biol. *22*, 4652–4660.

Decker, C.J., and Sollner-Webb, B. (1990). RNA editing involves indiscriminate U changes throughout precisely defined editing domains. Cell *61*, 1001–1011.

El-Sayed, N.M., Myler, P.J., Bartholomeu, D.C., Nilsson, D., Aggarwal, G., Tran, A.N., Ghedin, E., Worthey, E.A., Delcher, A.L., Blandin, G., *et al.* (2005). The genome sequence of *Trypanosoma cruzi*, etiologic agent of Chagas disease. Science *309*, 409–415.

Ernst, N.L., Panicucci, B., Igo, R.P., Jr., Panigrahi, A.K., Salavati, R., and Stuart, K. (2003). TbMP57 is a 3′ terminal uridylyl transferase (TUTase) of the *Trypanosoma brucei* editosome. Mol. Cell *11*, 1525–1536.

Etheridge, R.D., Aphasizheva, I., Gershon, P.D., and Aphasizhev, R. (2008). 3′ adenylation determines mRNA abundance and monitors completion of RNA editing in *T. brucei* mitochondria. EMBO J. 1596–1608.

Feagin, J.E., Abraham, J.M., and Stuart, K. (1988). Extensive editing of the cytochrome c oxidase III transcript in *Trypanosoma brucei*. Cell *53*, 413–422.

Fisk, J.C., Ammerman, M.L., Presnyak, V., and Read, L.K. (2008). TbRGG2, an essential RNA editing accessory factor in two *Trypanosoma brucei* life cycle stages. J. Biol. Chem. 23016–23025.

Fisk, J.C., Presnyak, V., Ammerman, M.L., and Read, L.K. (2009). Distinct and overlapping functions of MRP1/2 and RBP16 in mitochondrial RNA metabolism. Mol. Cell. Biol. *29*, 5214–5225.

Grams, J., McManus, M.T., and Hajduk, S.L. (2000). Processing of polycistronic guide RNAs is associated with RNA editing complexes in *Trypanosoma brucei*. EMBO J. *19*, 5525–5532.

Grams, J., Morris, J.C., Drew, M.E., Wang, Z., Englund, P.T., and Hajduk, S.L. (2002). A trypanosome mitochondrial RNA polymerase is required for transcription and replication. J. Biol. Chem. *277*, 16952–16959.

Guo, X., Carnes, J., Ernst, N.L., Winkler, M., and Stuart, K. (2012). KREPB6, KREPB7, and KREPB8 are important for editing endonuclease function in *Trypanosoma brucei*. RNA *18*, 308–320.

Hashimi, H., Zikova, A., Panigrahi, A.K., Stuart, K.D., and Lukes, J. (2008). TbRGG1, an essential protein involved in kinetoplastid RNA metabolism that is associated with a novel multiprotein complex. RNA *14*, 970–980.

Hashimi, H., Cicova, Z., Novotna, L., Wen, Y.Z., and Lukes, J. (2009). Kinetoplastid guide RNA biogenesis is dependent on subunits of the mitochondrial RNA binding complex 1 and mitochondrial RNA polymerase. RNA *15*, 588–599.

Hayman, M.L., and Read, L.K. (1999). *Trypanosoma brucei* RBP16 is a mitochondrial Y-box family protein with guide RNA binding activity. J. Biol. Chem. *274*, 12067–12074.

Hernandez, A., Madina, B.R., Ro, K., Wohlschlegel, J.A., Willard, B., Kinter, M.T., and Cruz-Reyes, J. (2010). REH2 RNA helicase in kinetoplastid mitochondria: ribonucleoprotein complexes and essential motifs for unwinding and guide RNA (gRNA) binding. J. Biol. Chem. *285*, 1220–1228.

Hernandez, A., Panigrahi, A., Cifuentes-Rojas, C., Sacharidou, A., Stuart, K., and Cruz-Reyes, J. (2008). Determinants for association and guide RNA-directed endonuclease cleavage by purified RNA editing complexes from *Trypanosoma brucei*. J. Mol. Biol. *381*, 35–48.

Hong, M., and Simpson, L. (2003). Genomic organization of *Trypanosoma brucei* kinetoplast DNA minicircles. Protist *154*, 265–279.

Huang, C.E., Cruz-Reyes, J., Zhelonkina, A.G., O'Hearn, S., Wirtz, E., and Sollner-Webb, B. (2001). Roles for ligases in the RNA editing complex of *Trypanosoma brucei*: band IV is needed for U-deletion and RNA repair. EMBO J. *20*, 4694–4703.

Ivens, A.C., Peacock, C.S., Worthey, E.A., Murphy, L., Aggarwal, G., Berriman, M., Sisk, E., Rajandream, M.A., Adlem, E., Aert, R., *et al.* (2005). The genome of the kinetoplastid parasite, Leishmania major. Science *309*, 436–442.

Kable, M.L., Seiwert, S.D., Heidmann, S., and Stuart, K. (1996). RNA editing: a mechanism for gRNA-specified uridylate insertion into precursor mRNA. Science *273*, 1182–1183.

Kala, S., and Salavati, R. (2010). OB-fold domain of KREPA4 mediates high-affinity interaction with guide RNA and possesses annealing activity. RNA *16*, 1951–1967.

Kang, X., Gao, G., Rogers, K., Falick, A.M., Zhou, S., and Simpson, L. (2006). Reconstitution of full-round uridine-deletion RNA editing with three recombinant proteins. Proc. Natl. Acad. Sci. U.S.A. *103*, 13944–13949.

Kao, C.Y., and Read, L.K. (2005). Opposing effects of polyadenylation on the stability of edited and unedited mitochondrial RNAs in *Trypanosoma brucei*. Mol. Cell. Biol. *25*, 1634–1644.

Kao, C.Y., and Read, L.K. (2007). Targeted depletion of a mitochondrial nucleotidyltransferase suggests the presence of multiple enzymes that polymerize mRNA 3′ tails in *Trypanosoma brucei* mitochondria. Mol. Biochem. Parasitol. *154*, 158–169.

Koslowsky, D.J., and Yahampath, G. (1997). Mitochondrial mRNA 3′ cleavage/polyadenylation and RNA editing in *Trypanosoma brucei* are independent events. Mol. Biochem. Parasitol. *90*, 81–94.

Koslowsky, D.J., Bhat, G.J., Read, L.K., and Stuart, K. (1991). Cycles of progressive realignment of gRNA with mRNA in RNA editing. Cell *67*, 537–546.

Koslowsky, D.J., Riley, G.R., Feagin, J.E., and Stuart, K. (1992). Guide RNAs for transcripts with developmentally regulated RNA editing are present in both life cycle stages of *Trypanosoma brucei*. Mol. Cell. Biol. *12*, 2043–2049.

Lee, M.G., and Van der Ploeg, L.H. (1997). Transcription of protein-coding genes in trypanosomes by RNA polymerase I. Annu. Rev. Microbiol. *51*, 463–489.

Leung, S.S., and Koslowsky, D.J. (2001). RNA editing in *Trypanosoma brucei*: characterization of gRNA U-tail interactions with partially edited mRNA substrates. Nucleic Acids Res. *29*, 703–709.

Li, C., and Englund, P.T. (1997). A mitochondrial DNA primase from the trypanosomatid *Crithidia fasciculata*. J. Biol. Chem. *272*, 20787–20792.

Li, F., Ge, P., Hui, W.H., Atanasov, I., Rogers, K., Guo, Q., Osato, D., Falick, A.M., Zhou, Z.H., and Simpson, L. (2009). Structure of the core editing complex (L-complex) involved in uridine insertion/deletion RNA editing in trypanosomatid mitochondria. Proc. Natl. Acad. Sci. U.S.A. *106*, 12306–12310.

Li, F., Herrera, J., Zhou, S., Maslov, D.A., and Simpson, L. (2011). Trypanosome REH1 is an RNA helicase involved with the 3′–5′ polarity of multiple gRNA-guided uridine insertion/deletion RNA editing. Proc. Natl. Acad. Sci. U.S.A. *108*, 3542–3547.

Liu, B., Liu, Y., Motyka, S.A., Agbo, E.E., and Englund, P.T. (2005). Fellowship of the rings: the replication of kinetoplast DNA. Trends Parasitol. *21*, 363–369.

Madej, M.J., Alfonzo, J.D., and Huttenhofer, A. (2007). Small ncRNA transcriptome analysis from kinetoplast mitochondria of Leishmania tarentolae. Nucleic Acids Res. *35*, 1544–1554.

Madej, M.J., Niemann, M., Huttenhofer, A., and Goringer, H.U. (2008). Identification of novel guide RNAs from the mitochondria of *Trypanosoma brucei*. RNA Biol. *5*, 84–91.

Madina, B.R., Kuppan, G., Vashisht, A.A., Liang, Y.H., Downey, K.M., Wohlschlegel, J.A., Ji, X., Sze, S.H., Sacchettini, J.C., Read, L.K., *et al.* (2011). Guide RNA biogenesis involves a novel RNase III family endoribonuclease in *Trypanosoma brucei*. RNA *17*, 1821–1830.

Maslov, D.A., and Simpson, L. (1992). The polarity of editing within a multiple gRNA-mediated domain is due to formation of anchors for upstream gRNAs by downstream editing. Cell *70*, 459–467.

Michelotti, E.F., Harris, M.E., Adler, B., Torri, A.F., and Hajduk, S.L. (1992). *Trypanosoma brucei* mitochondrial ribosomal RNA synthesis, processing and developmentally regulated expression. Mol. Biochem. Parasitol. *54*, 31–41.

Militello, K.T., Hayman, M.L., and Read, L.K. (2000). Transcriptional and post-transcriptional in organello labelling of *Trypanosoma brucei* mitochondrial RNA. Int. J. Parasitol. *30*, 643–647.

Miller, M.M., Halbig, K., Cruz-Reyes, J., and Read, L.K. (2006). RBP16 stimulates trypanosome RNA editing *in vitro* at an early step in the editing reaction. RNA *12*, 1292–1303.

Milman, N., Motyka, S.A., Englund, P.T., Robinson, D., and Shlomai, J. (2007). Mitochondrial origin-binding protein UMSBP mediates DNA replication and segregation in trypanosomes. Proc. Natl. Acad. Sci. U.S.A. *104*, 19250–19255.

Mingler, M.K., Hingst, A.M., Clement, S.L., Yu, L.E., Reifur, L., and Koslowsky, D.J. (2006). Identification of pentatricopeptide repeat proteins in *Trypanosoma brucei*. Mol. Biochem. Parasitol. *150*, 37–45.

Missel, A., Souza, A.E., Norskau, G., and Goringer, H.U. (1997). Disruption of a gene encoding a novel mitochondrial DEAD-box protein in *Trypanosoma brucei* affects edited mRNAs. Mol. Cell. Biol. *17*, 4895–4903.

Muller, U.F., Lambert, L., and Goringer, H.U. (2001). Annealing of RNA editing substrates facilitated by guide RNA-binding protein gBP21. EMBO J. *20*, 1394–1404.

Ochsenreiter, T., Cipriano, M., and Hajduk, S.L. (2007). KISS: the kinetoplastid RNA editing sequence search tool. RNA *13*, 1–4.

Palenchar, J.B., and Bellofatto, V. (2006). Gene transcription in trypanosomes. Mol. Biochem. Parasitol. *146*, 135–141.

Panigrahi, A.K., Allen, T.E., Stuart, K., Haynes, P.A., and Gygi, S.P. (2003a). Mass spectrometric analysis of the editosome and other multiprotein complexes in *Trypanosoma brucei*. J. Am. Soc. Mass Spectrom. *14*, 728–735.

Panigrahi, A.K., Schnaufer, A., Ernst, N.L., Wang, B., Carmean, N., Salavati, R., and Stuart, K. (2003b). Identification of novel components of *Trypanosoma brucei* editosomes. RNA *9*, 484–492.

Panigrahi, A.K., Ernst, N.L., Domingo, G.J., Fleck, M., Salavati, R., and Stuart, K.D. (2006). Compositionally and functionally distinct editosomes in *Trypanosoma brucei*. RNA *12*, 1038–1049.

Panigrahi, A.K., Zikova, A., Dalley, R.A., Acestor, N., Ogata, Y., Anupama, A., Myler, P.J., and Stuart, K.D. (2008). Mitochondrial complexes in *Trypanosoma brucei*: a novel complex and a unique oxidoreductase complex. Mol. Cell Proteomics *7*, 534–545.

Pays, E. (2005). Regulation of antigen gene expression in *Trypanosoma brucei*. Trends Parasitol. *21*, 517–520.

Pelletier, M., and Read, L.K. (2003). RBP16 is a multifunctional gene regulatory protein involved in editing and stabilization of specific mitochondrial mRNAs in *Trypanosoma brucei*. RNA *9*, 457–468.

Petersen-Mahrt, S.K., Estmer, C., Ohrmalm, C., Matthews, D.A., Russell, W.C., and Akusjarvi, G. (1999). The splicing factor-associated protein, p32, regulates RNA splicing by inhibiting ASF/SF2 RNA binding and phosphorylation. EMBO J. *18*, 1014–1024.

Pollard, V.W., Rohrer, S.P., Michelotti, E.F., Hancock, K., and Hajduk, S.L. (1990). Organization of minicircle genes for guide RNAs in *Trypanosoma brucei*. Cell *63*, 783–790.

Pusnik, M., Small, I., Read, L.K., Fabbro, T., and Schneider, A. (2007). Pentatricopeptide repeat proteins in *Trypanosoma brucei* function in mitochondrial ribosomes. Mol. Cell. Biol. *27*, 6876–6888.

Read, L.K., Stankey, K.A., Fish, W.R., Muthiani, A.M., and Stuart, K. (1994). Developmental regulation of RNA editing and polyadenylation in four life cycle stages of *Trypanosoma congolense*. Mol. Biochem. Parasitol. *68*, 297–306.

Reifur, L., Yu, L.E., Cruz-Reyes, J., Vanhartesvelt, M., and Koslowsky, D.J. (2010). The impact of mRNA structure on guide RNA targeting in kinetoplastid RNA editing. PLoS One *5*, e12235.

Riley, G.R., Corell, R.A., and Stuart, K. (1994). Multiple guide RNAs for identical editing of *Trypanosoma brucei* apocytochrome b mRNA have an unusual minicircle location and are developmentally regulated. J. Biol. Chem. *269*, 6101–6108.

Rogers, K., Gao, G., and Simpson, L. (2007). Uridylate-specific 3′ 5′-exoribonucleases involved in uridylate-deletion RNA editing in trypanosomatid mitochondria. J. Biol. Chem. *282*, 29073–29080.

Rusche, L.N., Cruz-Reyes, J., Piller, K.J., and Sollner-Webb, B. (1997). Purification of a functional enzymatic editing complex from *Trypanosoma brucei* mitochondria. EMBO J. *16*, 4069–4081.

Ryan, C.M., and Read, L.K. (2005). UTP-dependent turnover of *Trypanosoma brucei* mitochondrial mRNA requires UTP polymerization and involves the RET1 TUTase. RNA *11*, 763–773.

Ryan, C.M., Militello, K.T., and Read, L.K. (2003). Polyadenylation regulates the stability of *Trypanosoma brucei* mitochondrial RNAs. J. Biol. Chem. *278*, 32753–32762.

Schmid, B., Riley, G.R., Stuart, K., and Goringer, H.U. (1995). The secondary structure of guide RNA molecules from *Trypanosoma brucei*. Nucleic Acids Res. *23*, 3093–3102.

Schnaufer, A., Panigrahi, A.K., Panicucci, B., Igo, R.P., Jr., Wirtz, E., Salavati, R., and Stuart, K. (2001). An RNA ligase essential for RNA editing and survival of the bloodstream form of *Trypanosoma brucei*. Science *291*, 2159–2162.

Schnaufer, A., Ernst, N.L., Palazzo, S.S., O'Rear, J., Salavati, R., and Stuart, K. (2003). Separate insertion and deletion subcomplexes of the *Trypanosoma brucei* RNA editing complex. Mol. Cell *12*, 307–319.

Schumacher, M.A., Karamooz, E., Zikova, A., Trantirek, L., and Lukes, J. (2006). Crystal structures of *T. brucei* MRP1/MRP2 guide-RNA binding complex reveal RNA matchmaking mechanism. Cell *126*, 701–711.

Seiwert, S.D., and Stuart, K. (1994). RNA editing: transfer of genetic information from gRNA to precursor mRNA *in vitro*. Science *266*, 114–117.

Seiwert, S.D., Heidmann, S., and Stuart, K. (1996). Direct visualization of uridylate deletion *in vitro* suggests a mechanism for kinetoplastid RNA editing. Cell *84*, 831–841.

Simpson, L., Aphasizhev, R., Gao, G., and Kang, X. (2004). Mitochondrial proteins and complexes in Leishmania and Trypanosoma involved in U-insertion/deletion RNA editing. RNA *10*, 159–170.

Simpson, L., Aphasizhev, R., Lukes, J., and Cruz-Reyes, J. (2010). Guide to the nomenclature of kinetoplastid RNA editing: a proposal. Protist *161*, 2–6.

Simpson, L., and Maslov, D.A. (1999). Evolution of the U-insertion/deletion RNA editing in mitochondria of kinetoplastid protozoa. Ann. NY Acad. Sci. *870*, 190–205.

Simpson, L., and Shaw, J. (1989). RNA editing and the mitochondrial cryptogenes of kinetoplastid protozoa. Cell *57*, 355–366.

Simpson, L., Thieman, O.H., Savill, N.J., Alfonzo, J.D., and Maslov, D.A. (2000). Evolution of RNA editing in trypanosome mitochondria. Proc. Natl. Acad. Sci. U.S.A. *97*, 6986–6993.

Sogin, M.L. (1991). Early evolution and the origin of eukaryotes. Curr. Opin. Genet. Dev. *1*, 457–463.

Sprehe, M., Fisk, J.C., McEvoy, S.M., Read, L.K., and Schumacher, M.A. (2010). Structure of the *Trypanosoma brucei* p22 protein, a cytochrome oxidase subunit II-specific RNA editing accessory factor. J. Biol. Chem. *285*, 18899–18908.

Stuart, K., Brun, R., Croft, S., Fairlamb, A., Gurtler, R.E., McKerrow, J., Reed, S., and Tarleton, R. (2008). Kinetoplastids: related protozoan pathogens, different diseases. J. Clin. Invest. *118*, 1301–1310.

Stuart, K.D., Schnaufer, A., Ernst, N.L., and Panigrahi, A.K. (2005). Complex management: RNA editing in trypanosomes. Trends Biochem. Sci. *30*, 97–105.

Sturm, N.R., Maslov, D.A., Blum, B., and Simpson, L. (1992). Generation of unexpected editing patterns in Leishmania tarentolae mitochondrial mRNAs: misediting produced by misguiding. Cell *70*, 469–476.

Sturm, N.R., Yu, M.C., and Campbell, D.A. (1999). Transcription termination and 3′-End processing of the spliced leader RNA in kinetoplastids. Mol. Cell. Biol. *19*, 1595–1604.

Trotter, J.R., Ernst, N.L., Carnes, J., Panicucci, B., and Stuart, K. (2005). A deletion site editing endonuclease in *Trypanosoma brucei*. Mol. Cell *20*, 403–412.

Vanhamme, L., Perez-Morga, D., Marchal, C., Speijer, D., Lambert, L., Geuskens, M., Alexandre, S., Ismaili, N., Goringer, U., Benne, R., *et al.* (1998). *Trypanosoma brucei* TBRGG1, a mitochondrial oligo(U)-binding protein that co-localizes with an *in vitro* RNA editing activity. J. Biol. Chem. 273, 21825–21833.

Vickerman, K. (1985). Developmental cycles and biology of pathogenic trypanosomes. Br. Med. Bull. *41*, 105–114.

Vondruskova, E., van den Burg, J., Zikova, A., Ernst, N.L., Stuart, K., Benne, R., and Lukes, J. (2005). RNA interference analyses suggest a transcript-specific regulatory role for mitochondrial RNA-binding proteins MRP1 and MRP2 in RNA editing and other RNA processing in *Trypanosoma brucei*. J. Biol. Chem. *280*, 2429–2438.

Wang, Z., Morris, J.C., Drew, M.E., and Englund, P.T. (2000). Inhibition of *Trypanosoma brucei* gene expression by RNA interference using an integratable vector with opposing T7 promoters. J. Biol. Chem. *275*, 40174–40179.

Weng, J., Aphasizheva, I., Etheridge, R.D., Huang, L., Wang, X., Falick, A.M., and Aphasizhev, R. (2008). Guide RNA-binding complex from mitochondria of trypanosomatids. Mol. Cell 32, 198–209.

Will, C.L., and Luhrmann, R. (2011). Spliceosome structure and function. Cold Spring Harb. Perspect. Biol. 3, pii: a003707.

Xu, W., and Kimelman, D. (2007). Mechanistic insights from structural studies of beta-catenin and its binding partners. J. Cell. Sci. *120*, 3337–3344.

Yoshikawa, H., Komatsu, W., Hayano, T., Miura, Y., Homma, K., Izumikawa, K., Ishikawa, H., Miyazawa, N., Tachikawa, H., Yamauchi, Y., *et al.* (2011). Splicing factor 2-associated protein p32 participates in ribosome biogenesis by regulating the binding of Nop52 and fibrillarin to preribosome particles. Mol. Cell Proteomics *10*, M110 006148.

Zhang, H., Kolb, F.A., Jaskiewicz, L., Westhof, E., and Filipowicz, W. (2004). Single processing center models for human Dicer and bacterial RNase III. Cell *118*, 57–68.

Zhao, G., Li, G., Schindelin, H., and Lennarz, W.J. (2009). An Armadillo motif in Ufd3 interacts with Cdc48 and is involved in ubiquitin homeostasis and protein degradation. Proc. Natl. Acad. Sci. U.S.A. *106*, 16197–16202.

Zimmer, S.L., McEvoy, S.M., Li, J., Qu, J., and Read, L.K. (2011). A novel member of the RNase D exoribonuclease family functions in mitochondrial guide RNA metabolism in *Trypanosoma brucei*. J. Biol. Chem. *286*, 10329–10340.

Structural Studies of U-Insertion/ Deletion RNA Editing in Trypanosomes 5

Blaine H.M. Mooers

Abstract

We review the progress in the past decade in structural studies of the proteins and RNAs associated with the U-insertion/deletion RNA editing (or k-RNA editing) in the mitochondrion of trypanosomes. This review includes the electron microscopy studies of RNA editing complexes. Beyond the intellectual quest to understand the structural basis of RNA editing, these studies share the goal of using structures of essential proteins to design better inhibitors of RNA editing for medical and research purposes.

Introduction to U-deletion/ insertion editing

The pathogenic protozoans of the genera *Trypanosoma* and *Leishmania* threaten 600 million people with debilitating and sometimes fatal infections. New drugs to treat these infections are needed because the current drugs have toxic side-effects and are growing ineffective with the emergence of drug-resistant strains of parasites (Gehrig and Efferth, 2008; Wilkinson *et al.*, 2008; Wilkinson and Kelly, 2009; Chakravarty and Sundar, 2010; Paes *et al.*, 2011). The members of these two genera are in the order Kinetoplastida named after the kinetoplast – a granule containing DNA at the base of the cell's flagellum within the single mitochondrion. All kinetoplastids have a U-insertion/U-deletion RNA editing system (for recent reviews, see Chapter 4; see also Stuart and Panigrahi, 2002; Stuart *et al.*, 2005; Hajduk and Ochsenreiter, 2010) that is not found in humans, leading to interest in this system as a drug target (Salavati *et al.*, 2011). In *T. brucei* – which causes African sleeping sickness – RNA editing inserts 3583 uridylates (Us) and deletes 322 Us in 12 pre-mature mRNAs (pre-mRNA) in the mitochondrion in a post-transcription process that can double the length of the mRNA in the more extreme cases (Feagin *et al.*, 1988). The editing process is essential for making mature mRNAs that can be translated into functional proteins.

Large (~1.6 MDa) ribonucleoprotein complexes (Stuart *et al.*, 2005) execute the editing reactions under the direction of fragmented templates found in hundreds guide RNAs (gRNAs) (generally 50–80 nucleotides long) (Blum *et al.*, 1990) (Fig. 5.1). Three distinct types of editing complexes have been isolated (Carnes *et al.*, 2008). There are other multiprotein complexes associated with RNA editing that may have roles in processing the precursors of the mRNA substrate and the gRNA, for example the mitochondrial RNA-binding complex 1 (MRB1), has a core of six proteins that associate with several different subcomplexes through RNA-mediated interactions (Ammerman *et al.*, 2012).

The gRNAs bind antiparallel to their cognate mRNAs to form editing substrates (Blum *et al.*, 1990) (Fig. 5.1). The editing substrates have three functional domains: a 5′ anchor helix, a central template or guiding domain, and a 3′ U-helix (Blum and Simpson, 1990) (Fig. 5.1). The anchor helix matches the gRNA to its cognate mRNA just downstream from the first editing site. The template domain directs the insertion or deletion of Us in the mRNA. The U-helix holds the cleaved mRNA ends in proximity for re-ligation during editing.

The two parallel catalytic cascades begin with the endonucleolytic cleavage of the pre-mRNA by pathway specific RNA editing endonucleases (KREN1 and KREN2) (Fig. 5.1). Next, one or more Us are added to the 3′ end of the 5′ cleavage product by a RNA editing 3′-terminal uridylate

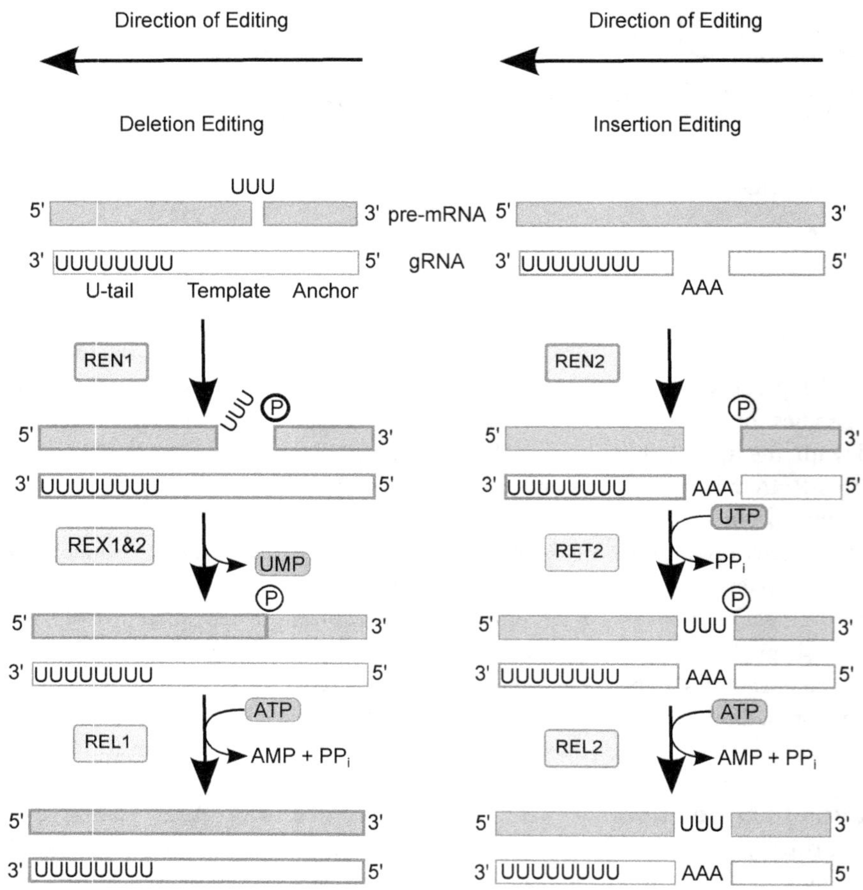

Figure 5.1 The two parallel RNA editing reaction cascades in the mitochondrion of trypanosomes (kinetoplastid RNA editing or k-RNA editing). The upper, grey band represents a mRNA strand and the lower, white band represents the gRNA strand. The oligo(U) tail or U-tail at the 3′ end of the gRNA strand is denoted by the run of Us. The gap in the mRNA strand represents the RNA editing site. The region below the gap is the template domain of the gRNA that directs the editing reaction. The circled P represents the 5′ phosphate at the end of the 3′-fragment of the mRNA after cleavage of the mRNA strand by a pathway-specific endonuclease. The phosphate is reincorporated.

transferase (TUTase 2 or RET2), or one or more Us are removed from the 3′ end of the 5′ cleavage product by a RNA editing 3′ exouridylylase activity (exoUase or KREX). Finally, the two mRNA fragments are ligated by ligases specific for U-insertion and U-deletion (Hajduk and Ochsenreiter, 2010). An RNA editing helicase (KREH1) activity modulates gRNA–mRNA interactions while a gRNA is still bound to the mRNA, or it displaces the gRNA after it has been used by the editing complex (Missel *et al.*, 1997; Li *et al.*, 2011). The editing reaction cycles create new anchor sequences upstream for the subsequent gRNAs with complementary sequences to bind; this enforces the 5′-to-3′ polarity of the editing reactions (Maslov and Simpson, 1992).

The secondary structure of the gRNA–mRNA duplex varies depending on the mRNA and gRNA involved and the stage of editing (Koslowsky *et al.*, 1991; Leung and Koslowsky, 2001a,b; Reifur and Koslowsky, 2008; Reifur *et al.*, 2010). Generally, the editing site is part of a three-helix junction (Reifur and Koslowsky, 2008). The editing site is flanked on the downstream side by the anchor helix and on the upstream side by the U-helix (Fig. 5.16). The editing site is opposite the end of the stem of a stem–loop formed by the template domain of the gRNA. Non-standard basepairs in the U-helix and near the editing sites can have profound effects on RNA editing efficiency (Cruz-Reyes *et al.*, 2001; Cifuentes-Rojas *et al.*, 2005, 2007), yet the 3D structures of the editing sites remain unknown.

However, structural studies of RNA editing complexes (Golas *et al.*, 2009; Li *et al.*, 2009), individual proteins from the editosome (Deng *et al.*, 2004, 2005; Park *et al.*, 2011; Wu *et al.*, 2011; Park *et al.*, 2012), and proteins and RNA associated with the editosome (Schumacher *et al.*, 2006; Sprehe *et al.*, 2010; Stagno *et al.*, 2010; Mooers and Singh, 2011) initiated in the past decade are yielding insights into the molecular basis of the RNA editing machinery. The structures of the RNA editing core complexes (RECC) are reviewed in the section 'EM studies of RNA editing core complexes' (Golas *et al.*, 2009; Li *et al.*, 2009). The structures of two editosome enzymes are discussed in 'Structures of enzymes from the editosome', below (Deng *et al.*, 2004, 2005). These enzymes have been the subject of homology modelling, molecular dynamics simulation, and drug design studies (Amaro *et al.*, 2007, 2008; Durrant *et al.*, 2010; Liang and Connell, 2010; Durrant and McCammon, 2011; Moshiri *et al.*, 2011; Demir and Amaro, 2012). The structures of three of a family of six OB-fold containing proteins that mediate protein–protein interactions in the core of the editosome are discussed in 'RNA editing complex structural proteins' (Park *et al.*, 2011, 2012; Wu *et al.*, 2011). The crystal structures of these proteins are a major breakthrough because these proteins have been very difficult to crystallize. Crystallization of these three proteins was enhanced by the use of single-domain antibodies from llamas (nanobodies). In 'Structures of the kRNA editing accessory factors', we discuss the crystal structures of macromolecules that associate with the editosome at least temporarily, act upon macromolecules that in turn interact with the editosome, or associate with a new editosome-like complex of unknown function (Schumacher *et al.*, 2006; Sprehe *et al.*, 2010; Stagno *et al.*, 2010; Mooers and Singh, 2011). This chapter focuses on the structures and their biological insights. See Chapter 4 for insights into the enzymology and molecular biology of kRNA editing.

The crystal structures discussed here are listed in Table 5.1. These are not necessarily the structures with the highest resolution for a particular protein. Often, the ligand-bound structures that we discuss have lower resolution limits than the apo structures. These structures are available in the Protein Data Bank or the Electron Microscopy Data Bank and were used to make the figures in this review.

There are at least five schemes of names for

Table 5.1 Selected X-ray crystal structures of *Trypanosoma brucei* mitochondrial RNA editing

Macromolecule	Resolution (Å)	PDB-ID	Reference
Editing complex enzymes			
TUTase 2	(RET2)	1.97 2B56	Deng *et al.* (2005)
RNA ligase 1 (REL1)	1.2	1XDN	Deng *et al.* (2004)
Editing complex structural proteins			
KREPA6	2.1	3K7U	Wu *et al.* (2011)
KREPA3^{OB} – A6	2.5	3STB	Park *et al.* (2011)
KREPA1$^{OB\Delta}$	1.97	4DKA	Park *et al.* (2012)
Editing-associated proteins, RNA and protein–RNA complexes			
MEAT1 + UTP + Mg^{2+}	2.3	3HIY	Stagno *et al.* (2010)
p22 protein	2.0	3JV1	Sprehe *et al.* (2010)
MRP1/MRP2/guide RNA	3.37	2GJE	Schumacher *et al.* (2006)
MRP1/MRP2	1.89	2GIA	Schumacher *et al.* (2006)
U-Helix RNA	1.37	3ND3	Mooers and Singh (2011)

k-RNA editing proteins in the literature (Simpson *et al.*, 2010). A 'Rosetta Stone' for translating these competing schemes is found in Table 1 of Simpson *et al.* (2010). Some RNA editing accessory proteins, like p22 ('Structures of the kRNA editing accessory factors'), are not in this table. We follow Simpson and colleagues' proposed naming convention, but we sometimes resort to two-letter abbreviations when it is more convenient (i.e. A1 for KREPA1). We use 'K' to represent 'kinetoplastid', and we use two-letter abbreviations to represent species, as in *Tb* for *Trypanosma brucei.*

EM studies of RNA editing core complexes

The editing reaction cycle is driven by enzymes found in the RNA editing core complexes (RECC). Native editing complexes isolated from mitochondrial vesicles have seven (Rusche *et al.*, 1997), 13 (Seiwert *et al.*, 1996) or 20 polypeptides (Aphasizhev *et al.*, 2003a). The low yields of these complexes suggest that either they have low steady-state concentrations or low kinetic or thermodynamic stabilities. RECCs have been recently isolated from transgenic trypanosomes in the insect stage that conditionally express tandem affinity tagged versions of editosome proteins (Panigrahi *et al.*, 2007). The tandem tagged proteins permit chromatographic purification of the editosomes under chemically moderate, native-like conditions. The purity of the editing complexes was high enough to permit visualization by transmission electron microscopy (TEM) and by cryo-EM (Golas *et al.*, 2009; Li *et al.*, 2009) and the subsequent generation of molecular models with resolutions of 13–20 Å. These studies have been recently reviewed (Goringer *et al.*, 2011).

Two distinct 20S editing complexes can be isolated in the absence of RNA. These complexes consist of 13 polypeptides that include the activities of the editing reaction cycle. One complex is specific for U-insertion editing activity, and one complex is specific for U-deletion editing activity (Schnaufer *et al.*, 2003). Each complex can accurately edit synthetic RNA substrates under the direction of the cognate gRNA (Igo *et al.*, 2000, 2002; Carnes and Stuart, 2007). The 20S editing complexes bind RNA editing substrates with nanomolar affinity; whereas the 35–40S complexes do not bind synthetic substrates because native substrates already occupy the editosome (Golas *et al.*, 2009). EM studies of the 20S complexes show that they form monodisperse, elongated structures 210 × 260 Å with two non-equivalent domains (Fig. 5.2A and B) (Golas *et al.*, 2009).

Each type of 20S complex (Fig. 5.2A and B) assembles around the RNA editing substrate to form the 35–40S complex (Fig. 5.2C). The molecular surface of the 35–40S complex from cryo negative staining EM was asymmetrical and up to 260 Å in diameter. The molecular surface enclosed a calculated molecular mass of 1.45 ± 0.15 MDa and gave a calculated sedimentation coefficient of 35–41S, which agrees well with the apparent sedimentation behaviour observed in isokinetic glycerol gradients (Pollard *et al.*, 1992; Peris *et al.*, 1997).

Modelling of the molecular envelopes of the 20S (Fig. 5.2A and B) complexes into the molecular envelopes of the 35–40S complexes suggest that the RNA is at the interface between the 20S complexes and that the enzymatic activities line the interfaces (Golas *et al.*, 2009; Goringer *et al.*, 2011). EM images of the 35–40S complex show diverse structures that vary in the size of the semispherical back element (Fig. 5.2B). This back element may be the location of the portion of the RNA substrate that is not in the active site. The large variation in the length of the pre-mRNAs may account for the variation in the size of the back element because the shortest mRNA is 60 kDa (unedited CR3) while the longest is 450 kDa (edited ND7) (Goringer *et al.*, 2011).

The averaged molecular images from both the small and large complexes are in the Electron Microscopy Data Bank (EMDB) (Lawson *et al.*, 2011). The structures are stored in ccp4-style electron density files that are readily displayed in most molecular graphics programs including *PyMOL*. The maps should be displayed at the recommended contour level which is usually much less than one sigma.

Another group did 3D structural analysis of the 20–25S RNA editing complexes [i.e. the L (Ligase) complex] isolated from *Leishmania*

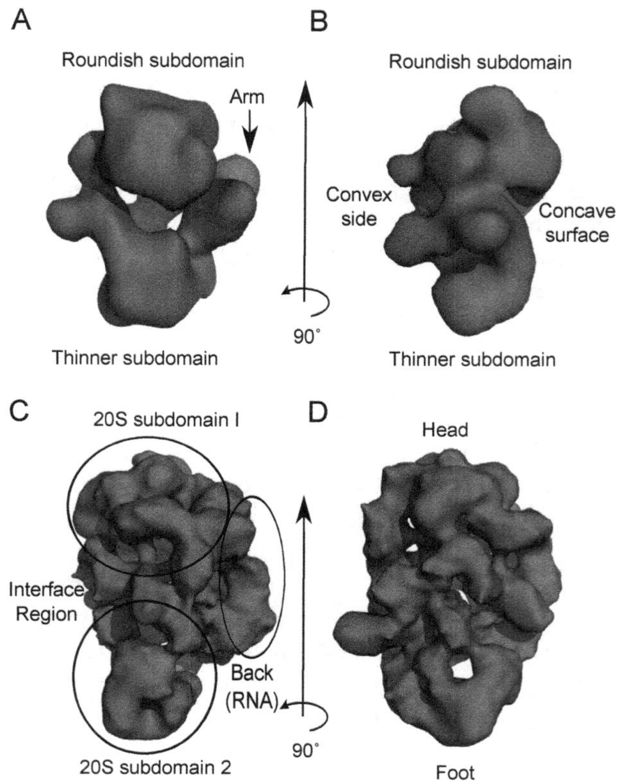

Figure 5.2 Electron micrographs of RNA editing complexes from *Trypanosoma brucei.* (A) Consensus model of the 20S editosome. (B) Same as (A) but rotated by 90° about the vertical axis. (C) Positions of the 20S complexes and (D) the prominent structural features in the consensus structure of the *Trypanosoma brucei* 35–40S RNA editing complex (not on the same scale as A and B). The isosurfaces were generated in *PyMOL* (Schrodinger, 2010) from accession 1595 (A and B) and 1594 (C and D) in the Electron Microscopy Data Bank (EMDB) (Lawson *et al.*, 2011). Contoured at the level of 0.0419 as recommended by EMDB for both surfaces.

tarentolae mitochondria (Li *et al.,* 2009). They used electron microscopy and tomography. The data were not deposited in the EMDB thereby inhibiting direst comparisons with the structures from *T. brucei.* Electron microscopy gave triangular shaped structures with dimensions of about 200 Å × 140 Å × 80 Å. Electron tomography generates images from single particles and does not require averaging. These unbiased images were consistent with the averaged structures from electron microscopy. In addition, modelling of the electron microscopy data starting with a Gaussian ball led to final models that were triangular in shape.

Some particles had an additional density extending from the central region. These particles may have an additional component.

Future directions

Further work in both systems with anti-REL1 IgGs and other anti-editosome protein antibodies can be expected to map by electron microscopy the location of editing enzymes within the editing complex.

Structures of enzymes from the editosome

The editing reactions are catalysed by U-deletion and U-insertion specific endonucleases KREN1 and KREN2 respectively. There are no structures of these endonucleases. Next, the Us are added to the 3′ end of the 5′ mRNA fragment by a RNA editing TUTase 2 (*Tb*RET2) for which there is a 1.8 Å crystal structure (Deng *et al.,* 2005)

('RNA editing TUTase 2'). [There are crystal structures of other TUTases from trypanosomes (see 'Structures of the kRNA editing accessory factors', below): *Tb*MEAT1 (Stagno *et al.*, 2010), and cytosolic *Tb*TUT4 (Stagno *et al.*, 2007a,b).] RNA editing exonuclease 1 (KREX1) and 2 (KREX2) remove Us from the 3′ end of the 5′ fragment in the U-deletion pathway. There are no crystal structures of these enzymes. The cleaved strands are ligated by pathway specific RNA editing ligases 1 and 2 (KREL1 and KREL2). There is a 1.2 Å resolution crystal structure of *Tb*REL1 (Deng *et al.*, 2004) [see 'RNA editing ligase I (KREL1)', below]. Finally, there is a RNA editing helicase 1 (KREH1) that removes the guide RNA from the mRNA. There is not a crystal structure of this enzyme.

RNA editing TUTase 2

Terminal uridylate transferases (TUTases) are a functionally and structurally diverse group of enzymes (Aphasizhev and Aphasizheva, 2008). They catalyse the template independent addition of Us to the 3′ ends of single-stranded RNA and to the cleaved ends of mRNA substrates in the gRNA-directed addition of Us in U-insertion editing pathway.

Terminal uridylate transferase 2 (TUTase 2) or RNA editing TUTase 2 (KRET2) is a monomeric protein that is an integral part of the 20S editing complex that is responsible for mRNA U-insertion editing (Aphasizhev *et al.*, 2003b; Ernst *et al.*, 2003). KRET2 catalyses the addition of as many Us to the 3′ end of the 5′ fragment in the cleaved pre-mRNA as specified by the gRNA. KRET2 is the only TUTase that is part of the editing complex. KRET2 interacts directly with the structural protein KREPA1 ('RNA editing complex structural proteins'). This interaction enhances the activity of RET2 (Ernst *et al.*, 2003).

In contrast, RET1 is a tetramer of 121-kDa monomers in *Leishmania tarentolae* that is also found in the mitochondrion and indirectly interacts with the editosome through its RNA substrates. KRET1 adds the oligoU tails to the 3′ end of gRNAs (Aphasizhev *et al.*, 2003b), and it adds 100s of Us to mRNA in the regulation of mRNA turnover (Ryan and Read, 2005). RET1 has 24% sequence identity with RET2 over a 554 amino acid region.

Both KRET1 and KRET2 are essential to the survival of the insect form of the parasite (Aphasizhev *et al.*, 2003b). KRET2 is also essential in the bloodstream form of the parasite (Deng *et al.*, 2005).

TUTases are members of the nucleotidyl transferase (NT) superfamily that is also called the DNA polymerase β or Pol β superfamily (Holm and Sander, 1995). Members of this family have a helix–turn–helix motif with the following amino acid sequence motif hG[G/S]X9–13Dh[D/E] h where h is a hydrophobic residue and X is any residue (Holm and Sander, 1995). A subfamily was identified that has conserved C-terminal sequences and that includes the TUTases and eukaryotic nuclear mRNA poly(A) polymerases (PAP). TUTases are most closely related to the non-canonical Trf4/5-type poly(A) polymerases (ncPAPs) found in the nucleus (Rogozin *et al.*, 2003).

The trypanosomes have a diverse array of TUTases and PAPs that may compensate for the lack of transcriptional control in the mitochondrial and nuclear genomes (Aphasizhev and Aphasizheva, 2008). There are two nuclear ncPAPs (ncPAP1 and ncPAP2) (Etheridge *et al.*, 2009), two mitochondrial poly(A) polymerases (KAP1 and KAP2) (Kao and Read, 2007; Etheridge *et al.*, 2009), two cytoplasmic TUTases (TUT3 and TUT4) (Aphasizhev *et al.*, 2004; Stagno *et al.*, 2007b), and three mitochondrial TUTases: RNA editing TUTase 1 (RET1) (Aphasizhev *et al.*, 2002), RNA editing TUTase 2 (RET2) (Aphasizhev *et al.*, 2003a), and mitochondrial editosome-like complex-associated TUTase 1 (MEAT1) (Panigrahi *et al.*, 2003). RET1 catalyses the uridylation of all classes of mitochondrial RNAs. RET1 adds tails of ~20 Us to guide RNAs and ribosomal RNAs. RET1 and KPAP2 acts in concert to add (A/U) heteropolymers to mRNAs (Aphasizheva and Aphasizhev, 2010).

TUTase1 is the only human TUTase (PDB-ID 3PQ1) homologue (Bai *et al.*, 2011). It has 20% sequence identity with *Tb*RET2, shares the overall folds of the C and N-terminal domains, and the position and identity of the catalytic residues. It polyadenylates a specific set of mRNAs associated with nuclear speckles, and it is a specific terminal uridylyltransferase for U6 snRNA *in vitro*.

The structure for *Tb*TUT4, a cytoplasmic TUTase, that shares 30% sequence identity with RET2, is now known (Stagno *et al.*, 2007a). Its crystal structure (PDB-ID 2IFK) has a root mean-square deviation (RMSD) of 1.6 Å with the crystal structure of RET2.

Reaction mechanism

Most members of the NT superfamily have two conserved acidic residues in the above-mentioned signature motif that form a metal binding triad with a third conserved acidic residue from outside of this motif. These residues coordinate two metal ions that are essential for catalysis (Pelletier and Sawaya, 1996). The chemistry of the nucleotidyl transfer reaction involves a metal coordinated in-line nucleophilic attack by the 3′ hydroxyl group of the nucleic acid substrate (the 5′ fragment of the pre-mRNA in this case) on the α-phosphate of the NTP without the formation of a covalent intermediate and with the release of pyrophosphate (Fig. 5.3). One metal cation (metal cation A) facilitates this reaction by lowering the affinity of the 3′ hydroxyl oxygen atom for its hydrogen atom. The other metal cation (metal cation B) helps to stabilize the pyrophosphate. This chemistry is conserved throughout the superfamily; however, there is variation in how the substrate specificity is achieved.

Domain organization

Most of the RET2 polypeptide chain was found in the electron density maps (Deng *et al.*, 2005). Only the first 10 residues from the N-terminus and the last 14 from the C-terminus were missing. The structure is wedge-shaped with overall

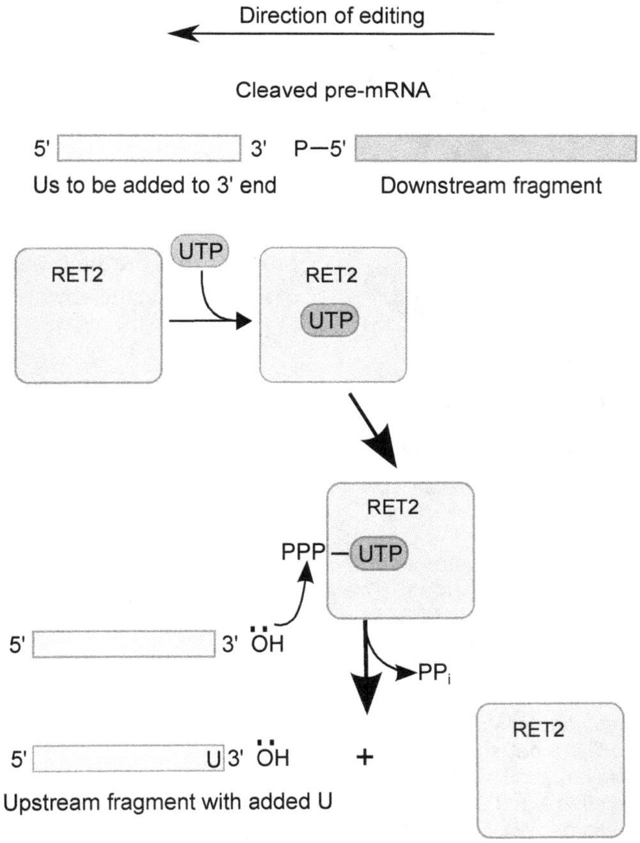

Figure 5.3 Cartoon of the three-step reaction mechanism of *Trypanosoma brucei* RNA editing terminal uridylate transferase 2 (*Tb*RET2).

dimensions of 80 Å long and 50 Å wide at the blunt end of the wedge (Fig. 5.4A). TbRET2 has three domains [N-terminal domain (NTD), middle domain (MD), and C-terminal domain (CTD)] arranged in a way that is unique to the NT superfamily. The CTD forms the base of the wedge and the MD forms the point of the wedge. The NTD is nestled between the CTD and MD.

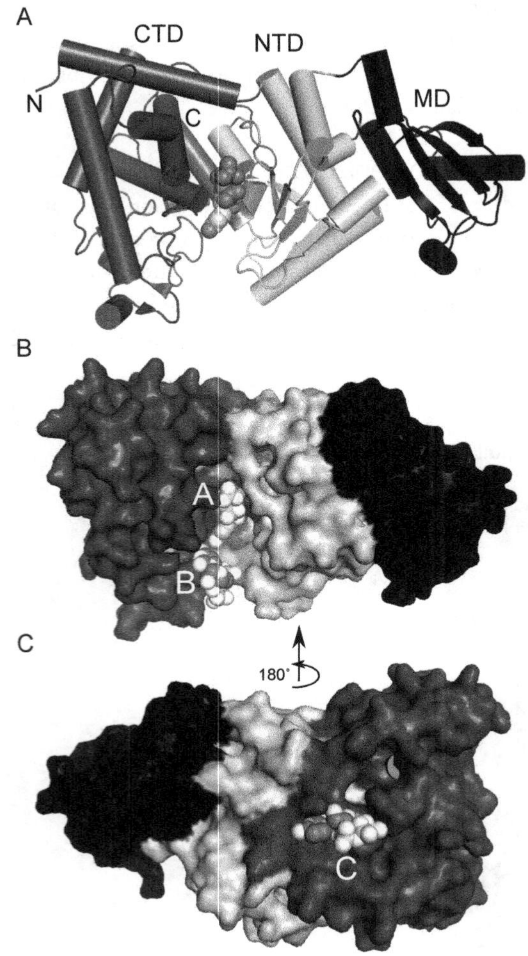

Figure 5.4 Global view of the crystal structure of *Tb*RET2 (PDB-ID 2B51). (A) Ribbon diagram. The three domains (the N-terminal domain, NTD; the middle domain, MD; and the C-terminal domain, CTD) are different shades of grey: NTD – light grey; MD – black; and CTD dark grey. The UTP in the A binding site is shown as van der Waals spheres. The UTPs at sites B and C are not shown. (B) The molecular surface representation coloured by domain. The UTP binding sites A and B are labelled. The UTPs are shown as van der Waals spheres. (C) The backside of the image in (B) shows the C-UTP binding site C.

The crystal structure of *Tb*RET2 (Deng *et al.*, 2005) demonstrates that it shares with other members of the superfamily a conserved N-terminal polymerase domain topology consisting of a five-stranded mixed β-sheet flanked by two or three α-helices (Fig. 5.4A). The NTD of RET2 has a five-strand antiparallel β-sheet and three α-helices (Fig. 5.4). The NTD is non-contiguous and is formed by residues 53–152 and 263–272. The NTD interacts extensively with the RNA binding C-terminal domain (i.e. bury 2923 Å^2 of molecular surface area). There is a deep cleft between the two domains. The cleft forms the UTP binding site. The UTP makes contacts with CTD and the Pol β signature sequence of the NTD.

The NTD is interrupted between strands β-4 and β-9 by an insertion of 110 residues (residues 153–262) (Fig. 5.4A) that form the MD. The insertion forms a compact middle domain that interacts only with the NTD (buries 2595 Å^2 of surface area) and that also extends out into solution. Deletion of the MD inactivates the enzyme. The MD is contiguous and contains six α-helices and a four-stranded antiparallel β-sheet.

The CTD (Fig. 5.4A) has seven contiguous α-helices, four β-strands, and the non-contiguous α-helix 1 from near the N-terminus. The CTD provides all of the direct interactions with the base of the UTP and thereby supplies the specificity for uracil.

UTP binding sites

At concentrations of 1 and 10 mM, UTP is found only in site A (Fig. 5.4B). At concentrations of 100 mM, UTP is also found in sites B and C (Fig. 5.4B) (Deng *et al.*, 2005). In 10 mM UTP and 100 mM UMP, UMPs occupy only sites B and C. In the presence of 10 mM ATP, GTP, or CTP, there is well-defined density only for the triphosphate and no density at sites B and C.

In site A, the Mg^{2+} is found with nearly perfect octahedral coordination. Three ligands are phosphate oxygen atoms from each phosphate in the triphosphate, two ligands are oxygen atoms in the conserved aspartates Asp-97 and Asp-99, and the sixth ligand is an oxygen atom in a water molecule.

This active site magnesium can be swapped with manganese (Deng *et al.*, 2005).

All three of the phosphates also form direct or water-mediated hydrogen bonds with residues from both the NTD and the CTD (Fig. 5.5). The beta phosphate forms a hydrogen bond with the backbone nitrogen of Ser-85 from the NTD. This serine is part of the NT superfamily signature motif.

The catalytic triad is formed by three aspartates in the NTD: Asp-97, Asp-99 and Asp-267. The first two aspartates coordinate one of the two Mg^{2+} expected in the active site (Fig. 5.5). The second Mg^{2+} was not visible. RET2 may require the RNA substrate to create the binding site for the second magnesium.

The side chain of Asn-277 is hydrogen bonded to the O2 oxygen atom of the base and the O2′ hydroxyl of the ribose ring (Fig. 5.5). The side chain of Ser-278 also forms hydrogen bonds to the O2′ hydroxyl and the O3′ hydroxyl oxygen atoms of the ribose ring. These two side chains are oriented by their interactions with the nucleoside to favour the binding of ribonucleotides over deoxyribonucleotides.

The specificity for a pyrimidine is provided by stacking of the base on the aromatic ring of Tyr-319 (Fig. 5.5). The specificity for a uracil over a cytosine is provided by a water molecule (HOH-633, see bottom of Fig. 5.5A) that is coordinated by two conserved carboxylates (Asp-421 and Glu-424) to be oriented and positioned to serve as a hydrogen bond acceptor on the N3 ring nitrogen atom which is a hydrogen bond donor in uracil in the common lactam tautomeric state. (The angle between D421:OD1, HOH633, and E424:OE1 is 109°, so the water's hydrogen atoms must point towards these two carboxylate oxygen atoms. This orients the water molecule to place one of the lone pair of electrons of the water's oxygen atom at a favourable distance and orientation for hydrogen bond formation with the uracil's N3 nitrogen atom. In contrast, the N3 ring nitrogen atom of cytosine is a hydrogen bond acceptor when in the normal tautomeric state of cytosine (i.e. N3 is unprotonated). This water molecule has a high residency at this site because it has a temperature factor of 24 Å^2 compared with the average temperature factor of 20 Å^2 for the base atoms. In addition, this water was found in all of 20 crystal structures of REL1. There are no specific interactions with the exocyclic O4 of the uracil that would distinguish it from the N4 of cytosine.

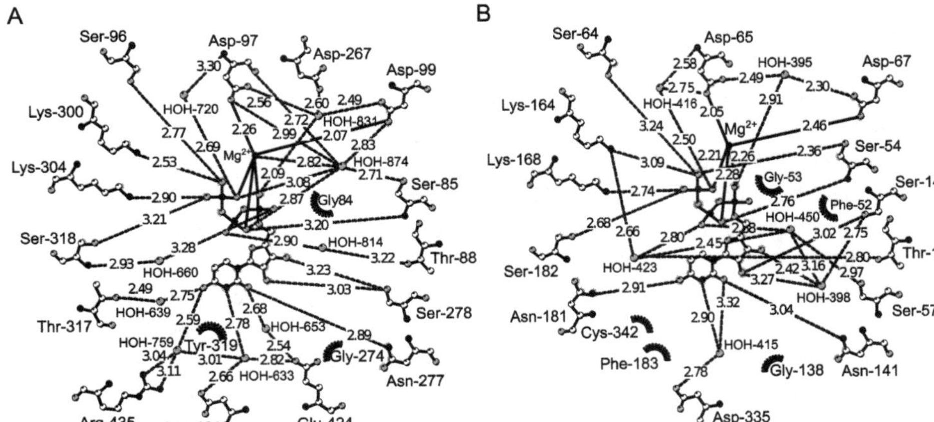

Figure 5.5 Comparison of (A) the UTP binding site A in *Trypanosoma brucei* RET2 (PDB-ID 2B4V) and in (B) the putative active site of *Tb*MEAT1 (PDB-ID 3HIY) in terms of the interactions between the UTP, Mg^{2+}, and the surrounding protein atoms. The 3D structure is projected in 2D, and the residues are repositioned for clarity. The short arcs of parallel lines represent atoms involved in hydrophobic contacts. The distances are in Å. The carbon atoms are coloured white, the oxygen atoms are grey, and all other atom types are black. The plot was made with *LIGPLOT+* (Laskowski and Swindells, 2011).

Future directions

There is interest in designing inhibitors that target KRET2 (Deng *et al.*, 2005; Demir and Amaro, 2012). However, pursuit of this goal may be challenging because of the presence of multiple TUTases in humans. Crystal structures of the uncharacterized human and trypanosomal TUTases are needed for more effective structure-based drug design campaigns.

RNA editing ligase I (KREL1)

The first crystal structure of a trypanosome RNA editing-related protein is the *Trypanosoma brucei* RNA editing ligase I (*Tb*REL1) (Deng *et al.*, 2004). *Tb*REL1 is one of two k-RNA editing-associated ligases. The other is *Tb*REL2. These two ligases share 41% sequence identity. *Tb*REL1 has a molecular weight of 52 kDa, and *Tb*REL2 has a molecular weight of 48 kDa (Panigrahi *et al.*, 2001a; Schnaufer *et al.*, 2001; Gao and Simpson, 2003). *Tb*REL1 associates with the U-deletion 20S editing complex while *Tb*REL2 associates with the U-insertion 20S complex. *Tb*REL1 can substitute for *Tb*REL2 but not vice-versa (Schnaufer *et al.*, 2001; Gao and Simpson, 2003). As a result, *Tb*REL1 is essential for the survival of both the insect and bloodstream stages of the parasite (Huang *et al.*, 2001; Panigrahi *et al.*, 2001a; Schnaufer *et al.*, 2001). The crystal structure has played a vital role in molecular simulation and homology modelling studies (Amaro *et al.*, 2007; Shaneh and Salavati, 2009) and in structure-based drug design efforts (Amaro *et al.*, 2008; Swift and Amaro, 2009; Swift *et al.*, 2009; Durrant *et al.*, 2010; Moshiri *et al.*, 2011; Salavati *et al.*, 2011) that would not have been possible or would have been inaccurate without this crystal structure.

REL1 and REL2 are members of the covalent nucleotidyl transferase superfamily (Huang *et al.*, 2001; Panigrahi *et al.*, 2001a; Rusche *et al.*, 2001; Schnaufer *et al.*, 2001; Ho and Shuman, 2002; Gao and Simpson, 2003). Members of this family share an overall fold, common evolutionary traces, and five well-conserved structural motifs that are responsible for the three-step catalytic reaction. At the level of the superfamily, the sequence identity is 10%. *Tb*REL1 has 15% sequence identity with DNA ligases and 20% sequence identity with bacteriophage T4 RNA ligase 2 (T4Rnl2) (Ho and Shuman, 2002; Yin *et al.*, 2003; Ho *et al.*, 2004) – its closest known relative.

The RNA-editing ligases from kinetoplastids are closely related (Palazzo *et al.*, 2003; Worthey *et al.*, 2003). *Tb*REL1 had 20% sequence identity with bacteriophage T4 RNA ligase 2, the closest non-trypanosomatid homologue, and 15% amino acid sequence identity with DNA ligases.

RNA ligation generally occurs in three steps (Palazzo *et al.*, 2003) (Fig. 5.6). First, a critical lysine – K87 in *Tb*REL1 – autoadenylates to form a REL1-AMP intermediate and releases pyrophosphate. Second, the AMP is transferred to the 5′ phosphate group of the 3′ RNA fragment (i.e. the donor RNA) and a 5′–5′ pyrophosphate link is formed. Third, the 3′ hydroxyl group of the 5′ fragment (i.e. the acceptor RNA) displaces the 5′-AMP and forms a new phosphodiester bond.

The crystal structure

The structure of *Tb*REL1 (PDB-ID 1XDN) (Deng *et al.*, 2005) was determined with X-ray diffraction data collected to an unusually high-resolution limit of 1.2 Å – the threshold for being called atomic resolution data. Only about one out of 50 protein crystal structures in Protein Data Bank have the same or higher resolution. The uncertainty in the atomic positions can be expected to be better than 0.04 Å. The very high precision of this structure is very important in reducing the uncertainties in atomic position dependent parameters derived from the crystal structure such as hydrogen bond lengths, molecular surface properties, and covalent and non-covalent energy terms. As a result, it provides an accurate starting structure for molecular dynamics simulations, homology modelling, and structure-based drug design.

The crystal structure is of the N-terminal catalytic domain (residues 52–316) with ATP and magnesium bound (Deng *et al.*, 2005). The full-length protein is 469 amino acids long. The first 50 residues are a mitochondrial import signal peptide that is absent after the protein is imported into the mitochondrion. The C-terminal domain is predicted to interact with other editosome proteins that have oligonucleotide binding (OB-fold) domains (Schnaufer *et al.*, 2003; Worthey *et al.*, 2003). For example, KREPA2 is thought

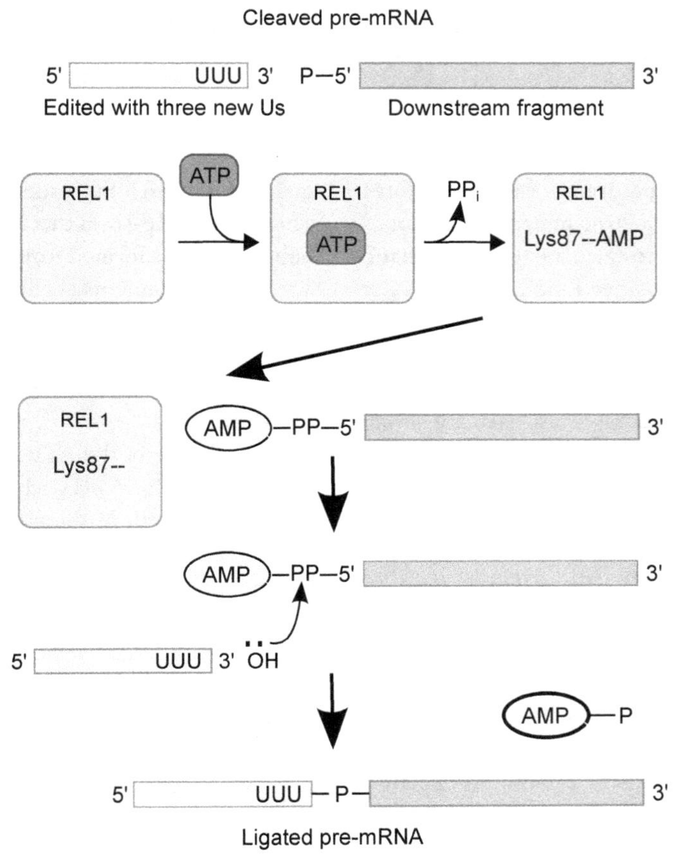

Figure 5.6 Cartoon of the three-step reaction mechanism of *Trypanosoma brucei* RNA editing ligase 1 (*Tb*REL1) which seals the cleaved pre-mRNA after either U-insertion or U-deletion RNA editing at the 3′ end of the 5′ fragment of the left. Lys-87 is adenylated by ATP with the release of pyrophosphate. The adenylated enzyme transfers AMP to the 5′ phosphate of the downstream fragment on the right to form a 5′–5′ pyrophosphate linkage. The 3′ hydroxyl of the edited pre-mRNA fragment attacks the anhydride bond of the pyrophosphate to release AMP while forming a phosphodiester bond with the 5′ end of the downstream fragment thereby joining together the two mRNA fragments.

to provide an oligonucleotide binding (OB-fold) domain *in trans* (Schnaufer *et al.*, 2003). In an analogous fashion, KREPA1 binds the C-terminal domain of REL2 *in trans* with KREPA1's second zinc-binding motif and enhances the ligation activity REL2 (Park *et al.*, 2012). Unfortunately, the full-length recombinant *Tb*REL1 did not produce enough protein for structural studies. However, the N-terminal domain of *Tb*REL1 is capable of RNA strand joining *in vitro* but with reduced activity and altered kinetics (Deng *et al.*, 2004).

The catalytic domain has a rectangular block shape with relatively flat faces. The rectangle has a pair of square faces ~ 50 × ~ 50 Å in dimension, and it is about 30 Å wide (Deng *et al.*, 2004) (Fig. 5.8A and B). The active site is located in a deep and narrow pocket that is near the middle of the square face with the flattest surface. The mouth of the pocket is open to the solvent in the crystal structure (i.e. it is not blocked by crystal packing contacts with neighbouring protein molecules). The pocket is about 20 Å long (Fig. 5.8C).

Putative RNA binding site

When the electrostatic potential was calculated with the computer program Delphi (Nicholls *et al.*, 1991) from the crystal structure coordinates and was mapped onto the solvent accessible surface, the pocket had a net positive charge whereas the outer surface were essentially neutral with only small patches of positive and negative electrostatic potential and without a large patch of positive electrostatic potential expected for a RNA binding surface (Deng *et al.*, 2004). However, when the ensemble average (4000 structures at 5 ps intervals from a 20 ns molecular dynamics simulation) was used to calculate the electrostatic isosurfaces, a large positive lobe emerges on the flat surface near the ATP binding pocket (Amaro *et al.*, 2007). This lobe overlaps with the location of the DNA in the co-crystal structure with the homologous human DNA ligase (Amaro *et al.*, 2007).

*Tb*REL1 has two loops (Tyr-165 to Lys-175 and Val-190 to Tyr-200) on the outer edge of the positive electrostatic surface near the ATP binding site (Fig. 5.7). These loops may be involved in RNA binding because each loop has a positive residue in the middle (Lys 166 and Arg194 respectively) (Deng *et al.*, 2004; Amaro *et al.*, 2007). These loops are absent in the homologue T4Rnl2 that, in the absence of its C-terminal domain, fails to catalyse the ligation reaction. This difference suggests that *Tb*REL1's catalytic domain has its own RNA binding motif and does not require its C-terminal domain for RNA binding.

The active site

The ATP binding pocket is flanked by two subdomains (Fig. 5.8A). Each subdomain is centred on a highly curved β-sheet that is flanked by α-helices. Subdomain 1 is formed from contiguous secondary structure elements (helices α3–α6 and strands 2–β10); on the other hand, subdomain 2 is formed from non-contiguous secondary structure elements (helices α1, α2, α7-α11 and strands β1, β11–13).

The adenine of the ATP projects halfway into the pocket (Fig. 5.8C) while the triphosphate blocks the mouth of the pocket. There are eight waters between the end of the adenine and the bottom of the pocket. In molecular dynamics simulations, these water molecules have high residency times (Amaro *et al.*, 2007). The presence of the water molecules suggests that the ATP and some of the water molecules could be replaced by a bulkier moiety in the form of an inhibitor (Deng *et al.*, 2004), although buried protein waters are not always easy to displace as intended (Mooers *et al.*, 2003).

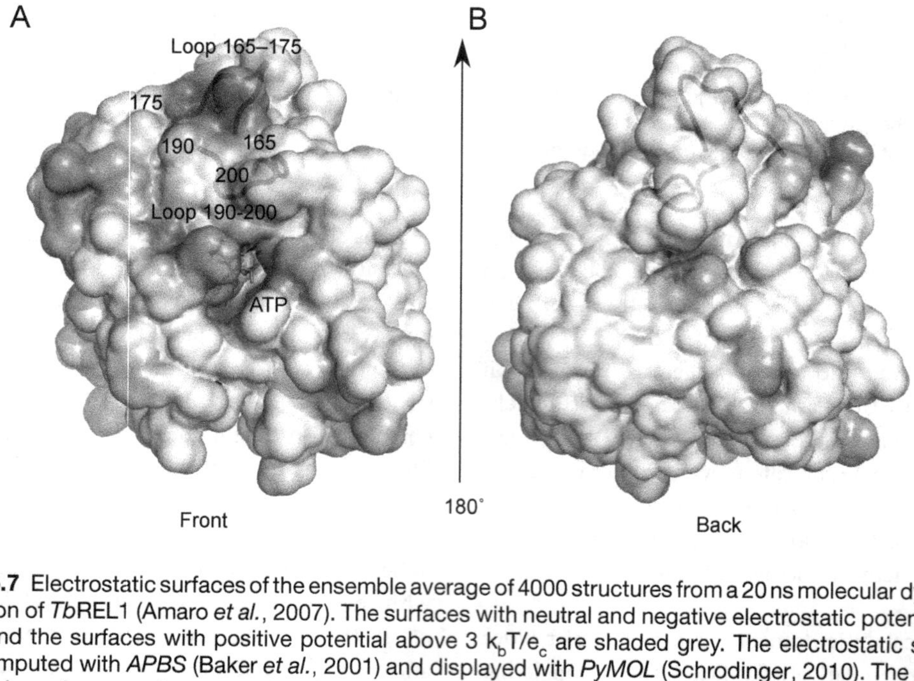

Figure 5.7 Electrostatic surfaces of the ensemble average of 4000 structures from a 20 ns molecular dynamics simulation of *Tb*REL1 (Amaro *et al.*, 2007). The surfaces with neutral and negative electrostatic potentials are white, and the surfaces with positive potential above 3 k_bT/e_c are shaded grey. The electrostatic surfaces were computed with *APBS* (Baker *et al.*, 2001) and displayed with *PyMOL* (Schrodinger, 2010). The solvent-accessible surfaces are transparent and reveal two ribbons that represent the backbones of two loops that are in the positively charged patch near the active site and that may have a role in RNA binding.

The adenine's faces are parallel to the β-strands of the two flanking β-sheets. The faces of the adenine are sandwiched between the side chains of Lys-87 and Val-286 on one side and the side chain of Phe-209 on the other side (Fig. 5.9A). There are no intervening waters between the faces of the adenine and these side chains. The adenine essentially plugs the pocket and blocks the exit of the buried waters on the bottom of the pocket (Fig. 5.8C). The adenine is twisted around into a high-energy *syn*-conformation while the ribose ring is in the 3′-endo conformation expected for a ribonucleotide.

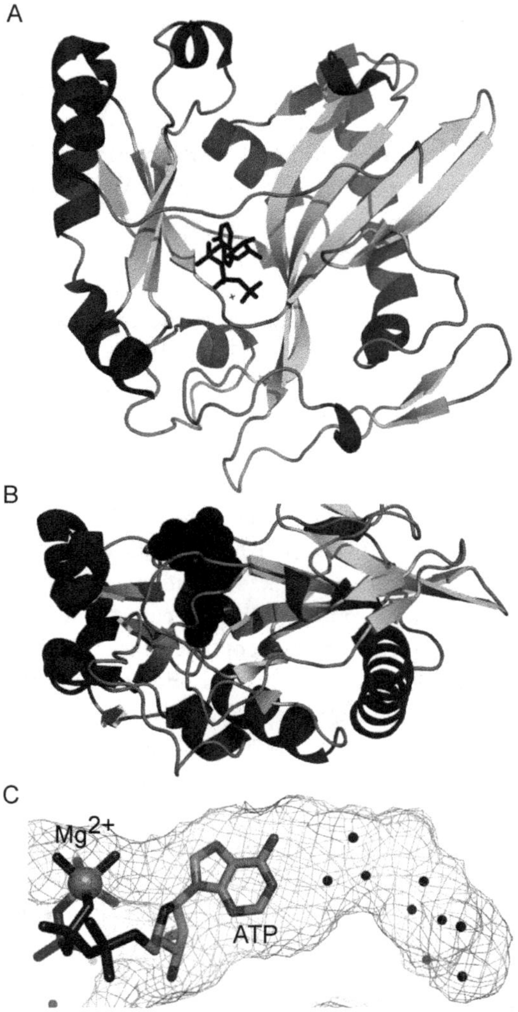

Figure 5.8 Crystal structure (PDB-ID 1XDN) of the catalytic domain of the *T. brucei* RNA editing Ligase I (*Tb*REL1). In A and B the cartoon is shaded by secondary structure: helix – dark grey; loops – medium grey; and strands – light grey. (A) View towards the active site. The triphosphate of the ATP in the ATP-binding pocket is closest to the viewer. (B) View in (A) rotated clockwise by 90° about the horizontal axis. The ATP represented as a surface to show how the adenine is buried in the core of the protein while the triphosphate remains near the mouth of the cleft. (C) Mesh rendering of the molecular surface of the ATP binding pocket. The mouth of the pocket is on the left and the bottom of the pocket is on the right. The view in (A) was rotated clockwise by 90°. The small spheres represent the oxygen atoms of water molecules.

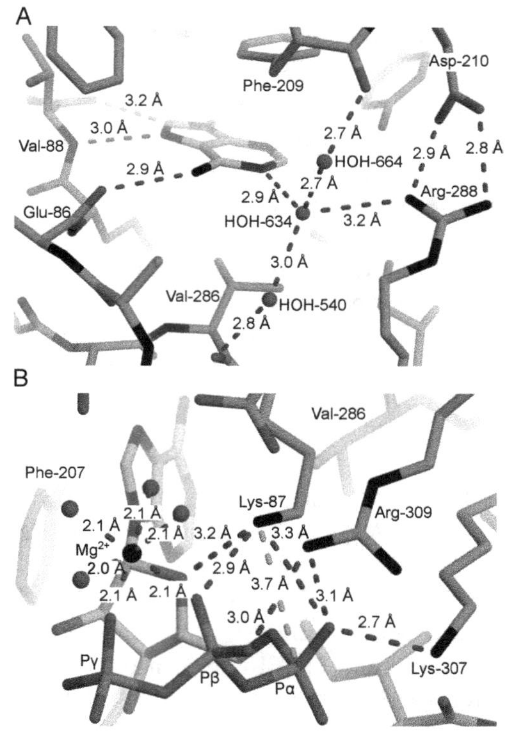

Figure 5.9 The ATP binding site in the crystal structure (PDB-ID 1XDN) of the N-terminal catalytic domain of *the T. brucei* RNA editing ligase I (*Tb*REL1). Stick model of the interactions between the ATP, Mg^{2+}, water molecules, and the surrounding protein atoms. (A) The interactions between the adenine base of the ATP and surrounding residues. The black dashed lines represent conventional hydrogen bonds. The white dashed line represents a CH–O hydrogen bond. (B) Some of the interactions between the backbone of the ATP, a hydrated Mg^{2+}, and the surrounding protein atoms. The white dashed line represents the distance between the catalytic Lys-87 and the target α-phosphorous atom.

The adenine's hydrogen bonding potential along its edges is not completely satisfied (Fig. 5.9A). The adenine's ring N3 nitrogen atom lacks a hydrogen bond, and its N6 amino group has one rather than the expected two hydrogen bond partners. The N1 ring nitrogen faces the bottom of the pocket and is hydrogen bonded to a water HOH-634 which in turn is hydrogen bonded to Arg-288:$N^{\eta 1}$ of motif IV and to two water molecules (HOH-540 and HOH-664) that are hydrogen bonded to the carbonyl oxygen atoms of the conserved protein residues Val-286 and Phe-209 (motif IIIa) respectively. This triad of waters may help stabilize the orientation of the adenine.

The major groove face of the base has several interactions with protein atoms (Fig. 5.9A). The exocyclic N6 nitrogen atom forms a hydrogen bond with the backbone carbonyl of Glu-86 of motif I. The ring N7 nitrogen atom forms a hydrogen bond with the backbone nitrogen of Val-88. The C8 ring carbon atom forms a CH–O hydrogen bond (distance of 3.16 Å) with the carbonyl oxygen atom of Val-88.

The ATP-binding pocket is also flanked by five highly conserved structural motifs of the covalent nucleotidyl transferase superfamily (Ho and Shuman, 2002): I, Lys-87 to Gly-90; II, Val-155 to Gly-162; III, Phe-207 to Phe-209; IV, Glu-283 to Ile-287; and V, Phe-262 to Ala-282 and Ile-305 to Arg-309. Essentially all of the ATP's atoms make direct and indirect interactions with the surrounding *Tb*REL1 atoms. The adenine's binding pocket is distinct, and there are no close homologues to *Tb*REL1 in humans. These features suggest that selective inhibitors of *Tb*REL1 will block this RNA editing pathway in these pathogens.

The triphosphate of ATP sits at the mouth of the pocket (Fig. 5.8C). The negative charge of the triphosphate is partially neutralized by inner shell covalent coordinate bonds with a divalent magnesium cation (Fig. 5.9B). The β – and γ-phosphate oxygen atoms of the ATP provide two ligands to the magnesium. Four water molecules provide the remaining free ligands (Fig. 5.9B). Three of these waters interact with Glu-236 of motif IV, Glu-156 of motif III, and His-89 and Gly-90 of motif I, so these water molecules may be highly conserved in trypanosome RNA ligases. The O3G oxygen atom of the terminal gamma phosphate forms a hydrogen bond with the $N^{\eta 2}$ of Arg-111. The O2B oxygen atom of the middle β-phosphate is hydrogen bonded to the catalytic residue Lys-87 of motif I, and its O1B oxygen is hydrogen bonded to Arg-309 of motif V (Fig. 5.9B). The O1A oxygen atom forms bifurcated hydrogen bonds with the side chains of Lys-307 and Arg-309 from motif V. The O2A oxygen atom forms a hydrogen bond with the backbone nitrogen of residue Ile-61.

The coordination of the triphosphate by the magnesium cation is unique in the crystal structures of RNA ligases (Fig. 5.9B). The magnesium in the mRNA capping enzyme was proposed to stabilize the α-phosphate as it is brought close to the catalytic lysine prior to adenylation of the lysine. Instead, the magnesium in *Tb*REL1 stabilizes the β – and γ-phosphates, and Lys-87 is not adenylated (Fig. 5.9B). The α-phosphorus to Lys-87:NZ distance is 3.74 Å, and the nitrogen is not oriented properly with respect to the leaving pyrophosphate for a nucleophilic attack. The Lys-87:NZ nitrogen atom instead interacts with the ribose O4′ and O5′ oxygen atoms and the O2B oxygen atom of the phosphate. A large part of the Lys-87 side chain packs against one face of the adenine base.

Lys-307 is highly conserved (motif V), close to the α-phosphorous atom (3.55 Å), and hydrogen bonded to the phosphate O1A oxygen atom (Fig. 5.9B). Lys-307 may play a role in the transfer of the α-phosphate to the canonical Lys-87 in the first reaction step. The Mg^{2+} may play multiple roles in *Tb*REL1. The Mg^{2+} may stabilize the β – and γ-phosphates in the initial Mg^{2+} –ATP–*Tb*REL1 complex, and then the Mg^{2+} may bind the α-phosphate and promote the formation of a covalent bond with Lys-87 in the second stage of the reaction.

Arg-309 is unique to REL1 ligases and may explain in part the difference in catalytic properties between REL1 and REL2. The guanidinium group of Arg-307 is close to the α – and β-phosphate groups of the ATP (Fig. 5.9B). Replacement of the homologous arginine with a lysine in T4Rnl2 abolished adenylation and strand joining activities (Ho *et al.*, 2004).

Structure-based drug design

Because *Tb*REL1 is an essential enzyme and because the crystal structure is determined with high precision at atomic resolution, the crystal structure is an excellent starting model for structure-based drug design. Virtual screening with *AUTODOCK* (Morris *et al.*, 1998; Huey *et al.*, 2007) found 14 potential inhibitors. Five of these inhibitors were experimentally validated *Tb*REL1 inhibitors with low micromolar IC_{50} (Amaro *et al.*, 2008). *AUTODOCK* was also able to distinguish weak from strong inhibitors. *AUTODOCK* was then used to find more naphthalene-based *Tb*REL1 inhibitors – one of which is effective against the whole parasite (Durrant *et al.*, 2010). The predicted binding modes of the active compounds were further evaluated using a flexible receptor model and computational fragment matching (Durrant and McCammon, 2011). This work found four low-micromolar inhibitors. Parts of three of these new inhibitors may bind a newly revealed cleft in the simulated protein structures that opens between Glu-60 and Arg-111 on one side of the ATP-binding pocket near the position occupied by the adenine in the crystal structure. Three of the four inhibitors have hydrolysable diazene linkers that compromise the effectiveness of the inhibitors in cells. The inhibitors need further optimization of their drug-like features (Lipinski *et al.*, 2001). Further experimental work on the high-throughput screening of inhibitors of RNA editing has been reported recently (Liang and Connell, 2010; Moshiri *et al.*, 2011) and has been reviewed (Salavati *et al.*, 2011).

Future research

Additional crystal structures of *Tb*REL1 would enhance our understanding of its unique catalytic mechanism and inhibition. First, the crystal structure of the adenylated Lys-87 is needed to better understand the role of the magnesium and surrounding residues in the second step in the catalytic mechanism. Second, the crystal structure of the *Tb*REL1:RNA substrate complex would reveal how the nicked substrate is recognized. Third, the crystal structure of *Tb*REL2 and its most promising inhibitors would probe the plasticity of the ATP binding pocket [e.g. do Glu-60 and Arg-111 separate as predicted (Durrant and McCammon, 2011)?] and the tenacity with which the buried water molecules at the bottom of the pocket remain in place. Fourth, the crystal structure of the *Tb*REL2 may shed light on why it can be substituted with *Tb*REL1. Fifth, the crystal structure of the full-length protein may reveal more about the function of the missing C-terminal domain. Sixth, the crystal structure of the *Tb*KREPA2 and *Tb*REL1 would reveal the role the OB-fold of *Tb*KREPA2 plays in RNA binding by *Tb*REL1. Seventh, structures of the homologues from *Trypanosoma cruzi* and *Leishmania major* would provide insights into (1) the evolution of *Tb*REL1, (2) the common features of their reaction mechanism, and (3) how an inhibitor developed against one *Tb*REL1 may or may not be effective inhibitors of other kinetoplastid REL1s. Finally, crystal structures of the complexes between REL1 and promising inhibitors are needed to get precise insights into how REL1 is being inhibited and to aid the design of more effective inhibitors.

RNA editing complex structural proteins

The core of the editing complex has six structural proteins – kinetoplastid RNA editing proteins A-type (KREPA1–KREPA6 or A1–A6) – that are members of the large superfamily of single-strand nucleic acid binding folds [or oligonucleotide binding (OB)], otherwise known as SSB domains. The A1–A6 proteins share an OB-fold near their C-terminus (Worthey *et al.*, 2003), so they may play a structural role, although some members of this group are known to bind RNA (Tarun *et al.*, 2008). All six *Tb*KREPAs are members of the three known editing complexes (Carnes *et al.*, 2005, 2008; Panigrahi *et al.*, 2006). Their stoichiometry in the editing complexes remains unclear due to the large uncertainty in the molecular weights of the editing complexes. Nonetheless, they are essential to functioning of the RNA editing core complex (Drozdz *et al.*, 2002; Huang *et al.*, 2002; Schnaufer *et al.*, 2003; Worthey *et al.*, 2003; Kang *et al.*, 2004; Brecht *et al.*, 2005; Panigrahi *et al.*, 2006; Salavati *et al.*, 2006; Law *et al.*, 2007, 2008; Guo *et al.*, 2008; Kala and Salavati, 2010). A1–A3 are the larger proteins of the six proteins. They contain two zinc finger motifs followed by a C-terminal OB-fold (Panigrahi *et al.*, 2001b; Worthey *et al.*, 2003). A4–A6 lack the zinc fingers but still have the C-terminal OB domain (Worthey *et al.*, 2003). Each KREPA protein has unique characteristics and its own set of interacting partners (Schnaufer *et al.*, 2010).

The OB-fold consists of a five-stranded antiparallel β-sheet. There is usually one OB-fold per polypeptide chain. In other systems, the OB-fold is found sometimes as a monomer but more often as a dimer. To dimerize, the first β-strand of the OB-fold hybridize antiparallel to the first β-strand of a second OB-fold to form an extended β-sheet. A pair of these OB dimers can dimerize to form a tetramer with *D*2 symmetry. The relative orientation of the two dimers varies to a large extent. In mitochondria and prokaryotes, SSB proteins are involved in RNA transcription and DNA repair, recombination, and replication (Chase and Williams, 1986; Lohman *et al.*, 1988; Meyer and Laine, 1990; Murzin, 1993; Bogden *et al.*, 1999; Arcus, 2002; Theobald *et al.*, 2003; Worthey *et al.*, 2003; Eggington *et al.*, 2004).

The isolated A1–A6 proteins are difficult to crystalize by conventional crystallization methods (Park *et al.*, 2011; Wu *et al.*, 2011; Park *et al.*, 2012). The Hol laboratory at the University of Washington recently made progress by crystallizing the KREPA proteins with antibodies from llamas. They have reported the crystal structures of A6 homodimers (Wu *et al.*, 2011), A3–A6 heterodimers (Park *et al.*, 2011), and A1 homodimer (Park *et al.*, 2012). These crystal structures should enable the generation of more accurate predictions of how the components of the editing complexes assemble. These predictions in turn can be tested *in vitro* against low-resolution structural data such as from electron microscopy studies (Golas *et al.*, 2009; Li *et al.*, 2009).

KREPA6 homodimer

Although KREPA6 (A6) varies in length from 164 to 229 residues across trypanosome species, it interacts with A1, A2, A3, and A4 (Schnaufer *et al.*, 2003, 2010) and with single-stranded RNA (Tarun *et al.*, 2008). As a result, A6 is central to the structural integrity of the RECC (Tarun *et al.*, 2008).

Recombinant A6 was recalcitrant to crystallization in thousands of crystallization solutions. The standard strategies were used – homologues from different species, length variants, and mutations designed to improve protein folding and solubility (Wu *et al.*, 2011). Success was achieved by using antibodies as 'crystallization chaperones'. The nanobody adds new surfaces that promote intermolecular interactions required for crystal lattice formation and reduces the conformational heterogeneity of the protein sample by binding proteins in one specific conformation (Lah *et al.*, 2003; Loris *et al.*, 2003; Tereshko *et al.*, 2008). Camelid antibodies lack the light chain and the C_H domain of conventional IgGs while retaining the antibody repertoire of the antigen binding heavy-chain variable domain (V_HH) (Hamers-Casterman *et al.*, 1993). Although these nanobodies are small (~ 125 residues), they still bind their protein targets with high affinity (Muyldermans, 2001; Muyldermans *et al.*, 2001). Nanobodies derived from llamas had been successfully applied recently to other difficult to crystallize proteins in the Hol laboratory (Korotkov *et al.*, 2009; Lam *et al.*, 2009).

Purified recombinant A6 was used to obtain 17 anti-A6 nanobodies (Nb1–Nb17) from a llama. Fourteen of these nanobodies were successfully expressed and purified. Four of the purified KREPA6:nanobody complexes gave promising crystals and two (A6:Nb5 and A6:Nb15) gave 2.3 and 2.1 Å resolution X-ray data respectively. These two nanobodies bound to A6 at the same sites and gave very similar structures of A6 (RMSD of 0.6 Å).

In parallel, a semi-synthetic library of 10^9 V_HH domains (Goldman *et al.*, 2006) was explored by phage display. These synthetic antibodies are also single domain antibodies (Abs). After four rounds of selection, 27 phages gave a positive signal. The phage DNA was sequenced, and two clones with the most frequently enriched V_HH complementarity-determining region 3 (CDR3) were selected for expression and co-crystallization trials. The two binders (Ab1 and Ab2) were expressed, purified, and mixed with purified A6. Purified heterotetramers of A6:Ab2 crystallized and gave low-resolution X-ray data (3.4 Å). The Ab2 antibody bound a different site on A6 compared with the Nb5 and Nb15.

The stoichiometry of A6 alone and in combination with the antibodies was determined in solution. Isolated A6 forms a homotetramer. This tetramer separates into dimers upon the addition of each antibody. The antibodies bind the A6 dimers to form $(A6:V_HH)_2$ heterotetramers in solution. This stoichiometry agrees with that found in the crystal structures.

The folds of both A6 and the antibodies were similar to known structures, so molecular replacement was successful in the determination of the structures. The structure of A6:Nb15 was determined using as search models the antibody (PDB-ID 1I3V) (Spinelli *et al.*, 2001) and *E. coli* SSB (PDB-ID 1EQQ) (Matsumoto *et al.*, 2000), a protein homologous to A6. The structure of the A6:Nb15 was then used to determine the structures of the two remaining complexes.

The A6 monomers have the five β-strands of the OB-fold (Fig. 5.10). An alpha helix is between strands β-3 and β-4. The loop L_{12} is in well-defined electron density, but loop L_{23} and L_{45} are missing electron density hence the breaks in the cartoon (Fig. 5.10). The N-terminal residues 19–20 and C-terminal residues 135–164 are also missing electron density.

The crystal structure of the A6:Nb15 complex had one heterodimer in an asymmetrical unit. A crystallographic 2-fold axis generates the second heterodimer that makes up the heterotetramer, so both halves of the heterotetramer are identical (Fig. 5.10A). The buried solvent accessible surface area between the two heterodimers is extensive (2030 Å^2 with a calculated $\Delta G°$ of interaction of −6.2 kcal/mol according to the *PISA* server (Krissinel and Henrick, 2007). The A6 β–1strand interacts antiparallel to the corresponding β-1 strand (β-1′) from the symmetry related copy of A6. This leads to an extended anti-parallel β-sheet of six-strands (β-3, β-2, β-1, β-1′, β-2′, β-3′) where the two copies of A6 interact. The 2-fold crystallographic axis that relates the two copies of A6 is perpendicular to this extended β-sheet. This axis is called the '*P*-axis' where it is found in other SSB proteins (Saikrishnan *et al.*, 2003). Twenty-two of the 25 residues at the interface are the same in all trypanosomes, so this interface is highly

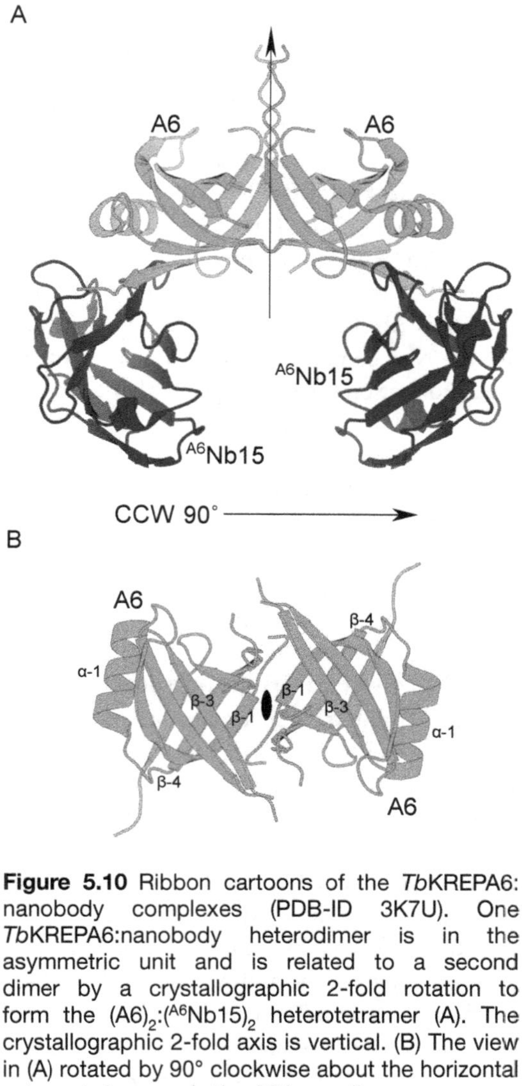

Figure 5.10 Ribbon cartoons of the *Tb*KREPA6: nanobody complexes (PDB-ID 3K7U). One *Tb*KREPA6:nanobody heterodimer is in the asymmetric unit and is related to a second dimer by a crystallographic 2-fold rotation to form the $(A6)_2:(^{A6}Nb15)_2$ heterotetramer (A). The crystallographic 2-fold axis is vertical. (B) The view in (A) rotated by 90° clockwise about the horizontal axis and shows only the A6 homodimer.

conserved. The loop residue types and lengths are also very similar across trypanosome species, so the A6 dimer structure is likely the same across all trypanosome species.

The A6 dimer is in the centre of the heterotetramer and is flanked by two copies Nb15. The Nb15 nanobodies do not interact directly with each other. A parallel pair of β-strands forms the interface between Nb15 and A6 (β6 from A6 and βC″ from Nb15). Seventeen residues from A6 and 17 residues from Nb15 interact and bury 1390 Å^2 of molecular surface area. According to the PISA server, the $\Delta G°$ of interaction is −8.0 kcal/mol.

The crystal structure A6:Nb5 had one tetramer

in the asymmetric unit. Both dimers superimpose with a RMSD of 0.6 Å for 217 C$^{\alpha}$ atoms, so they are essentially identical. The A6 dimer superimposes on the A6 dimer from the crystal structure of A6:Nb15 with a RMSD of 0.6 Å for 196 equivalent C$^{\alpha}$ atoms. The Ab2 antibody from the synthetic antibody screen is 65.6% identical to Nb15 with large differences in the CDR1, CDR2, and CDR3. The crystal structure the A6:Ab2 also had a (A6:Ab2)$_2$ tetramer in the asymmetric unit like the Nb5 complex. The Ab2 domains contact the A6 dimer on the face opposite that contacted in the crystal structures of the Nb5 and Nb15 complexes. The L45 loop interacts with the CDR3 residues of Ab2 to a larger extent than in the other two complexes. Interactions with the CDR3 bury 438 Å^2 (in contrast to ~ 130 Å^2 in the crystal structures of the other two complexes) and interactions with the framework 3 bury 191 Å^2.

KREPA3^{OB}–KREPA6 heterodimer

One of A6's interacting partners is A3. A3 binds ssRNA as well as dsRNA (Brecht *et al.*, 2005). Its C-terminal OB-fold interacts with A6 and the editosome protein B5 (Schnaufer *et al.*, 2010). The heterodimer of A3 and A6 forms by interacting through their OB-folds. This interaction was captured in a recent crystal structure of the heterotetramer formed by the OB-folds of A3 and A6 and two copies of the anti-A3 nanobody ^{A3}Nb14 (Park *et al.*, 2011). The nanobody binds to both A3 and A6 in spite of only 40% sequence identity between A3 and A6. The crystal structures of one antibody bound to two different proteins are rare. The previous examples involved proteins with higher sequence identity (Lescar *et al.*, 1995; Igonet *et al.*, 2007). The structure of the A3^{OB}: A6 heterodimer may be the first structure of a heterodimer formed by OB-folds. A3 and A6 usually form homodimers. However, the crystal structure shows that the A3^{OB}:A6 interface buries 50% more surface area than the A6:A6 interface and this may explain the favouring of the A3^{OB}:A6 heterodimer over (A3^{OB})$_2$ and (A6)$_2$ homodimers.

A ternary complex of A2^{OB} (residues 474–587):A3^{OB} (residues 245–393):A6 (residues 20–164). The A2^{OB} protein disassociated from the complex over time and left behind the A3^{OB}–A6 binary complex. An immune library was generated in a llama against full-length A3, and a set of 14 nanobodies was initially isolated for further evaluation. According to electromobility shift analysis, two nanobodies ^{A3}Nb14 and ^{A3}Nb8 bind A3^{OB}. The complex with ^{A3}Nb14 gave crystals of the native protein and of the complex containing A3^{OB} with selenomethionine in place of methionine. The initial electron density maps showed one OB-fold with three selenomethiones which had to be A3^{OB} and a second OB-fold without selenomethionines which had to be A6 (Fig. 5.11). The native crystals gave 2.5 Å resolution X-ray data.

Superposition of the A3^{OB} onto A6 using 91 equivalent C$^{\alpha}$ atoms gives a RMSD of 1.2 Å. Most of the largest differences were located in the loop L23 and the β4–β5 hairpin. The two monomers interact by hydrogen bonding between their antiparallel β-1 strand as in the structure of the A6 homodimer. The monomers are also related by a pseudo-*P*-axis that is perpendicular to the extended β-sheet. The ^{A3}Nb14 molecules are related by the same pseudodyad.

The surface residues have an asymmetric distribution. The 'front' and 'top' surface is dominated by positively charged residues, the 'back' surface is dominated by negatively charged residues, and the 'bottom' or β-surface is dominated by aliphatic residues with branched side chains. The β-surface faces a symmetry related β-surface to form a heterotetramer of OB-folds: (A3^{OB}–A6)$_2$. The interface between the heterodimers in a neighbouring asymmetrical unit buries ~ 2000 Å^2 (Park *et al.*, 2012).

The crystal structures of the A6 homodimer and the A3^{OB}:A6^{OB} heterodimer were used in a molecular modelling exercise to develop a 'shifted heterotetramer' model for the cluster of OB-folds in the core of the editosome (Park *et al.*, 2011). The model has been subsequently updated in the publication about the KREP protein that will be discussed next.

KREPA1$^{OB\Delta}$ homodimer

KREPA1 (or A1) is the largest of the six interaction proteins in the RNA editing core complex. A1 has two zinc finger motifs followed by one OB-fold near the C-terminus. The OB-fold of A1 interacts with A6. A6 in turn interacts with

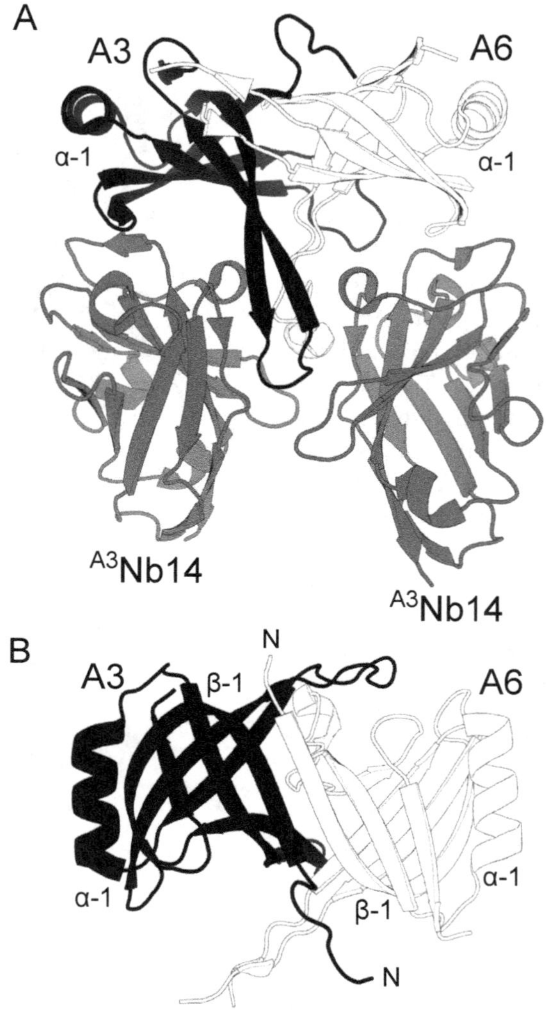

Figure 5.11 Ribbon cartoons of the *T. brucei* KREPA3^{OB}–KREPA6–(^{A3}Nb14)$_2$ heterotetramer from the 2.5 Å resolution crystal structure (PDB-ID 3STB). KREPA3^{OB} is coloured dark grey, KREPA6 is coloured white, and the anti-A3 nanobodies are coloured medium grey. (A) View perpendicular to the pseudodyad that relates the KREP/nanobody dimers. (B) View down the pseudodyad that relates KREPA3^{OB} and KREPA6. The pseudodyad axis is represented by the open version of the symbol for a 2-fold rotation axis.

A2, A3, and A4. This suggests that A1 is part of a central complex that contains five different kinds of OB-folds (Panigrahi *et al.*, 2001a; Schnaufer *et al.*, 2003, 2010; Park *et al.*, 2011). A1 also interacts with the ligase REL2 (Gao and Simpson, 2003). A series of pulldown experiments with truncation mutants showed that the first zinc-binding site on A1 (residues 196–331) interacts with REL2. On the other hand, the C-terminal domain of REL2 (residues 308–416) was shown to interact with A1 (Park *et al.*, 2012). The binding of *Tb*REL2 by A1 *in trans* enhanced the ligation of nicked mRNA by *Tb*REL2 (Park *et al.*, 2012). The second zinc-binding site on A1 (residues 396–482) binds *Tb*RET2 (Schnaufer *et al.*, 2010). In summary, A1 binds directly to REL2, RET2, and KREPA6.

Crystals of the A1^{OB} domain were obtained after replacing the 38-residue loop L_{23} (residues 658–695) with the linker GASG to create A1$^{OB\Delta}$ and after mixing with the anti-A1$^{OB\Delta}$ nanobody ^{A1}Nb10 (Park *et al.*, 2012). A stable ternary complex of (A1$^{OB\Delta}$:A6:^{A1}Nb10) was used in the crystallization trials. The complex crystallized in three different crystal lattices, but only the

($A1^{OB\Delta}$:^{A1}Nb10) dimer was found in the electron density maps. Apparently, A6 dissociated from the ternary complex prior to crystallization. The crystal structures were determined and proved to be similar. The authors deposited all three structures. Only the crystal structure (3DKA) with the highest resolution diffraction data (2.0 Å) is discussed here.

The crystal structure of the $A1^{OB\Delta}$ monomer has 29% sequence identity with A6 at 80 equivalent sites and a RMSD of 1.6 Å with the corresponding C^{α} carbon atoms. The structure of $A1^{OB\Delta}$ shares the OB-fold with a β-barrel of six β-strands. One very striking difference was the absence of electron density in the $A1^{OB\Delta}$ structure where the α-1 helix between strands β-3 and β-4 is expected to be. The residues in this segment of the polypeptide chain must be very flexible. Modelling of the OB-fold core in the RECC with the crystal structure of $A1^{OB\Delta}$ and the crystal structures of the previously discussed OB-fold proteins suggest that the α–1 helix remains exposed in the core and may be stabilized by the binding of another protein. As in the crystal structures of A3 and A6, antiparallel N-terminal β-strands are at the centre of the extended β3–β2–β1–β1′–β2′–β3′ β-sheet. The homodimers in crystal forms I and II were similar to each other but differed from the dimer in crystal form three in that the second subunit was rotated by 9° and the shifted by 2 Å. There might be multiple types of monomers in solutions.

The heterotetramer was also unusual in its broken molecular symmetry (Fig. 5.12A). Two nanobodies bound to one $A1^{OB\Delta}$, and the two nanobodies were not related to each other by a 2-fold rotation axis (Fig. 5.12A). However, the two $A1^{OB\Delta}$ proteins were related by a molecular pseudo 2-fold (Fig. 5.12B).

$A1^{OB\Delta}$ also had an asymmetric distribution of electrostatic potential on its surface (Fig. 5.12C and D). In particular, seven positively charged side chains are clustered on the surface of the β3, β4, β4′, and β5′ strands (Fig. 5.12C). Five of these residues are highly conserved (Arg-703, Lys-715, Arg-731, Arg-734, and Arg-742). The *T. brucei* protein has two more positively charged residues, Lys-719 and Lys-741. Arginine to glutamate single mutations at sites 703, 731, and 734 abolish

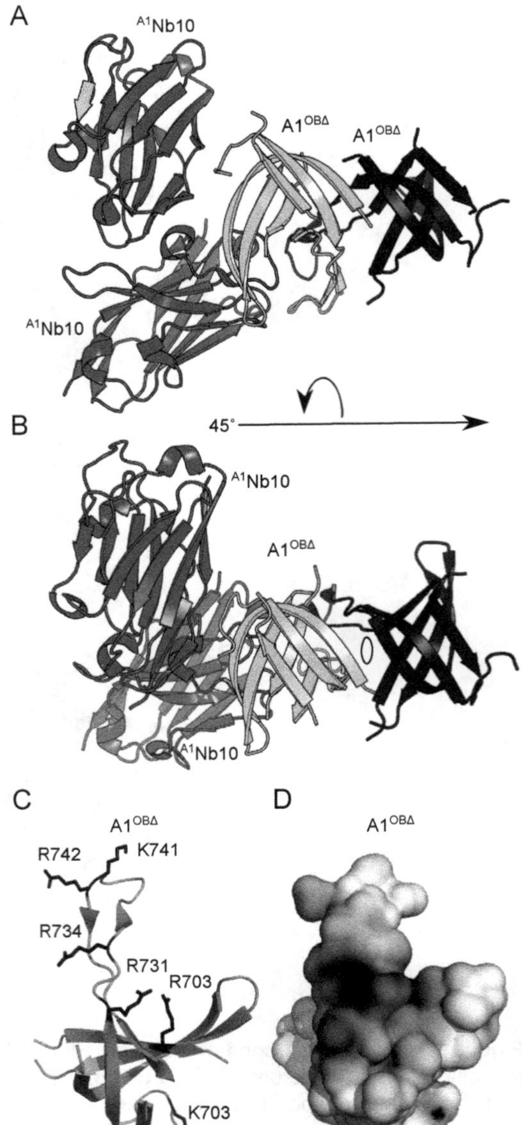

Figure 5.12 Ribbon cartoons of the *T. brucei* KREPA1$^{OB\Delta}$–(^{A1}Nb10)$_2$ heterotetramer from the 1.97 Å resolution crystal structure (PDB-ID 3DKA). KREPA1$^{OB\Delta}$ (A1 $^{OB\Delta}$) from Chain D is white, $A1^{OB\Delta}$ from Chain C is dark grey, and the ^{A1}Nb10 nanobodies are medium grey. (A) View showing how the two nanobodies interact with one $A1^{OB\Delta}$ perpendicular to the dyad that relates the A1/nanobody homodimers. (B) View down the $A1^{OB\Delta}$ homodimer pseudodyad. (C) View of the positively charged face on $A1^{OB\Delta}$ that is thought to bind to RNA. Seven charged side chains are shown as stick models in black. (D) The electrostatic potential mapped onto the solvent accessible surface using APBS (Baker *et al.*, 2001) with the regions above 5 k_bT/e_c coloured dark grey.

dsRNA binding (Park *et al.*, 2012). This positively charged surface remains exposed in the updated 'shifted heterotetramer' model of the OB-fold core of the editing complex (Park *et al.*, 2012).

Future directions

The crystal structures of $A1^{OB\Delta}$, $A3^{OB}$, and A6 provide glimpses into the OB-folds at the core of the RNA editing complexes. We now have the structures of $A1^{OB\Delta}$:$A1^{OB\Delta}$ and A6:A6 homodimers and $A3^{OB}$:A6 heterodimer. The heterodimer is a big surprise. It suggests that there are 120 possible dimers. Nanobodies will likely continue to play a vital role in the crystallization of the remaining OB-folds. It takes several years to develop and screen the nanobodies, but the investment clearly improves the chances of success at crystallization.

Structures of the kRNA editing accessory factors

This section starts with one protein that is part of a RNA editing-like complex of undetermined function. Then we discuss the structures of proteins associated with the RNA editing core complex (p22 and MRP1/MRP2). Finally, we discuss the structure of a fragment of the U-helix domain of a gRNA/mRNA duplex that is probably formed outside of the RECC and is latter part of the RECC during RNA editing.

Mitochondrial editosome-like complex-associated TU*Tase 1* (MEAT1)

The *T. brucei* mitochondrial editosome-like complex-associated terminal uridylate transferase 1 (MEAT1) interacts with an editosome-like complex in the mitochondrion and is exclusively U-specific (Aphasizheva *et al.*, 2009). The function of this editosome-like complex is presently unclear (Aphasizheva *et al.*, 2009). The RNA editing TUTase *Tb*RET2 interacts with a protein partner KREPA1. However, KREPA1 is missing from this editing-like complex. *Tb*MEAT1 is essential for the survival of insect and bloodstream forms of *Trypanosoma brucei,* so it is an attractive drug target (Aphasizheva *et al.*, 2009). MEAT1's function is not fully understood yet. *Tb*MEAT1 has limited similarity with the editing TUTases. Like the cytosolic TUTase4 (Aphasizhev *et al.*, 2004), it lacks the middle domain found in *Tb*RET2 (Deng *et al.*, 2005) (Section 2.1). However, unlike the cytosolic TUTase4 that adds both UTP and CTP, *Tb*MEAT1 is highly specific for UTP, a feature that it shares with *Tb*RET2 (Aphasizheva *et al.*, 2009). Although MEAT1's relation to U-insertion/deletion editing is murky at this time, we included this TUTase and compare it with *Tb*RET2.

The crystal structures of the apo *Tb*MEAT1 were determined with X-ray data with a resolution of 1.56 Å (PDB-ID 3HJ4), *Tb*MEAT1–UTP complex to a resolution of 1.95 Å (PDB-ID 3HJI), and *Tb*MEAT1-UTP-Mg^{2+} to a resolution of 2.25 Å (Stagno *et al.*, 2010). *Tb*MEAT1 shows no large conformational change upon the binding of UTP (0.3 Å RMSD for superposition of *apo* and UTP complex structure), and it does not require the RNA substrate to select UTP. *Tb*MEAT1 has a bridging domain described below that holds the two domains in a fixed position relative to each other and that may provide a surface for protein–protein interactions with its protein partner in the editosome-like complex.

MEAT1 adopts a spherical bidomain architecture similar to that found in *Tb*TUT2 and *Tb*TUT4 in spite of low sequence identity (12% and 14% respectively) (Stagno *et al.*, 2010). The deep cleft between the two domains harbours both the catalytic site and the nucleotide-binding site. The NTD domain is highly conserved and contains three highly conserved carboxylates (Asp-65, Asp-67, and Asp-130) that bind metal cations. The NTD also has other residues that interact with the triphosphate moiety. The NTD adopts the DNA polymerase β-like catalytic fold (Pol β). It contains a five-stranded β-sheet flanked on both sides by α-helices. Asp-65 and Asp-67 are in the β-sheet.

The CTD resembles the ATP cone domain and is conserved among UTPases (Stagno *et al.*, 2010). It contains an insertion that folds into two antiparallel alpha helices that bridge the CTD and NTD. This bridging domain (BD) buries about 650 $Å^2$ of molecular surface area and creates a channel above the active site cleft between the NTD and CTD (Fig. 5.13B). The site of this

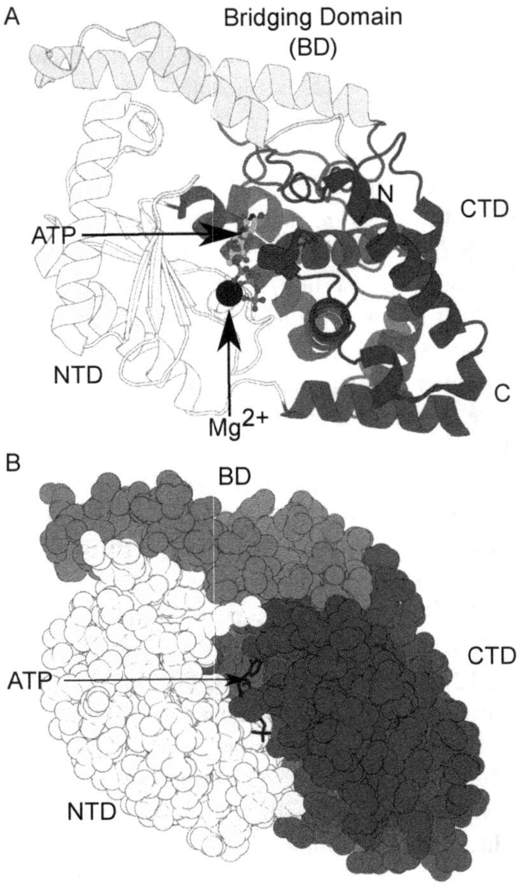

Figure 5.13 Crystal structure of *T. brucei* mitochondrial editosome-like complex-associated terminal uridylate terminase 1 (MEAT1) (PDB-ID 3HIY). (A) Ribbon cartoon with the N-terminal domain (NTD) in white and the C-terminal domain (CTD) in black and grey. The bridging domain in grey is part of the CTD. The ATP molecule in the active site cleft is shown as a stick model while one Mg^{2+} cation is shown as a white sphere. (B) Van der Waals sphere representation of MEAT1 with the same view and shading scheme as in (A).

insertion is occupied by unstructured loops in *Tb*RET2 and *Tb*TUT4.

Crystals of the apo protein were soaked with the other nucleoside triphosphates (Stagno *et al.*, 2010). The electron densities around the ribose ring and the triphosphate are consistent between structures, but the electron density it is not well defined around the position of the base suggesting that different nucleotides are present in different unit cells in the crystal (Stagno *et al.*, 2010).

The binding of the sugar–phosphate appears conserved among TUTases. The conserved aspartates Asp-65 and Asp-67 and single non-bridging phosphate oxygen atoms from all three phosphates coordinate a single Mg^{2+} or a water molecule in the absence of metal cations (Fig. 5.5B). The other side of the triphosphate is hydrogen bonded to Ser-54, Lys-164, Lys1–68, and Ser-182 (Stagno *et al.*, 2010) (Fig. 5.5B).

UTP-binding specificity

Both residues conserved among TUTases and MEAT1 specific residues contribute to MEAT1's exclusive specificity for UTP (Aphasizheva *et al.*, 2009). The chain side nitrogen atom of Asn-141 makes a hydrogen bond with the exocyclic O6 oxygen atom of the uracil (Fig. 5.5B). MEAT1 has an Asn-181 at the position that is a serine in other trypanosomal TUTases. The Asn-181 side-chain carbonyl oxygen makes a hydrogen bond with the backbone amide and side-chain hydroxyl of Thr-184 (Fig. 5.5B). These interactions lock the side chain of Asn-181 into an orientation that allows the amide nitrogen to donate a hydrogen bond to O4 oxygen atom of the uracil base. This interaction distinguishes uracil from cytosine. A network of water-mediated hydrogen bonds also contributes to the specificity for uracil (Fig. 5.5B).

Structure of P22 accessory factor

P22 is a non-editosome associated k-RNA editing accessory factor. A specific accessory factor affects the extent to which a specific mRNA is edited. P22 is specific for the cytochrome oxidase subunit II (COII) mRNA that is one of two mRNAs edited in the procyclic form found in the insect stage of the trypanosome life cycle. The COII mRNA is minimally edited by the insertion of four Us. The editing is directed by an unusual cis-acting guide RNA (most gRNAs are *trans*-acting) that is encoded in the 3′-end of the COII precursor mRNA (Golden and Hajduk, 2005). One of the three known editing complexes is specific for the editing of COII mRNA (Carnes *et al.*, 2008). This editing complex has at least two COII specific proteins: a ribonuclease (KREN3) and an adaptor protein (KREPB6). P22 contributes to cell growth in the procyclic form of *T. brucei* (Sprehe *et al.*, 2010).

The overall shape of p22 resembles the shape

of the crystal structure of the human protein p32 with the exception that p32 has the central channel rather than the chamber. Interestingly, p32 has a number of functions including the regulation of RNA splicing (Petersen-Mahrt *et al.*, 1999) and the maintenance of oxidative phosphorylation (Muta *et al.*, 1997). Both proteins are members of the mitochondrial acidic matrix protein (Mam33p) family (Seytter *et al.*, 1998).

P22 is encoded in the nucleus with an open reading frame for a 227 amino acid pre-protein. A 46 amino acid signal peptide is removed upon entry into the mitochondrion to give the mature protein with 181 residues.

The crystal structure of recombinant p22 (PDB-ID 3JV1) shows that three mature p22 proteins form a trimer (Fig. 5.14A). The molecular triad axis is superposed on a crystallographic 3-fold rotation axis in the crystal structure, so one polypeptide chain is found in the asymmetric unit. As a result, all three chains have identical structures.

The trimer has a triangular shape and is about twice (~75 Å) as wide as it is thick (~35 Å) (Fig. 5.14C and D). One surface is closed and has a mix of positive and negative electrostatic potential (Fig. 5.14E). The opposite surface contains a partially covered chamber (Fig. 5.14F and G) that is about 30 Å deep. This surface has a net negative electrostatic potential (Fig. 5.14C). This is a common feature in the proteins of Mam33p family (Jiang *et al.*, 1999).

The p22 polypeptide chain forms a six-stranded antiparallel β-sheet flanked by five α-helices. Usually, β-sheets associate in multimers by edge-on-edge interactions between the terminal β-strands to extend the β-sheet region. In contrast, the β-sheets in p22 are isolated from each other, and the interactions between subunits occur between the α-helices. The third α-helix from each monomer interacts by forming several hydrogen bonds in the centre of the flat face (Fig. 5.14A). Helices α-2 and α-3 form the floor of the central chamber (Fig. 5.14G). These two helices are absent in human p32 that has an open channel along the molecular 3-fold axis at the position occupied by the three copies of helices α-2 and α-3 in p22. The α-1 and α-5 are long and almost antiparallel. These two helices provide the outer borders of each subunit and place the N – and C-termini near each other. The deletion of α-1 disrupts the interaction of p22 with TbRGG2 – a RNA editing accessory factor (Sprehe *et al.*, 2010). P22 also interacts with the editosome, so p22 may act as an adaptor protein (Sprehe *et al.*, 2010).

Structures of guide RNA matching-making proteins MRP1/MRP2

The RNA editing-associated proteins mitochondrial RNA-binding proteins 1 and 2 (previously named gBP21 and gBP25 respectively; Köller *et al.*, 1997) promote the matching of guide RNA with its cognate mRNA (Müller *et al.*, 2001; Müller and Göringer, 2002). Free gRNA is folded into a two stem–loop structure in the four gRNA characterized to date (Schmid *et al.*, 1995; Schmid *et al.*, 1996). MRP1 unfolds the first stem–loop in gRNA. This first stem–loop often sequesters part of the anchor sequence by intramolecular hydrogen bonding. The unfolding of this first stem–loop occurs without the expenditure of ATP. Meanwhile, MRP2 binds the stem of the second stem–loop that contains the template sequence of the guide RNA that directs RNA editing.

MRP1 and MRP2 are difficult to express individually in a soluble forms in bacteria (Schumacher *et al.*, 2006). MRP1 solubility is improved by genetic truncation of the last 30 residues from the C-terminus. Both proteins can be expressed together for improved solubility as members of a $(MRP1)_2$:$(MRP2)_2$ heterotetramer.

The native protein crystallized with one tetramer in the asymmetric unit (PDB-ID 2GIA) and gave high-resolution X-ray diffraction data (1.89 Å) (Schumacher *et al.*, 2006). These X-ray data are available in the Protein Data Bank. Crystals soaked with mercury acetate were used to collect diffraction data at multiple wavelengths around the mercury edge for MAD phasing. Selenomethionine containing protein crystallized with two heterotetramers in the asymmetric unit (PDB-ID 2GID) and diffracted to low resolution (3.35 Å). The X-ray data for 2GID are not available in the Protein Data Bank. The selenomethione containing $(MRP1/MRP2)_2$ heterotetramer crystallized with a 44-nucleotide version of the guide RNA gDN7-506 (the oligoU tail was missing)

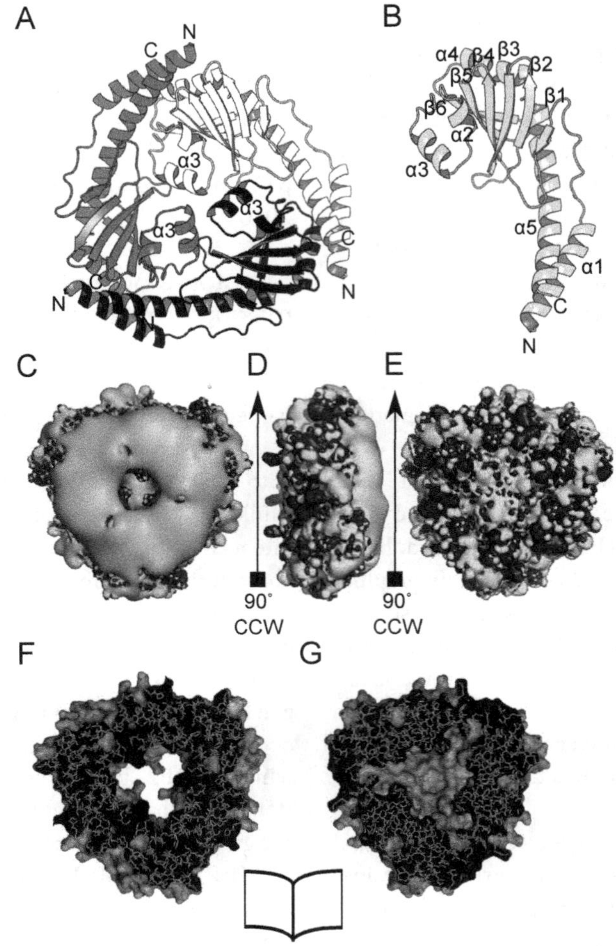

Figure 5.14 Representations of the 2.0 Å resolution crystal structure of *Trypanosoma brucei* p22 protein (PDB-ID 3JV1). (A) Ribbon cartoon of p22 trimer. Each subunit is coloured a different shade of grey. The view is down the molecular 3-fold axis. (B) Ribbon diagram of an isolated subunit from the trimer in (A). The secondary structure elements labelled. (C) The electrostatic potential computed from the crystal structure coordinates using APBS (Baker *et al.*, 2001). The regions with potential below −3 k_bT/e_c are coloured light grey. The regions with potential above 3 k_bT/e_c are coloured dark grey. (D) The view in (C) rotated counterclockwise (CCW) by 90°. (E) The view in (D) rotated CCW by 90°. (F and G) The p22 protein represented by its molecular surface and by sticks representing bonds between atoms. The molecular surface has been sliced in half and opened like a book. (F) The top half with the view from the middle of the central chamber and out of the hole in the top surface. (G) View from the middle of the central chamber and down towards the bottom surface of the chamber.

(PDB-ID 2JGE) (Schumacher *et al.*, 2006). The crystals of this RNA–protein complex gave X-ray diffraction data for only 3 days after the crystals appeared which reflected the labile nature of these crystals. The resolution is reported to be 3.37 Å. The X-ray data for 2GJE are not available in the Protein Data Bank. No mention is made of attempts at co-crystallizing the gRNA with the native protein that had diffracted so well without RNA. This suggests that this system was still suboptimal. The high-resolution limit of the apo structure suggests that the addition of RNA added disorder to the system and is responsible for the lower-quality crystals and diffraction data. The structure was solved by molecular replacement using the high-resolution structure of MRP1 and

MRP2 described above as the search model. An omit map shows the presence of the RNA (Schumacher *et al.*, 2006). The position of U43 was verified with 3.5 Å diffraction data collected from a crystal with the U43 replaced with 5-iodouracil. These data are also not available in the PDB. In spite of the lack of publicly available X-ray data, we focus on the structure of the complex with gRNA (Fig. 5.15).

MRP1 and MRP2 share the 'Whirly' transcription factor fold from plants (Desveaux *et al.*, 2002). They share a β-β-β-β-α-β-β-β-β-α topology. They have very low sequence identity (18%) but remarkably high structural homology. They superpose with a RMSD of 1.38 Å for 126 corresponding C^{α} positions (Schumacher *et al.*, 2006). They assemble into a dimer of heterodimers or really a pseudo homotetramer with pseudocyclic

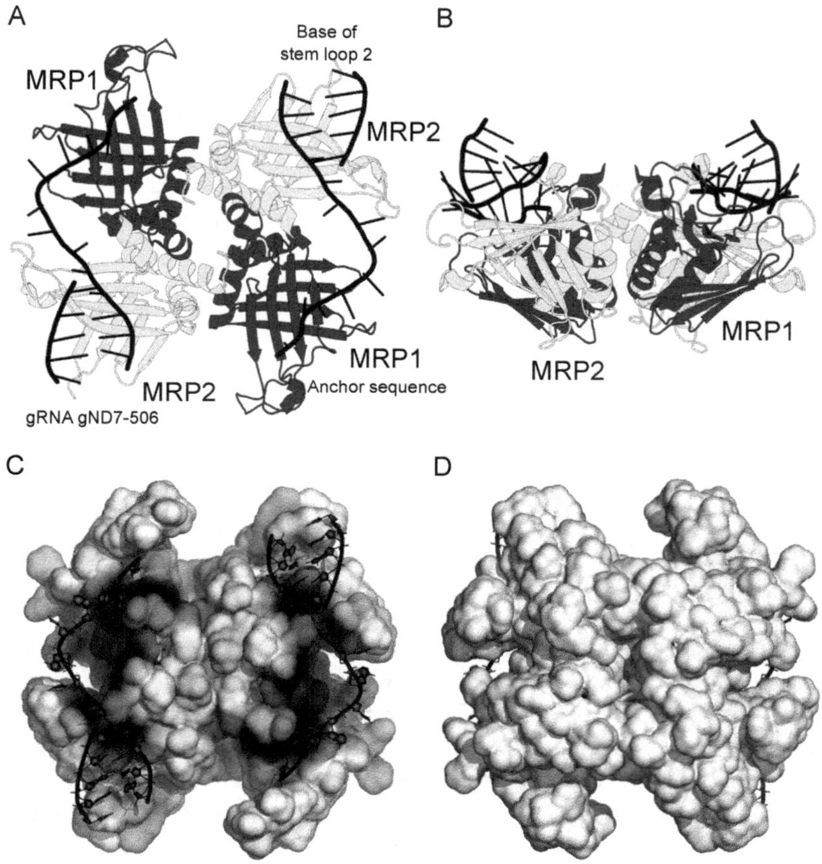

Figure 5.15 Ribbon cartoons of the crystal structure of guide RNA gND7–506 (minus the U-tail) bound to the MRP1/MRP2 heterotetramer (PDB-ID 2GJE, 3.37 Å resolution). MRP1 (dark grey) and MRP2 (light grey) superimpose very well (126 corresponding C^{α} atoms superimpose with a RMSD of 1.38 Å) in spite of only 18% sequence identity (Schumacher *et al.*, 2006). They form a dimer of heterodimers with pseudo 4-fold symmetry. The gRNA is represented by a ladder cartoon were each rung represents a base and the thick tube represents the phosphate backbone. Top view (A) and view in (A) rotated 90° about the horizontal axis in the counterclockwise direction (B). About 40% of the RNA was visible in the electron density. The β-sheets form a platform for the unfolding and binding of the gRNA backbone. (C and D) The electrostatic potential mapped out the solvent accessible surface. The regions with electrostatic potential above 5 k_bT/e_c are dark grey. The regions with electrostatic potential below −5 k_bT/e_c are light grey. The neutral surfaces are white. The calculations were done in the absence of the RNA with APBS (Baker *et al.*, 2001). (C) Same view as in (A) shows the shape and charge complementarity between the RNA backbone and the RNA binding surface of MRP1 and MRP2. (D) View of the electrostatic potential on the backside of the heterotetramer that was largely neutral.

C4 symmetry when considering the high structural similarity between MRP1 and MRP2 (Fig. 5.15A and B). All eight termini of the four subunits associate in a central coil of helices. This coil has the lowest temperature factors in the structure although 35 residues at the C-terminus of MRP2 were not visible in the electron density. Their absence may be due to the C-terminal truncation of the MRP1. This central coil is surrounded by a disc of β-sheets that are almost coplanar and provides a relatively flat surface for non-specific binding of guide RNA (Fig. 5.15A and B). The β-sheets present a positively charged surface that is extensive enough for the binding of two guide RNA molecules per heterotetramer (Fig. 5.15C). The β-sheets have intermediate temperature factors. The outer edge of the tetramer is flanked by loops. These loops have the temperature factors that are 2–3 times higher than in the central region of the tetramer. The outer edges of the heterotetramer also have a negative electrostatic potential, and the side opposite the RNA binding surface has nearly neutral electrostatic potential (Fig. 5.15D).

About 40% of the RNA was built into the electron density map (Fig. 5.15A). The RNA has a double-stranded stem bound to the surface of MRP2. This stem–loop is stem–loop II in the folded gRNA that contains the template sequence that directs RNA editing. The loop region is missing in the electron density suggesting that it is very mobile or disordered. In low-resolution electron density maps, the bases can be distinguished from the backbone, but purine bases are hard to distinguish from pyrimidine bases. Watson–Crick basepairing rules and the position of the brominated U43 provide constraints on the likely RNA sequence in this part of the model. The remainder of the RNA is a single-stranded, and its backbone is bound to the surface of MRP1. This region of the RNA represents the anchor sequence with its bases unpaired and exposed for pairing with the complementary sequence in the mRNA. (Later work shows that MRP1/MRP2 can pair gRNA with non-cognate mRNA (Zikova *et al.*, 2008), so other factors may contribute to the specificity of the interaction.) The temperature factors of the RNA had an average of 136.9 Å^2. In contrast, MRP1 had an average temperature factor of 60.7 Å^2, and MRP2 had an average temperature factor of 34.0 Å^2. The scaling of low-resolution data can be difficult, and this can lead to unrealistically high temperature factors. In addition, with low-resolution data, it is very difficult to accurately refine the temperature factors for individual atoms due to problems with overfitting the X-ray diffraction data. The high B-factors of the RNA suggest several possibilities: (1) the RNA is very mobile on the surface of MRP1 and MRP2, (2) the RNA is not present in all of the unit cells of the crystal (~25% relative to MRP2), (3) the RNA is statically disordered with stem–loop II sometimes present on MRP1 as well as on MRP2, or (4) some combination of 1–3. We conclude that there is crystallographic evidence for the RNA being bound to the surface of MRP1 and MRP2, but the evidence is not very precise.

Future directions

There are many hundreds of gRNAs with very different template sequences (Ochsenreiter *et al.*, 2007; Hajduk and Ochsenreiter, 2010). One low-resolution crystal structure (2JGE) does not adequately represent all gRNA/MRP1/MRP2 structures. More structural work on gRNA/MRP1/MRP2 complexes is needed.

U-helix crystal structure

Most of the structural biology work in the trypanosome RNA editing field is focused on the proteins in the editosome or associated with it in part because these proteins have low homology with human proteins and thus make good drug targets. Less attention is given to the structures of the gRNAs, pre-mRNA, and their complexes (Homann, 2008). The gRNAs do not to appear to be attractive targets for structural studies due to the low thermal stability of the gRNAs characterized to date. However, studies by Koslowsky and colleagues show that some of the pre-mRNAs form stable structures as pre-mRNA/gRNA complexes (Leung and Koslowsky, 2001a,b; Reifur and Koslowsky, 2008; Reifur *et al.*, 2010). These latter complexes often have three structural domains: the simple helical domain of largely complementary basepairs in anchor helix, the three-way junction of the editing site/template domain, and the non-specific helical domain of a poly(U)/largely polypurine tract in the U-helix (Fig. 5.16A). The last domain was studied by X-ray crystallography (Mooers and Singh, 2011) (Fig. 5.16B).

The trans-acting gRNAs have a oligo(U) tail at their 3′ end (the U-tail for brevity) that is added by RET1 after transcription in the guide RNA processing complex (Aphasizheva and Aphasizhev, 2010). The U-tail has an average length of 15 Us and varies from 4 to 25 Us (Blum and Simpson, 1990). Of the four most common nucleotides in RNA, Us are the most promiscuous in their basepairing in double-stranded RNA and can form basepairs with Gs, As, and Cs that do not dramatically change the width of the double helix. Homopolymers of Us do not form single-stranded helices at room temperature (Inners and Felsenfeld, 1970; Chen *et al.*, 2012), unlike homopolymers of Gs, As, and Cs. Instead, they form random coils, perhaps due to unfavourable base dipole–dipole interactions that disfavour the stacking of uracil bases on top of each other. A large decrease in conformational entropy is expected to occur when the U-tail binds to the pre-mRNA contributing to a lowering of the melting temperature of the U-helix compared with a double helix of the same length composed of Watson–Crick basepairs. The low thermal stability of the U-helix eases the displacement of the U-tail from the pre-mRNA when the RNA editing machinery is moving from one editing site to the next.

The U-helix was targeted for structural studies because it is a ubiquitous feature of gRNA/pre-mRNA complexes (Mooers and Singh, 2011). Attempts to crystallize the heteroduplexes of oligo(U) RNA and polypurines failed, probably due to the propensity of U-tails to form worm-like random coils at room temperature. Low temperature crystallization trials were not explored due to limited RNA quantities. An alternative approach of fusing two RNAs head-to-head was ultimately adopted (Fig. 5.16A). The fused RNA could form double-stranded RNA by self-annealing which simplified the preparation of RNA for crystallization trials. The polypurine tract in the 5′ half of the RNA reassembles the 3′ end of the template domain in the guide RNA and functions as an anchoring point for the nucleating of the double

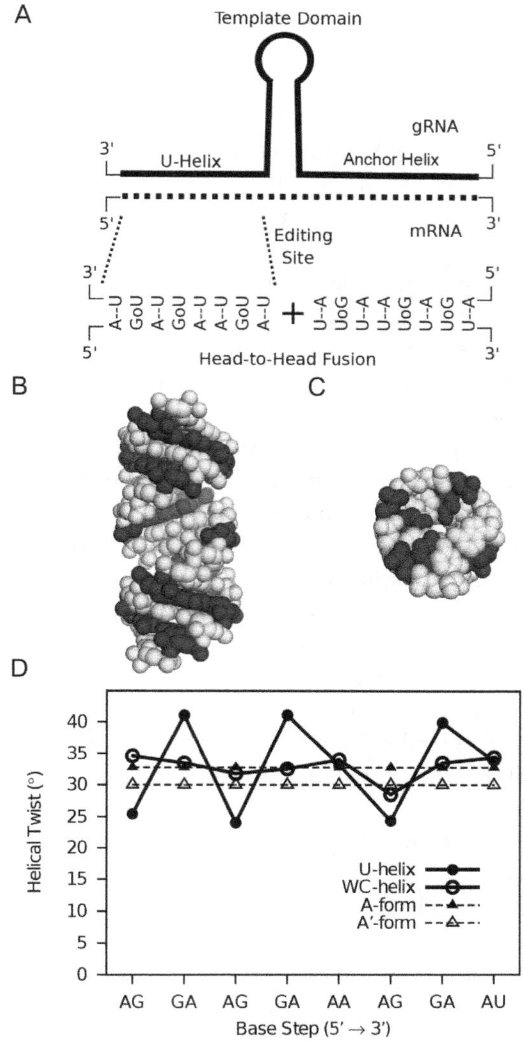

Figure 5.16 Crystal structure of the RNA fusion of two eight basepair fragments from the U-helix domain of the A6 mRNA/gA6[14] gRNA duplex (PDB-ID 3ND3). The domains are not to scale. (A) Cartoon of the U-helix design. Two copies of the U-helix fragment were fused head-to-head. The resulting RNA duplex was assembled from a single RNA species and had 2-fold molecular symmetry. (B and C) Van der Waals sphere representation of the crystal structure designed in (A). The G·U wobble basepairs are in dark grey while that Watson–Crick basepairs are in light grey. The view is towards the deep major groove in the middle in (B) while in (C) the view is down the helical axis. (D) Variation in helical twist by base step for the first eight unique base steps of the U-helix crystal structure shown in (B) and for the crystal structure of a control RNA with all the G·U wobble basepairs replaced with Watson–Crick G·C basepairs (PDB-ID 3ND4).

helix structure formation between the U-tail and the purine-rich tract in the pre-mRNA.

A crystal of a RNA fusion of two U-helices eight basepairs in length for the A6–14 gRNA/A6 pre-mRNA duplex diffracted X-rays to 1.37 Å (Mooers and Singh, 2011). This RNA has three isolated G·U wobble basepairs in each eight basepair fragment (Fig. 5.16B). The RNA crystallized with one strand in the asymmetric unit, so only half of the duplex provides crystallographically independent information. The G·U wobble basepairs did not cause variation in the width of the double helix, but they did introduce asymmetry in the geometry of the surfaces of the minor and major grooves due to the shift of the Gs towards the minor groove and the shift of the Us towards the major groove. This shifting of the bases relative to each other in the wobble basepair is due to the breaking of the pseudo dyad symmetry found in Watson–Crick basepairs. This change in basepair geometry introduces apparent low helical twist before the G·U wobble basepair and a high twist angle after the G·U basepair (Fig. 5.16C). The variation in helical twist is strongly correlated with the shifting of the bases in the G·U wobble basepairs (Mooers and Singh, 2011). A crystal structure (PDB-ID 3ND4) of the control sequence with the G·U wobble basepairs replaced with G·C basepairs showed little variation in helical twist except at the site of a magnesium cation that bridged symmetry-related helices and induced a change in helical twist at the sixth base step (Fig. 5.16C). This control structure is the longest crystal structure reported to date of a double-stranded RNA with only Watson–Crick basepairs.

Future directions

Many of the sequence motifs found on U-helices are absent in the RNA structural databases. For example, we do not know the structural effects of three contiguous G·U wobble basepairs. The study of some of these motifs can be pursued in the above 16-basepair RNA fusion system, but longer RNA fusions would enable the study of longer motifs in the absence of crystal packing artefacts introduced by end effects. The one drawback of longer RNAs is that the corresponding control sequences may form very stable stems that favour stem–loop formation over duplex formation even at the high strand concentrations used in structural studies. This would hinder the structure determination of the double-stranded RNA form of the control sequences.

Future trends

It is safe to anticipate that the additional crystal structures of the RNA editing core complex proteins and accessory proteins will be determined in the near future. Additional editing core proteins will probably be determined with the aid of nanobodies. It is uncertain if the crystal structure of one of the known RNA editing complexes from *T. brucei* will ever be obtained. A stable editing complex that can be crystallized may require extensive protein engineering and/or the right combination of nanobodies. Alternatively, someone may discover a more stable RNA editing complex in one of the many poorly characterized kinetoplastid species. Nonetheless, the crystal structures of the essential RNA editing enzymes have been and will continue to be important in the structure-based virtual screening and structure-based optimization of potential inhibitors for research and ultimately for drugs that provide safer and more effective therapies to treat people with debilitating and sometime fatal infections by the pathogenic kinetoplastids.

Acknowledgements

The author thanks the structural biologists who deposited their data in public databases and thereby greatly facilitated this independent review of their work. The author also thanks Drs Young-Jun Park and Wim Hol for kindly sending the coordinates of the KREPA1 A1$^{OB\Delta}$:nanobody complex before their release to the public. The author apologizes to colleagues whose important work is not cited because of space limitations or his unfortunate negligence. This work was supported by the OUHSC College of Medicine Alumni Association, the Presbyterian Health Foundation (PHF#1545-Mooers), the Oklahoma Center for the Advancement of Science and Technology (OCAST HR08–138) and National Institutes of Health R01 AI088011.

References

Amaro, R.E., Swift, R.V., and McCammon, J.A. (2007). Functional and structural insights revealed by molecular dynamics simulations of an essential RNA editing ligase in *Trypanosoma brucei*. PLoS Negl. Trop. Dis. *1*, e68.

Amaro, R.E., Schnaufer, A., Interthal, H., Hol, W., Stuart, K.D., and McCammon, J.A. (2008). Discovery of drug-like inhibitors of an essential RNA editing ligase in *Trypanosoma brucei*. Proc. Natl. Acad. Sci. U.S.A. *105*, 17278–17283.

Ammerman, M.L., Downey, K.M., Hashimi, H., Fisk, J.C., Tomasello, D.L., Faktorova, D., Kafkova, L., King, T., Lukes, J., and Read, L.K. (2012). Architecture of the trypanosome RNA editing accessory complex, MRB1. Nucleic Acids Res. *40*, 5637–5650.

Aphasizhev, R., and Aphasizheva, I. (2008). Terminal RNA uridylyltransferases of trypanosomes. Biochim. Biophys. Acta *1779*, 270–280.

Aphasizhev, R., Aphasizheva, I., Nelson, R.E., Gao, G., Simpson, A.M., Kang, X., Falick, A.M., Sbicego, S., and Simpson, L. (2003a). Isolation of a U-insertion/deletion editing complex from Leishmania tarentolae mitochondria. EMBO J. *22*, 913–924.

Aphasizhev, R., Aphasizheva, I., and Simpson, L. (2003b). A tale of two TUTases. Proc. Natl. Acad. Sci. U.S.A. *100*, 10617–10622.

Aphasizhev, R., Aphasizheva, I., and Simpson, L. (2004). Multiple terminal uridylyltransferases of trypanosomes. FEBS Lett. *572*, 15–18.

Aphasizhev, R., Sbicego, S., Peris, M., Jang, S.H., Aphasizheva, I., Simpson, A.M., Rivlin, A., and Simpson, L. (2002). Trypanosome mitochondrial 3′ terminal uridylyl transferase (TUTase): the key enzyme in U-insertion/deletion RNA editing. Cell *108*, 637–648.

Aphasizheva, I., and Aphasizhev, R. (2010). RET1-catalyzed uridylylation shapes the mitochondrial transcriptome in *Trypanosoma brucei*. Mol. Cell. Biol. *30*, 1555–1567.

Aphasizheva, I., Ringpis, G.E., Weng, J., Gershon, P.D., Lathrop, R.H., and Aphasizhev, R. (2009). Novel TUTase associates with an editosome-like complex in mitochondria of *Trypanosoma brucei*. RNA *15*, 1322–1337.

Arcus, V. (2002). OB-fold domains: a snapshot of the evolution of sequence, structure and function. Curr. Opin. Struct. Biol. *12*, 794–801.

Bai, Y., Srivastava, S.K., Chang, J.H., Manley, J.L., and Tong, L. (2011). Structural basis for dimerization and activity of human PAPD1, a noncanonical poly(A) polymerase. Mol. Cell *41*, 311–320.

Baker, N.A., Sept, D., Joseph, S., Holst, M.J., and McCammon, J.A. (2001). Electrostatics of nanosystems: application to microtubules and the ribosome. Proc. Natl. Acad. Sci. U.S.A. *98*, 10037–10041.

Blum, B., and Simpson, L. (1990). Guide RNAs in kinetoplastid mitochondria have a nonencoded 3′ oligo(U) tail involved in recognition of the preedited region. Cell *62*, 391–397.

Blum, B., Bakalara, N., and Simpson, L. (1990). A model for RNA editing in kinetoplastid mitochondria: 'guide' RNA molecules transcribed from maxicircle DNA provide the edited information. Cell *60*, 189--198.

Bogden, C.E., Fass, D., Bergman, N., Nichols, M.D., and Berger, J.M. (1999). The structural basis for terminator recognition by the Rho transcription termination factor. Mol. Cell *3*, 487–493.

Brecht, M., Niemann, M., Schluter, E., Muller, U.F., Stuart, K., and Goringer, H.U. (2005). TbMP42, a protein component of the RNA editing complex in African trypanosomes, has endo-exoribonuclease activity. Mol. Cell *17*, 621–630.

Carnes, J., and Stuart, K.D. (2007). Uridine insertion/deletion editing activities. Methods Enzymol. *424*, 25–54.

Carnes, J., Trotter, J.R., Ernst, N.L., Steinberg, A., and Stuart, K. (2005). An essential RNase III insertion editing endonuclease in *Trypanosoma brucei*. Proc. Natl. Acad. Sci. U.S.A. *102*, 16614–16619.

Carnes, J., Trotter, J.R., Peltan, A., Fleck, M., and Stuart, K. (2008). RNA editing in *Trypanosoma brucei* requires three different editosomes. Mol. Cell. Biol. *28*, 122–130.

Chakravarty, J., and Sundar, S. (2010). Drug resistance in leishmaniasis. J. Glob. Infect. Dis. *2*, 167–176.

Chase, J.W., and Williams, K.R. (1986). Single-stranded DNA binding proteins required for DNA replication. Annu. Rev. Biochem. *55*, 103–136.

Chen, H., Meisburger, S.P., Pabit, S.A., Sutton, J.L., Webb, W.W., and Pollack, L. (2012). Ionic strength-dependent persistence lengths of single-stranded RNA and DNA. Proc. Natl. Acad. Sci. U.S.A. *109*, 799–804.

Cifuentes-Rojas, C., Halbig, K., Sacharidou, A., Nova-Ocampo, M.D., and Cruz-Reyes, J. (2005). Minimal pre-mRNA substrates with natural and converted sites for full-round U insertion and U deletion RNA editing in trypanosomes. Nucleic Acids Res. *33*, 6610–6620.

Cifuentes-Rojas, C., Pavia, P., Hernandez, A., Osterwisch, D., Puerta, C., and Cruz-Reyes, J. (2007). Substrate determinants for RNA editing and editing complex interactions at a site for full-round U insertion. J. Biol. Chem. *282*, 4265–4276.

Cruz-Reyes, J., Zhelonkina, A., Rusche, L., and Sollner-Webb, B. (2001). Trypanosome RNA editing: simple guide RNA features enhance U deletion 100-fold. Mol. Cell. Biol. *21*, 884–892.

Demir, O., and Amaro, R.E. (2012). Elements of nucleotide specificity in the *Trypanosoma brucei* mitochondrial RNA editing enzyme RET2. J. Chem. Inf. Model *52*, 1308–1318.

Deng, J., Schnaufer, A., Salavati, R., Stuart, K.D., and Hol, W.G. (2004). High resolution crystal structure of a key editosome enzyme from *Trypanosoma brucei*: RNA editing ligase 1. J. Mol. Biol. *343*, 601–613.

Deng, J., Ernst, N.L., Turley, S., Stuart, K.D., and Hol, W.G. (2005). Structural basis for UTP specificity of RNA editing TUTases from *Trypanosoma brucei*. EMBO J. *24*, 4007–4017.

Desveaux, D., Allard, J., Brisson, N., and Sygusch, J. (2002). A new family of plant transcription factors displays a novel ssDNA-binding surface. Nat. Struct. Biol. *9*, 512–517.

Drozdz, M., Palazzo, S.S., Salavati, R., O'Rear, J., Clayton, C., and Stuart, K. (2002). TbMP81 is required for RNA editing in *Trypanosoma brucei*. EMBO J. *21*, 1791–1799.

Durrant, J.D., and McCammon, J.A. (2011). Towards the development of novel *Trypanosoma brucei* RNA editing ligase 1 inhibitors. BMC Pharmacol *11*, 9.

Durrant, J.D., Hall, L., Swift, R.V., Landon, M., Schnaufer, A., and Amaro, R.E. (2010). Novel naphthalene-based inhibitors of *Trypanosoma brucei* RNA editing ligase 1. PLoS Negl. Trop. Dis. *4*, e803.

Eggington, J.M., Haruta, N., Wood, E.A., and Cox, M.M. (2004). The single-stranded DNA-binding protein of *Deinococcus radiodurans*. BMC Microbiol 4, 2.

Ernst, N.L., Panicucci, B., Igo, R.P., Jr., Panigrahi, A.K., Salavati, R., and Stuart, K. (2003). TbMP57 is a 3′ terminal uridylyl transferase (TUTase) of the *Trypanosoma brucei* editosome. Mol. Cell *11*, 1525–1536.

Etheridge, R.D., Clemens, D.M., Gershon, P.D., and Aphasizhev, R. (2009). Identification and characterization of nuclear non-canonical poly(A) polymerases from *Trypanosoma brucei*. Mol. Biochem. Parasitol. *164*, 66–73.

Feagin, J.E., Abraham, J.M., and Stuart, K. (1988). Extensive editing of the cytochrome c oxidase III transcript in *Trypanosoma brucei*. Cell 53, 413–422.

Gao, G., and Simpson, L. (2003). Is the *Trypanosoma brucei* REL1 RNA ligase specific for U-deletion RNA editing, and is the REL2 RNA ligase specific for U-insertion editing? J. Biol. Chem. *278*, 27570–27574.

Gehrig, S., and Efferth, T. (2008). Development of drug resistance in *Trypanosoma brucei* rhodesiense and *Trypanosoma brucei* gambiense. Treatment of human African trypanosomiasis with natural products (Review). Int. J. Mol. Med. *22*, 411–419.

Golas, M.M., Bohm, C., Sander, B., Effenberger, K., Brecht, M., Stark, H., and Goringer, H.U. (2009). Snapshots of the RNA editing machine in trypanosomes captured at different assembly stages *in vivo*. EMBO J. *28*, 766–778.

Golden, D.E., and Hajduk, S.L. (2005). The 3′-untranslated region of cytochrome oxidase II mRNA functions in RNA editing of African trypanosomes exclusively as a cis guide RNA. RNA *11*, 29–37.

Goldman, E.R., Anderson, G.P., Liu, J.L., Delehanty, J.B., Sherwood, L.J., Osborn, L.E., Cummins, L.B., and Hayhurst, A. (2006). Facile generation of heat-stable antiviral and antitoxin single domain antibodies from a semisynthetic llama library. Anal. Chem. *78*, 8245–8255.

Goringer, H.U., Katari, V.S., and Bohm, C. (2011). The structural landscape of native editosomes in African trypanosomes. Wiley Interdiscip. Rev. RNA *2*, 395–407.

Guo, X., Ernst, N.L., and Stuart, K.D. (2008). The KREPA3 zinc finger motifs and OB-fold domain are essential for RNA editing and survival of *Trypanosoma brucei*. Mol. Cell. Biol. *28*, 6939–6953.

Hajduk, S., and Ochsenreiter, T. (2010). RNA editing in kinetoplastids. RNA Biol. *7*, 229–236.

Hamers-Casterman, C., Atarhouch, T., Muyldermans, S., Robinson, G., Hamers, C., Songa, E.B., Bendahman, N., and Hamers, R. (1993). Naturally occurring antibodies devoid of light chains. Nature *363*, 446–448.

Ho, C.K., and Shuman, S. (2002). Bacteriophage T4 RNA ligase 2 (gp24.1) exemplifies a family of RNA ligases found in all phylogenetic domains. Proc. Natl. Acad. Sci. U.S.A. *99*, 12709–12714.

Ho, C.K., Wang, L.K., Lima, C.D., and Shuman, S. (2004). Structure and mechanism of RNA ligase. Structure *12*, 327–339.

Holm, L., and Sander, C. (1995). DNA polymerase beta belongs to an ancient nucleotidyltransferase superfamily. Trends Biochem. Sci. *20*, 345–347.

Homann, M. (2008). Editing reactions from the perspective of RNA structure. In RNA Editing Nucleic Acids and Molecular Biology 20, Goringer, H.U., ed. (Springer-Verlag, Berlin), pp. 1–32.

Huang, C.E., Cruz-Reyes, J., Zhelonkina, A.G., O'Hearn, S., Wirtz, E., and Sollner-Webb, B. (2001). Roles for ligases in the RNA editing complex of *Trypanosoma brucei*: band IV is needed for U-deletion and RNA repair. EMBO J. *20*, 4694–4703.

Huang, C.E., O'Hearn, S.F., and Sollner-Webb, B. (2002). Assembly and function of the RNA editing complex in *Trypanosoma brucei* requires band III protein. Mol. Cell. Biol. *22*, 3194–3203.

Huey, R., Morris, G.M., Olson, A.J., and Goodsell, D.S. (2007). A semiempirical free energy force field with charge-based desolvation. J. Comput. Chem. *28*, 1145–1152.

Igo, R.P., Jr., Palazzo, S.S., Burgess, M.L., Panigrahi, A.K., and Stuart, K. (2000). Uridylate addition and RNA ligation contribute to the specificity of kinetoplastid insertion RNA editing. Mol. Cell. Biol. *20*, 8447–8457.

Igo, R.P., Jr., Weston, D.S., Ernst, N.L., Panigrahi, A.K., Salavati, R., and Stuart, K. (2002). Role of uridylate-specific exoribonuclease activity in *Trypanosoma brucei* RNA editing. Eukaryot. Cell *1*, 112–118.

Igonet, S., Vulliez-Le Normand, B., Faure, G., Riottot, M.M., Kocken, C.H., Thomas, A.W., and Bentley, G.A. (2007). Cross-reactivity studies of an anti-*Plasmodium vivax* apical membrane antigen 1 monoclonal antibody: binding and structural characterisation. J. Mol. Biol. *366*, 1523–1537.

Inners, L.D., and Felsenfeld, G. (1970). Conformation of polyribouridylic acid in solution. J. Mol. Biol. *50*, 373–389.

Jiang, J., Zhang, Y., Krainer, A.R., and Xu, R.M. (1999). Crystal structure of human p32, a doughnut-shaped acidic mitochondrial matrix protein. Proc. Natl. Acad. Sci. U.S.A. *96*, 3572–3577.

Kala, S., and Salavati, R. (2010). OB-fold domain of KREPA4 mediates high-affinity interaction with guide RNA and possesses annealing activity. RNA *16*, 1951–1967.

Kang, X., Falick, A.M., Nelson, R.E., Gao, G., Rogers, K., Aphasizhev, R., and Simpson, L. (2004). Disruption of the zinc finger motifs in the Leishmania tarentolae LC-4 (=TbMP63) L-complex editing protein affects the stability of the L-complex. J. Biol. Chem. *279*, 3893–3899.

Kao, C.Y., and Read, L.K. (2007). Targeted depletion of a mitochondrial nucleotidyltransferase suggests the presence of multiple enzymes that polymerize mRNA 3′ tails in *Trypanosoma brucei* mitochondria. Mol. Biochem. Parasitol. *154*, 158–169.

Köller, J., Müller, U.F., Schmid, B., Missel, A., Kruft, V., Stuart, K., and Göringer, H.U. (1997). *Trypanosoma brucei* gBP21. An arginine-rich mitochondrial protein that binds to guide RNA with high affinity. J. Biol. Chem. *272*, 3749–3757.

Korotkov, K.V., Pardon, E., Steyaert, J., and Hol, W.G. (2009). Crystal structure of the N-terminal domain of the secretin GspD from ETEC determined with the assistance of a nanobody. Structure *17*, 255–265.

Koslowsky, D.J., Bhat, G.J., Read, L.K., and Stuart, K. (1991). Cycles of progressive realignment of gRNA with mRNA in RNA editing. Cell *67*, 537–546.

Krissinel, E., and Henrick, K. (2007). Inference of macromolecular assemblies from crystalline state. J. Mol. Biol. *372*, 774–797.

Lah, J., Marianovsky, I., Glaser, G., Engelberg-Kulka, H., Kinne, J., Wyns, L., and Loris, R. (2003). Recognition of the intrinsically flexible addiction antidote MazE by a dromedary single domain antibody fragment. Structure, thermodynamics of binding, stability, and influence on interactions with DNA. J. Biol. Chem. *278*, 14101–14111.

Lam, A.Y., Pardon, E., Korotkov, K.V., Hol, W.G., and Steyaert, J. (2009). Nanobody-aided structure determination of the EpsI:EpsJ pseudopilin heterodimer from Vibrio vulnificus. J. Struct. Biol. *166*, 8–15.

Laskowski, R.A., and Swindells, M.B. (2011). LigPlot+: multiple ligand–protein interaction diagrams for drug discovery. J. Chem. Inf. Model *51*, 2778–2786.

Law, J.A., O'Hearn, S., and Sollner-Webb, B. (2007). In *Trypanosoma brucei* RNA editing, TbMP18 (band VII) is critical for editosome integrity and for both insertional and deletional cleavages. Mol. Cell. Biol. *27*, 777–787.

Law, J.A., O'Hearn, S.F., and Sollner-Webb, B. (2008). *Trypanosoma brucei* RNA editing protein TbMP42 (band VI) is crucial for the endonucleolytic cleavages but not the subsequent steps of U-deletion and U-insertion. RNA *14*, 1187–1200.

Lawson, C.L., Baker, M.L., Best, C., Bi, C., Dougherty, M., Feng, P., van Ginkel, G., Devkota, B., Lagerstedt, I., Ludtke, S.J., *et al.* (2011). EMDataBank.org: unified data resource for CryoEM. Nucleic Acids Res. *39*, D456–464.

Lescar, J., Pellegrini, M., Souchon, H., Tello, D., Poljak, R.J., Peterson, N., Greene, M., and Alzari, P.M. (1995). Crystal structure of a cross-reaction complex between Fab F9.13.7 and guinea fowl lysozyme. J. Biol. Chem. *270*, 18067–18076.

Leung, S.S., and Koslowsky, D.J. (2001a). Interactions of mRNAs and gRNAs involved in trypanosome mitochondrial RNA editing: structure probing of an mRNA bound to its cognate gRNA. RNA *7*, 1803–1816.

Leung, S.S., and Koslowsky, D.J. (2001b). RNA editing in *Trypanosoma brucei*: characterization of gRNA U-tail interactions with partially edited mRNA substrates. Nucleic Acids Res. *29*, 703–709.

Li, F., Ge, P., Hui, W.H., Atanasov, I., Rogers, K., Guo, Q., Osato, D., Falick, A.M., Zhou, Z.H., and Simpson, L. (2009). Structure of the core editing complex (L-complex) involved in uridine insertion/deletion RNA editing in trypanosomatid mitochondria. Proc. Natl. Acad. Sci. U.S.A. *106*, 12306–12310.

Li, F., Herrera, J., Zhou, S., Maslov, D.A., and Simpson, L. (2011). Trypanosome REH1 is an RNA helicase involved with the 3′–5′ polarity of multiple gRNA-guided uridine insertion/deletion RNA editing. Proc. Natl. Acad. Sci. U.S.A. *108*, 3542–3547.

Liang, S., and Connell, G.J. (2010). Identification of specific inhibitors for a trypanosomatid RNA editing reaction. RNA *16*, 2435–2441.

Lipinski, C.A., Lombardo, F., Dominy, B.W., and Feeney, P.J. (2001). Experimental and computational approaches to estimate solubility and permeability in drug discovery and development settings. Adv. Drug Deliv. Rev. *46*, 3–26.

Lohman, T.M., Bujalowski, W., and Overman, L.B. (1988). E. coli single strand binding protein: a new look at helix-destabilizing proteins. Trends Biochem. Sci. *13*, 250–255.

Loris, R., Marianovsky, I., Lah, J., Laeremans, T., Engelberg-Kulka, H., Glaser, G., Muyldermans, S., and Wyns, L. (2003). Crystal structure of the intrinsically flexible addiction antidote MazE. J. Biol. Chem. *278*, 28252–28257.

Maslov, D.A., and Simpson, L. (1992). The polarity of editing within a multiple gRNA-mediated domain is due to formation of anchors for upstream gRNAs by downstream editing. Cell *70*, 459–467.

Matsumoto, T., Morimoto, Y., Shibata, N., Kinebuchi, T., Shimamoto, N., Tsukihara, T., and Yasuoka, N. (2000). Roles of functional loops and the C-terminal segment of a single-stranded DNA binding protein elucidated by X-ray structure analysis. J. Biochem. *127*, 329–335.

Meyer, R.R., and Laine, P.S. (1990). The single-stranded DNA-binding protein of *Escherichia coli*. Microbiol. Rev. *54*, 342–380.

Missel, A., Souza, A.E., Nörskau, G., and Göringer, H.U. (1997). Disruption of a gene encoding a novel mitochondrial DEAD-box protein in *Trypanosoma brucei* affects edited mRNAs. Mol. Cell. Biol. *17*, 4895–4903.

Mooers, B.H., and Singh, A. (2011). The crystal structure of an oligo(U):pre-mRNA duplex from a trypanosome RNA editing substrate. RNA *17*, 1870–1883.

Mooers, B.H., Datta, D., Baase, W.A., Zollars, E.S., Mayo, S.L., and Matthews, B.W. (2003). Repacking the Core of T4 lysozyme by automated design. J. Mol. Biol. *332*, 741–756.

Morris, G.M., Goodsell, D.S., Halliday, R.S., Huey, R., Hart, W.E., Belew, R.K., and Olson, A.J. (1998). Automated docking using a Lamarckian genetic algorithm and an empirical binding free energy function. J. Comput. Chem. *19*, 1639–1662.

Moshiri, H., Acoca, S., Kala, S., Najafabadi, H.S., Hogues, H., Purisima, E., and Salavati, R. (2011). Naphthalene-based RNA editing inhibitor blocks RNA editing activities and editosome assembly in *Trypanosoma brucei*. J. Biol. Chem. *286*, 14178–14189.

Müller, U.F., and Göringer, H.U. (2002). Mechanism of the gBP21-mediated RNA/RNA annealing reaction: matchmaking and charge reduction. Nucleic Acids Res. *30*, 447–455.

Müller, U.F., Lambert, L., and Göringer, H.U. (2001). Annealing of RNA editing substrates facilitated by guide RNA-binding protein gBP21. EMBO J. *20*, 1394–1404.

Murzin, A.G. (1993). OB(oligonucleotide/oligosaccharide binding)-fold: common structural and functional solution for non-homologous sequences. EMBO J. *12*, 861–867.

Muta, T., Kang, D., Kitajima, S., Fujiwara, T., and Hamasaki, N. (1997). p32 protein, a splicing factor 2-associated protein, is localized in mitochondrial matrix and is functionally important in maintaining oxidative phosphorylation. J. Biol. Chem. *272*, 24363–24370.

Muyldermans, S. (2001). Single domain camel antibodies: current status. J. Biotechnol. *74*, 277–302.

Muyldermans, S., Cambillau, C., and Wyns, L. (2001). Recognition of antigens by single-domain antibody fragments: the superfluous luxury of paired domains. Trends Biochem. Sci. *26*, 230–235.

Nicholls, A., Sharp, K.A., and Honig, B. (1991). Protein folding and association: insights from the interfacial and thermodynamic properties of hydrocarbons. Proteins *11*, 281–296.

Ochsenreiter, T., Cipriano, M., and Hajduk, S.L. (2007). KISS: the kinetoplastid RNA editing sequence search tool. RNA *13*, 1–4.

Paes, L.S., Mantilla, B.S., Barison, M.J., Wrenger, C., and Silber, A.M. (2011). The uniqueness of the *Trypanosoma cruzi* mitochondrion: opportunities to target new drugs against Chagas' disease. Curr. Pharma. Design *17*, 2074–2099.

Palazzo, S.S., Panigrahi, A.K., Igo, R.P., Salavati, R., and Stuart, K. (2003). Kinetoplastid RNA editing ligases: complex association, characterization, and substrate requirements. Mol. Biochem. Parasitol. *127*, 161–167.

Panigrahi, A.K., Ernst, N.L., Domingo, G.J., Fleck, M., Salavati, R., and Stuart, K.D. (2006). Compositionally and functionally distinct editosomes in *Trypanosoma brucei*. RNA *12*, 1038–1049.

Panigrahi, A.K., Gygi, S.P., Ernst, N.L., Igo, R.P., Jr., Palazzo, S.S., Schnaufer, A., Weston, D.S., Carmean, N., Salavati, R., Aebersold, R., *et al.* (2001a). Association of two novel proteins, TbMP52 and TbMP48, with the *Trypanosoma brucei* RNA editing complex. Mol. Cell. Biol. *21*, 380–389.

Panigrahi, A.K., Schnaufer, A., Carmean, N., Igo, R.P., Jr., Gygi, S.P., Ernst, N.L., Palazzo, S.S., Weston, D.S., Aebersold, R., Salavati, R., *et al.* (2001b). Four related proteins of the *Trypanosoma brucei* RNA editing complex. Mol. Cell. Biol. *21*, 6833–6840.

Panigrahi, A.K., Schnaufer, A., Ernst, N.L., Wang, B., Carmean, N., Salavati, R., and Stuart, K. (2003). Identification of novel components of *Trypanosoma brucei* editosomes. RNA *9*, 484–492.

Panigrahi, A.K., Schnaufer, A., and Stuart, K.D. (2007). Isolation and compositional analysis of trypanosomatid editosomes. Methods Enzymol. *424*, 3–24.

Park, Y.J., Pardon, E., Wu, M., Steyaert, J., and Hol, W.G. (2011). Crystal structure of a heterodimer of editosome interaction proteins in complex with two copies of a cross-reacting nanobody. Nucleic Acids Res. *40*, 1828–1840.

Park, Y.J., Budiarto, T., Wu, M., Pardon, E., Steyaert, J., and Hol, W.G. (2012). The structure of the C-terminal domain of the largest editosome interaction protein and its role in promoting RNA binding by RNA editing ligase L2. Nucleic Acids Res. *40*, 6966–6977.

Pelletier, H., and Sawaya, M.R. (1996). Characterization of the metal ion binding helix-hairpin-helix motifs in human DNA polymerase beta by X-ray structural analysis. Biochemistry *35*, 12778–12787.

Peris, M., Simpson, A.M., Grunstein, J., Liliental, J.E., Frech, G.C., and Simpson, L. (1997). Native gel analysis of ribonucleoprotein complexes from a Leishmania tarentolae mitochondrial extract. Mol. Biochem. Parasitol. *85*, 9–24.

Petersen-Mahrt, S.K., Estmer, C., Ohrmalm, C., Matthews, D.A., Russell, W.C., and Akusjarvi, G. (1999). The splicing factor-associated protein, p32, regulates RNA splicing by inhibiting ASF/SF2 RNA binding and phosphorylation. EMBO J. *18*, 1014–1024.

Pollard, V.W., Harris, M.E., and Hajduk, S.L. (1992). Native mRNA editing complexes from *Trypanosoma brucei* mitochondria. EMBO J. *11*, 4429–4438.

Reifur, L., and Koslowsky, D.J. (2008). *Trypanosoma brucei* ATPase subunit 6 mRNA bound to gA6-14 forms a conserved three-helical structure. RNA *14*, 2195–2211.

Reifur, L., Yu, L.E., Cruz-Reyes, J., Vanhartesvelt, M., and Koslowsky, D.J. (2010). The impact of mRNA structure on guide RNA targeting in kinetoplastid RNA editing. PLoS ONE *5*, e12235.

Rogozin, I.B., Aravind, L., and Koonin, E.V. (2003). Differential action of natural selection on the N and C-terminal domains of 2′–5′ oligoadenylate synthetases and the potential nuclease function of the C-terminal domain. J. Mol. Biol. *326*, 1449–1461.

Rusche, L.N., Cruz-Reyes, J., Piller, K.J., and Sollner-Webb, B. (1997). Purification of a functional enzymatic editing complex from *Trypanosoma brucei* mitochondria. EMBO J. *16*, 4069–4081.

Rusche, L.N., Huang, C.E., Piller, K.J., Hemann, M., Wirtz, E., and Sollner-Webb, B. (2001). The two RNA ligases of the *Trypanosoma brucei* RNA editing complex: cloning the essential band IV gene and identifying the band V gene. Mol. Cell. Biol. *21*, 979–989.

Ryan, C.M., and Read, L.K. (2005). UTP-dependent turnover of *Trypanosoma brucei* mitochondrial mRNA requires UTP polymerization and involves the RET1 TUTase. RNA *11*, 763–773.

Saikrishnan, K., Jeyakanthan, J., Venkatesh, J., Acharya, N., Sekar, K., Varshney, U., and Vijayan, M. (2003). Structure of Mycobacterium tuberculosis single-stranded DNA-binding protein. Variability in quaternary structure and its implications. J. Mol. Biol. *331*, 385–393.

Salavati, R., Ernst, N.L., O'Rear, J., Gilliam, T., Tarun, S., Jr., and Stuart, K. (2006). KREPA4, an RNA binding protein essential for editosome integrity and survival of *Trypanosoma brucei*. RNA *12*, 819–831.

Salavati, R., Moshiri, H., Kala, S., and Shateri Najafabadi, H. (2011). Inhibitors of RNA editing as potential chemotherapeutics against trypanosomatid pathogens. Int. J. Parasitol. Drugs Drug Resistance 2, 35–46.

Schmid, B., Riley, G.R., Stuart, K., and Göringer, H.U. (1995). The secondary structure of guide RNA molecules from *Trypanosoma brucei*. Nucleic Acids Res. 23, 3093–3102.

Schmid, B., Read, L.K., Stuart, K., and Göringer, H.U. (1996). Experimental verification of the secondary structures of guide RNA-pre-mRNA chimaeric molecules in *Trypanosoma brucei*. Eur. J. Biochem. *240*, 721–731.

Schnaufer, A., Panigrahi, A.K., Panicucci, B., Igo, R.P., Jr., Wirtz, E., Salavati, R., and Stuart, K. (2001). An RNA ligase essential for RNA editing and survival of the bloodstream form of *Trypanosoma brucei*. Science *291*, 2159–2162.

Schnaufer, A., Ernst, N.L., Palazzo, S.S., O'Rear, J., Salavati, R., and Stuart, K. (2003). Separate insertion and deletion subcomplexes of the *Trypanosoma brucei* RNA editing complex. Mol. Cell *12*, 307–319.

Schnaufer, A., Wu, M., Park, Y.J., Nakai, T., Deng, J., Proff, R., Hol, W.G., and Stuart, K.D. (2010). A protein–protein interaction map of trypanosome ~20S editosomes. J. Biol. Chem. *285*, 5282–5295.

Schrodinger, LLC (2010). The PyMOL Molecular Graphics System, Version 1.3r1.

Schumacher, M.A., Karamooz, E., Zikova, A., Trantirek, L., and Lukes, J. (2006). Crystal structures of *T. brucei* MRP1/MRP2 guide-RNA binding complex reveal RNA matchmaking mechanism. Cell *126*, 701–711.

Seiwert, S.D., Heidmann, S., and Stuart, K. (1996). Direct visualization of uridylate deletion *in vitro* suggests a mechanism for kinetoplastid RNA editing. Cell *84*, 831–841.

Seytter, T., Lottspeich, F., Neupert, W., and Schwarz, E. (1998). Mam33p, an oligomeric, acidic protein in the mitochondrial matrix of Saccharomyces cerevisiae is related to the human complement receptor gC1q-R. Yeast *14*, 303–310.

Shaneh, A., and Salavati, R. (2009). Kinetoplastid RNA editing ligases 1 and 2 exhibit different electrostatic properties. J. Mol. Model *16*, 61–76.

Simpson, L., Aphasizhev, R., Lukes, J., and Cruz-Reyes, J. (2010). Guide to the nomenclature of kinetoplastid RNA editing: a proposal. Protist *161*, 2–6.

Spinelli, S., Tegoni, M., Frenken, L., van Vliet, C., and Cambillau, C. (2001). Lateral recognition of a dye hapten by a llama VHH domain. J. Mol. Biol. *311*, 123–129.

Sprehe, M., Fisk, J.C., McEvoy, S.M., Read, L.K., and Schumacher, M.A. (2010). Structure of the *Trypanosoma brucei* p22 protein, a cytochrome oxidase subunit II-specific RNA editing accessory factor. J. Biol. Chem. *285*, 18899–18908.

Stagno, J., Aphasizheva, I., Aphasizhev, R., and Luecke, H. (2007a). Dual role of the RNA substrate in selectivity and catalysis by terminal uridylyl transferases. Proc. Natl. Acad. Sci. U.S.A. *104*, 14634–14639.

Stagno, J., Aphasizheva, I., Rosengarth, A., Luecke, H., and Aphasizhev, R. (2007b). UTP-bound and Apo structures of a minimal RNA uridylyltransferase. J. Mol. Biol. *366*, 882–899.

Stagno, J., Aphasizheva, I., Bruystens, J., Luecke, H., and Aphasizhev, R. (2010). Structure of the mitochondrial editosome-like complex associated TUTase 1 reveals divergent mechanisms of UTP selection and domain organization. J. Mol. Biol. *399*, 464–475.

Stuart, K., and Panigrahi, A.K. (2002). RNA editing: complexity and complications. Mol. Microbiol. *45*, 591–596.

Stuart, K.D., Schnaufer, A., Ernst, N.L., and Panigrahi, A.K. (2005). Complex management: RNA editing in trypanosomes. Trends Biochem. Sci. *30*, 97–105.

Swift, R.V., and Amaro, R.E. (2009). Discovery and design of DNA and RNA ligase inhibitors in infectious microorganisms. Expert Opin. Drug Discov. *4*, 1281–1294.

Swift, R.V., Durrant, J., Amaro, R.E., and McCammon, J.A. (2009). Toward understanding the conformational dynamics of RNA ligation. Biochemistry *48*, 709–719.

Tarun, S.Z., Jr., Schnaufer, A., Ernst, N.L., Proff, R., Deng, J., Hol, W., and Stuart, K. (2008). KREPA6 is an RNA-binding protein essential for editosome integrity and survival of *Trypanosoma brucei*. RNA *14*, 347–358.

Tereshko, V., Uysal, S., Koide, A., Margalef, K., Koide, S., and Kossiakoff, A.A. (2008). Toward chaperone-assisted crystallography: protein engineering enhancement of crystal packing and X-ray phasing capabilities of a camelid single-domain antibody (VHH) scaffold. Protein Sci. *17*, 1175–1187.

Theobald, D.L., Mitton-Fry, R.M., and Wuttke, D.S. (2003). Nucleic acid recognition by OB-fold proteins. Annu. Rev. Biophys. Biomol. Struct. *32*, 115–133.

Wilkinson, S.R., and Kelly, J.M. (2009). Trypanocidal drugs: mechanisms, resistance and new targets. Expert Rev. Mol. Med. *11*, e31.

Wilkinson, S.R., Taylor, M.C., Horn, D., Kelly, J.M., and Cheeseman, I. (2008). A mechanism for cross-resistance to nifurtimox and benznidazole in trypanosomes. Proc. Natl. Acad. Sci. U.S.A. *105*, 5022–5027.

Worthey, E.A., Schnaufer, A., Mian, I.S., Stuart, K., and Salavati, R. (2003). Comparative analysis of editosome proteins in trypanosomatids. Nucleic Acids Res. *31*, 6392–6408.

Wu, M., Park, Y.J., Pardon, E., Turley, S., Hayhurst, A., Deng, J., Steyaert, J., and Hol, W.G. (2011). Structures of a key interaction protein from the *Trypanosoma brucei* editosome in complex with single domain antibodies. J. Struct. Biol. *174*, 124–136.

Yin, S., Ho, C.K., and Shuman, S. (2003). Structure-function analysis of T4 RNA ligase 2. J. Biol. Chem. *278*, 17601–17608.

Zikova, A., Kopecna, J., Schumacher, M.A., Stuart, K., Trantirek, L., and Lukes, J. (2008). Structure and function of the native and recombinant mitochondrial MRP1/MRP2 complex from *Trypanosoma brucei*. Int. J. Parasitol. *38*, 901–912.

RNA Editing and Small Regulatory RNAs

6

Bjorn-Erik Wulff and Kazuko Nishikura

Abstract

Adenosine-to-inosine (A-to-I) editing of double-stranded RNA (dsRNA) is catalysed by members of the adenosine deaminase acting on RNA (ADAR) family, which is conserved from man to sea anemone. It has recently become clear that the most common substrates of ADARs are non-coding RNAs, including small regulatory RNAs like microRNAs (miRNAs), short interfering RNAs (siRNAs) and endogenous siRNAs (esiRNAs). These mediate post-transcriptional gene silencing (PTGS) by basepairing to complementary transcripts. This review discusses the effects ADARs exert on small regulatory RNA pathways and the resulting biological consequences. This discussion includes how ADAR substrate specificity is controlled, how ADARs both edit and sequester substrates, how ADARs affect the miRNA pathway by editing miRNA targets and efforts to discover novel edited adenosines affecting small regulatory RNA pathways.

Introduction

Double-stranded RNAs (dsRNAs) are bound by dsRNA-binding domains (dsRBDs) with little sequence specificity (Ryter and Schultz, 1998). Various dsRNA-binding proteins (dsRBPs) therefore have many substrates in common and consequently compete with one another. A complete explanation of their function requires consideration of this competition. This review discusses the interplay between two biological systems that make heavy use of dsRBPs: adenosine-to-inosine (A-to-I) editing by adenosine deaminases acting on RNA (ADARs) and post-transcriptional regulation by small regulatory RNAs.

A-to-I RNA editing by ADARs

Conserved from man to sea anemones (Jin *et al.*, 2009), all ADARs contain a number of dsRBDs and a deaminase domain. They bind dsRNAs, in which they deaminate adenosines to form inosines.

Family members differ in the number and spacing of their dsRBDs and the occasional presence of additional domains (Fig. 6.1a). Mammals have three ADARs – ADAR1, ADAR2 and ADAR3. ADAR1 has two isoforms – the constitutively expressed ADAR1p110 and the interferon-inducible ADAR1p150. Any enzymatic function of ADAR3 remains to be identified. *C. elegans* has two ADARs, ADR-1 and ADR-2. *Drosophila* has one ADAR, dADAR.

Inosine preferentially basepairs with cytidine, which makes inosine and guanosine functionally equivalent for the translation, reverse transcription and splicing machineries, among others. For example, editing of Na^{1+} channel subunit GluR-B in a CAG (glutamine) codon causes it to be read as CGG (arginine). This critically lowers the conductivity of the channel (Higuchi *et al.*, 1993). Furthermore, the ADAR2 protein self-edits ADAR2 mRNAs in an AA nucleotide. This generates an AG 3′ splice site that leads to translation of a non-functional protein. This negative feedback regulates the expression of functional ADAR2 (Rueter *et al.*, 1999). ADARs also edit precursors of microRNAs (miRNAs), short interfering RNA (siRNA) and endogenous siRNAs (esiRNAs). In this way, they exert a major influence on small regulatory RNAs.

Post-transcriptional regulation by small RNAs

Post-transcriptional regulation can be carried out by proteins that bind specific transcripts. However,

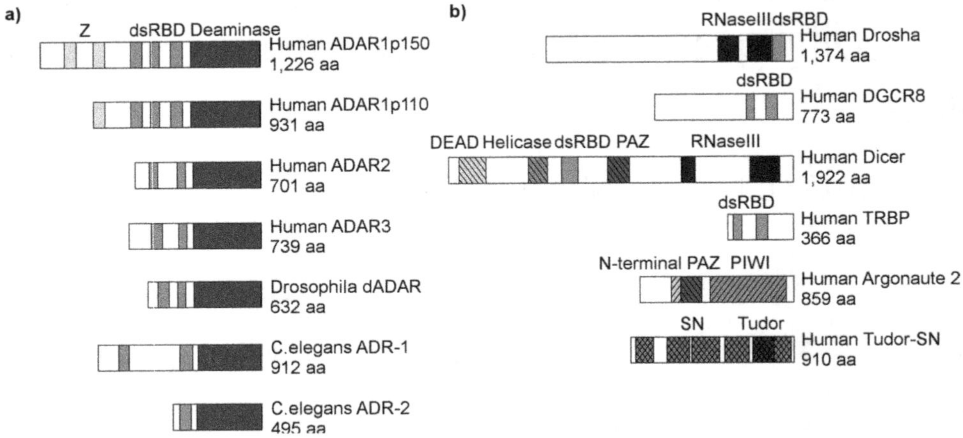

Figure 6.1 Domain structure of ADARs and proteins relevant to small RNAs. (a) ADARs have a variable number of dsRBDs and a catalytic deaminase domain. Mammalian ADAR1 also contains Z domains that can bind left-handed (Z) dsRNA or DNA. (b) Drosha has a dsRBD and two RNase III domains that each cleaves one strand of a pri-miRNA. DGCR8 uses its two dsRBDs to help Drosha bind its substrates. Dicer has a helicase domain, a DEAD domain, a PAZ domain that binds certain dsRNA ends, a dsRBD and two RNase III domains that can each cleave one strand of a dsRNA. TRBP uses its two dsRBDs to help Dicer bind its substrates. Argonaute 2 is a member of the Argonaute family, which can form the core of RISC. Argonaute 2 has an N-terminal domain, a PAZ domain and a catalytic PIWI domain that cleaves mRNAs complementary to the Argonaute's bound small RNA. Tudor-SN degrades inosine-containing dsRNA. It contains five staphylococcal nuclease (SN) domains and one Tudor domain.

it is faster, energetically cheaper and takes up less cellular space to use a regulatory RNA. Such an RNA could be as short as ~13 nucleotides (nt) and still exclusively identify a unique target by basepairing. While this makes small RNAs good for target recognition, they are poor effectors. They are therefore used as guides of non-specific effector machineries like the RNA-induced silencing complex (RISC).

RISC is a multiprotein complex that contains a member of the Argonaute protein family (Fig. 6.1b). The Argonaute binds the small RNA and orients it to basepair with target transcripts. Complimentary target transcripts can get cleaved by Argonaute's own catalytic activity or be acted upon other proteins recruited to RISC. This usually leads to sequestration or degradation of the target.

Natural small RNAs are produced in a form not yet ready for RISC loading. They can take various maturation pathways that generally involve Drosha and Dicer (Fig. 6.1b). Both of these RNase III family members cleave dsRNAs, leaving a 5′ phosphate and a ~2-nt 3′ overhang. Drosha is commonly aided by dsRBP DGCR8 and Dicer by dsRBP TRBP. Based on the pathway taken and the RNAs origin, the small RNA is usually classified as a miRNAs, siRNA, esiRNA or secondary siRNA, although other small RNA classes also exist.

Primary (pri-)miRNAs form hairpins in transcripts usually synthesized by RNA polymerase II (RNAPII). About half reside in precursor (pre-) mRNA introns, while the rest reside in intergenic regions. Drosha cleaves these out of the surrounding transcript to form ~70-nt hairpin precursor (pre-)miRNAs. Exportin-5 exports the pre-miRNA to the cytoplasm, where it gets cleaved again by Dicer. The resulting ~19-bp miRNA duplex gets loaded onto RISC, where one strand serves as a guide, while the other is discarded (Fig. 6.2a). The strand that was 5′-most in the pri-miRNA is referred to as the 5p strand, the other as the 3p strand.

siRNAs are the basis of the RNA interference (RNAi) response. They are produced from long dsRNAs in the cytoplasm, whether they originate from dsRNA viruses, scientific experimentation or other sources. Dicer repeatedly cleaves these dsRNAs to form multiple ~19-bp

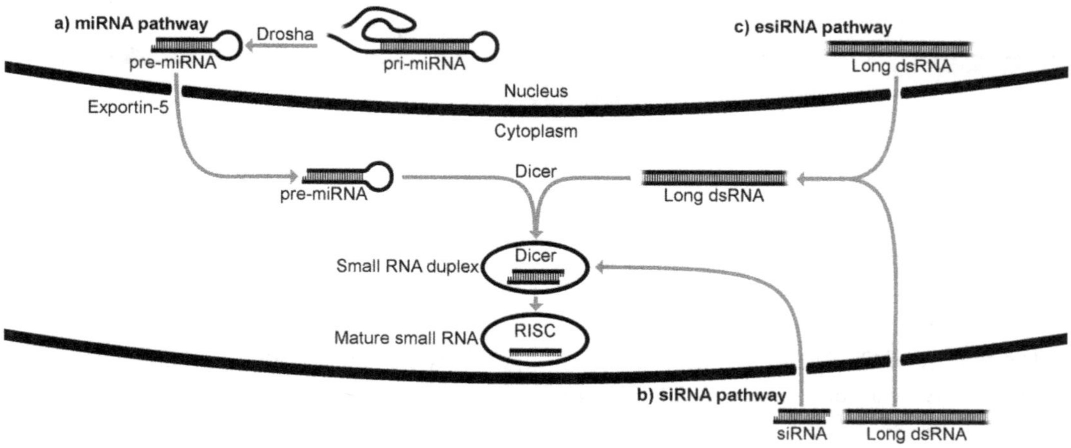

Figure 6.2 Small regulatory RNA biogenesis. (a) The miRNA pathway usually begins with RNAPII transcripts containing pri-miRNA hairpins. These are cleaved by Drosha with help from its partner DGCR8. The resulting pre-miRNAs are exported to the cytoplasm by Exportin-5. In the cytoplasm, the pre-miRNAs are cleaved by Dicer with help from its partner TRBP. Dicer hands the resulting miRNA duplexes to RISC, which discards one strand and retains the other. (b) siRNAs can originate as exogenous long dsRNA. These are cleaved repeatedly by Dicer-TRBP to form siRNA duplexes. Alternatively, these duplexes can be provided to the cell directly. Dicer hands the siRNA duplexes to RISC, which discards one strand and retains the other. (c) esiRNAs begin as long dsRNAs transcribed from the genome. These can for example be formed by intramolecular basepairing in transcripts from repetitive regions. In such a case, the two strands would be joined by a loop not shown in this figure. Repetitive cleavage by Dicer forms esiRNA duplexes, which can be loaded onto RISC. RISC discards one strand and retains the other.

siRNA duplexes that can be loaded onto RISC. One strand becomes the guide of RISC, while the other is discarded (Fig. 6.2b). To silence a transcript for experimental purposes, cells are commonly given siRNA duplexes already in these ~19-bp forms (Fig. 6.2b). These are less prone to trigger the interferon response than long dsRNAs.

esiRNAs follow a maturation pathway similar to that of siRNAs, but differ in origin. Long dsRNAs transcribed in the nucleus get cleaved repeatedly by Dicer and loaded onto RISC (Fig. 6.2c). The targets of the mature esiRNA could for example be transcripts of a gene overlapping the esiRNA's genomic origin.

Secondary siRNAs are not found in mammals, but play an important role in *C. elegans*. When a transcript is targeted by RISC, an RNA-dependent RNA polymerase (RdRP) can use the targeted transcript as a template for new siRNAs (Fig. 6.3). This serves to amplify and perpetuate the RNAi response. The secondary siRNAs differ structurally from the primary ones by being somewhat shorter and having 5′ triphosphates in place of monophosphates.

Whereas mature small regulatory RNAs are too small to be substrates of ADARs, their precursors are commonly not. This enables ADARs to exert major influence on small regulatory RNA biogenesis and function. Additionally, ADARs are also able to sequester certain small RNA duplexes (Yang *et al.*, 2005). A full understanding of small regulatory RNAs therefore demands consideration of ADARs.

Modulation of siRNA and esiRNA biogenesis and function by ADARs

siRNA and esiRNA are commonly generated from long dsRNAs that are potential ADAR substrates. By editing these RNAs, ADARs influence their biogenesis and function. Additionally, ADAR1p150 can sequester siRNA duplexes (Yang *et al.*, 2005). Both editing and sequestration allow ADARs to antagonize RNAi.

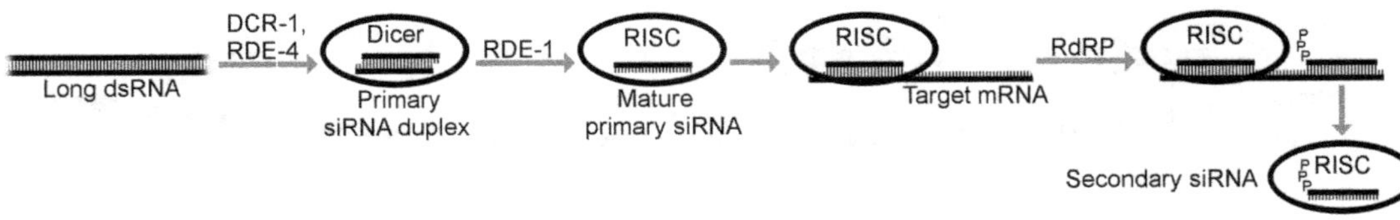

Figure 6.3 Secondary siRNA biogenesis. Secondary siRNAs are found in *C. elegans*, but not mammals or *Drosophila*. Following normal biogenesis of the primary siRNA or esiRNA, RISC recruits an RdRP to transcribe secondary siRNAs off its target mRNAs. This process amplifies and perpetuates the RNAi response.

Antagonism of RNAi by editing

Long dsRNAs can be processed into siRNA or esiRNA duplexes by Dicer. They can also be edited indiscriminately so that random A·U basepairs are replaced by less stable I·U wobble pairs. Editing continues until the double-stranded structure is so destabilized that it is no longer a suitable ADAR substrate, at which point ~ 50% of adenosines have been converted to inosines.

This affects RNAi in two ways. Firstly, the destabilized dsRNA is not a good substrate for Dicer. Secondly, any small RNA duplexes still produced have sequences no longer perfectly complementary to homologous transcripts. In both ways, editing antagonizes RNAi. This has been the conclusion of experiments studying RNAi *in vitro* and *in vivo* with both dsRNA of transgenic and endogenous origins.

Editing prevents dsRNA from triggering RNAi *in vitro*

An *in vitro* RNAi assay can be carried out by mixing *Drosophila* extract, a dsRNA and a homologous target RNA. Processing of the dsRNA to mature small RNAs and degradation of the target RNA indicates efficient RNAi (Scadden and Smith, 2001a).

The effect of editing on this system has been tested by treating the dsRNA with an ADAR prior to the assay. Increasing numbers of A-to-I conversions increasingly inhibited RNAi. Editing of 43% of adenosines by ADAR2 rendered the RNA ineffective in generating siRNAs. Lower levels of editing allowed some siRNAs to be produced, in which inosine could be detected (Zamore *et al.*, 2000; Scadden and Smith, 2001a).

Editing prevents transgenic dsRNA from triggering RNAi

Transgenic arrays in *C. elegans* commonly contain inverted copies of the same gene and produce unintended dsRNA upon read-through. This dsRNA could trigger RNAi and silence all copies of the transgene. Their expression therefore depends on prevention of the RNAi response by editing of the dsRNA.

This has been demonstrated for a green fluorescent protein (GFP)-expressing array (Knight and Bass, 2002). The non-coding strand lacked a promoter of its own and should therefore only be transcribed upon read-through between inverted copies. Reverse transcription, PCR amplification and sequencing of this strand showed that it is both transcribed and edited.

Knockout of ADR-1 and ADR-2 not only prevented detection of these editing events, it prevented any detection of the non-coding strand at all. This is consistent with a scenario where the dsRNA gets processed into siRNAs in the absence of editing. These siRNAs triggered an RNAi response directed at the transgenic GFP transcripts. Specifically, GFP expression was prevented by the ADAR knockouts, but rescued upon additional knockout of RNAi proteins RDE-1 or DCR-1 (Fig. 6.3).

Editing prevents endogenous dsRNA from triggering RNAi

Knockout of ADR-1 and ADR-2 disrupts chemotaxis in *C. elegans*, but leaves the worms otherwise normal (Tonkin and Bass, 2003). Interestingly, chemotaxis is rescued by additional knockout of RNAi proteins RDE-1 or RDE-4 (Fig. 6.3), even though these knockouts do not affect chemotaxis on their own (Tonkin and Bass, 2003). This implies that ADARs edit endogenous RNAs that would otherwise be processed in an RDE-1 and RDE-4-dependent manner into small regulatory RNAs triggering RNAi.

One search for these small RNAs has recently been carried out (Wu *et al.*, 2011). Short RNAs with 5′ monophosphates and 3′ hydroxyl groups were isolated from wild-type or ADR-1, ADR-2 double-knockout *C. elegans* at the embryonic or L4 stage. High-throughput sequencing revealed small RNAs up-regulated by knockout. Most of these were 23–24 nt long, consistent with the length of *C. elegans* primary siRNAs.

These small RNAs clustered to 545 loci that each spanned 0.1–9 kb. Together, they add up to 407 kb, or 0.4% of the *C. elegans* genome. The loci showed significant overlap with low-to-medium copy number regions. Of the few in annotated transcribed regions, over 60% were in introns.

This localization and clustering suggests that the small RNAs are processed from long dsRNAs, which would classify them as esiRNAs. Knockout of RDE-4 or RDE-1 (Fig. 6.3) in addition to ADR-1 and ADR-2 confirmed this classification. RDE-4 cleaves long dsRNAs into siRNAs. Its knockout lowered expression of the small RNAs to wild-type levels. RDE-1 cleaves the small RNA passenger strand following RISC loading, effectively halving the number of short RNAs in the cell. Its knockout nearly doubled expression of the small RNAs.

These esiRNA loci are presumably edited in wild-type worms. Small RNA sequencing initially failed to detect such editing, likely because of difficulties in aligning edited transcripts. An alignment algorithm that allowed A-to-G transitions overcame this problem and identified ~15,000 putative editing sites. About half of these clustered to ~130 of the 545 esiRNA loci. Editing at the remaining loci may still evade identification due to the difficulty of aligning edited small RNA sequences to their true genomic origins. Alternatively, editing might induce their degradation and thereby pre-empt their detection.

132 esiRNA loci overlapped annotated transcripts. Their esiRNAs could potentially template secondary siRNAs off these transcripts. Because of their 5′ triphosphates, secondary siRNAs can be detected by sequencing small RNA independent of their 5′ phosphorylation status. This identified additional small RNAs antisense to 35 of the annotated transcripts. These were mostly 21–22 nt long, common for *C. elegans* secondary siRNAs. ADR-1, ADR-2 double knockout increased putative secondary siRNA expression for all 35 transcripts, >2-fold for nine of them. Additional knockout of either RDE-1 or RDE-4 reverted expression of these small RNAs to wild-type levels, consistent with the notion that they are secondary siRNAs (Fig. 6.3).

The 35 transcripts able to template secondary siRNAs might be silenced by the esiRNAs produced in the absence of ADARs. Indeed, for the five transcripts for which secondary siRNA levels increased >4-fold upon ADR-1, ADR-2 knockout, captured mRNA counts also decreased significantly. Counts decreased by 89% for Y46D2A.2, by 70% for F39E9.22, by 65% for Y46D2A.5, by 61% for F39E9.1 and by 29% for R12C12.7. For the remaining transcripts with lesser secondary siRNA up-regulation, there were lesser decreases in mRNA counts.

The esiRNAs were also up-regulated by ADR-1 and ADR-2 single knockouts. Up-regulation was more severe for ADR-2 knockout than ADR-1 knockout, but most severe for the double knockout. Interestingly, this severity trend is the same as that observed for the effect of ADAR knockouts on *C. elegans* chemotaxis (Tonkin and Bass, 2003). The parallel between the effects of ADAR and RNAi knockouts on esiRNA expression and on chemotaxis suggests a causal relationship. However, no identified esiRNA locus overlaps a known chemotaxis-related gene. It is possible that the concurrent silencing of all of these transcripts has a complex effect on chemotaxis or that chemotaxis is affected by some yet undiscovered substrate common to ADARs and RNAi (Wu *et al.*, 2011).

Another outstanding question is whether these esiRNA loci serve a biologically relevant regulatory function. Perhaps some *C. elegans* cells lack strong ADAR function at certain developmental stages, at which point these esiRNA loci would specifically down-regulate overlapping genes. Alternatively, ADAR editing might constitutively guard the cell from sporadic dsRNA producing destructive esiRNA.

Subcellular localization of long dsRNA editing

The subcellular distribution of ADARs controls what dsRNAs it can encounter. Mammalian ADAR1p110 and ADAR2 localize mostly to the nucleus, though ADAR1p110 can also function in the cytoplasm (Fritz *et al.*, 2009). The interferon-inducible ADAR1p150 shuttles between nucleus and cytoplasm. *C. elegans* ADR-1 and ADR-2 reside in the nucleus, and *Drosophila* dADAR localizes mainly to the nucleus (Bhogal *et al.*, 2011; Jepson *et al.*, 2011).

Because editing of a long dsRNA can prevent its Dicer processing, and because Dicer processing leaves a dsRNA too short to be edited, which enzyme a long dsRNA encounters first can determine its fate. Studies in *C. elegans* suggest that dsRNA originating in the cytoplasm is primarily processed by Dicer, while dsRNA originating in the nucleus is primarily edited by ADARs.

Cytoplasmic long dsRNA is processed by Dicer

dsRNA injected into *C. elegans* or expressed by bacteria eaten by *C. elegans* enters the cytoplasm, where it is processed by Dicer and induces RNAi. Because ADR-1 and ADR-2 localize to the nucleus, they cannot efficiently compete with cytoplasmic Dicer for this substrate. Accordingly, the efficiency of the RNAi response is unaffected by ADR-1 and ADR-2 knockout (Knight and Bass, 2002).

This situation might be different in mammals, because of their interferon-inducible cytoplasmic ADAR1p150. How efficiently ADAR1p150 can compete for Dicer substrates remains unknown.

Nuclear long dsRNA is edited by ADAR

Long dsRNA is efficiently edited in the *C. elegans* nucleus by ADARs, which prevents it from triggering RNAi (Morse *et al.*, 2002). Only upon knockout of ADARs does an RNAi response take place. This has been shown for various endogenous transcripts (Wu *et al.*, 2011) and transcripts from a transgenic GFP-expressing construct (Knight and Bass, 2002).

The endogenous transcripts originated from regions with low to medium copy numbers of inverted repeats. The transgenic GFP line was generated by injecting DNA into *C. elegans* syncytial gonad. This can create extrachromosomal arrays consisting of hundreds of copies of the transgene arranged in tandem repeats. In both cases, read-through between two inverted copies of one sequence can produce a long dsRNA.

In wild-type worms, sequencing showed that both the GFP and the endogenous transcripts are extensively edited. The GFP was highly expressed, and sequencing of small RNAs showed that nearly no small RNAs were produced from the endogenous transcripts.

In ADR-1 and ADR-2 knockout worms, the situation was different. GFP was barely expressed, and Sanger sequencing could not identify the GFP transcript (Knight and Bass, 2002). Some unedited long dsRNAs were presumably processed into esiRNAs that silenced the remaining GFP transcripts. Sequencing of small RNAs showed that a high number of siRNAs were produced from the endogenous long dsRNAs, and genes overlapping the dsRNA loci were consequently silenced (Wu *et al.*, 2011).

Whereas most ADARs localize to the nucleus, Dicer mainly localizes to the cytoplasm (Provost *et al.*, 2002). While the presence of Dicer in the nucleus is of interest (Ando *et al.*, 2011), it is probably outnumbered there by ADARs. Additionally, there is evidence that ADARs can act co-transcriptionally (Laurencikiene *et al.*, 2006; Morlando *et al.*, 2008). This probably gives ADARs the opportunity to act upon dsRNAs that, in the cytoplasm, would have predominantly been substrates of Dicer.

Beyond *C. elegans*

As described above, *C. elegans* has provided valuable information on the non-specific editing of long dsRNAs by ADARs, and will likely continue to do so. A general conclusion from the above sections is that editing prevents dsRNA originating in the nucleus from erroneously triggering RNAi. One must wonder whether this will hold true for other organisms. Known edited long dsRNAs in humans could well produce esiRNAs were they not edited (Morse *et al.*, 2002). Additionally, putatively edited esiRNAs have been identified in *Drosophila* (Kawamura *et al.*, 2008). However, avoiding improper RNAi might not be as important for these species, as they cannot perpetuate an erroneous RNAi response via secondary siRNAs.

Unfortunately, investigations like those carried out in *C. elegans* can be difficult in other organisms, where the ADAR knockout phenotypes are generally far more severe. *C. elegans* lacking ADR-1 and ADR-2 show defective chemotaxis, but are otherwise viable and fertile (Tonkin *et al.*, 2002). By contrast, mice lacking ADAR1 show defective erythropoiesis and die at embryonic day (E)12 (Wang *et al.*, 2000, 2004; Hartner *et al.*, 2004, 2009). Mice lacking ADAR2 die from seizures shortly after birth (Higuchi *et al.*, 2000). *Drosophila* lacking dADAR show adult-stage extreme un-coordination, seizures, temperature-sensitive paralysis and a complete lack of courtship (Jepson and Reenan, 2009, 2010). Neither un-coordination nor temperature-sensitive paralysis can be rescued by additional mutations in the RNAi machinery, suggesting that these phenotypes do not result from improper RNAi (Jepson and Reenan, 2009).

However, these severe phenotypes should not prevent similar studies in animals beyond *C. elegans*. Tractable hypomorphic dADAR *Drosophila* strains have been made (Jepson *et al.*, 2011), but have yet to yield small RNA-related results. Furthermore, a single A-to-G mutation in the CAG editing site of GluR-B rescues the ADAR2 phenotype in mice (Higuchi *et al.*, 2000). Such mice, or mice below the age at which ADAR deficiencies cause death, could be used to examine small RNA editing in mammals (Vesely *et al.*, 2012). These kinds of studies could do much to establish how far beyond *C. elegans* the above results extend. For now, one hint of a similarly antagonistic ADAR–RNAi relationship in mammals may be the up-regulation of mouse ADAR1 by siRNAs duplexes, as explained below.

PU.1-dependent up-regulation of ADAR1 by siRNAs

siRNAs silence target transcripts. Curiously, higher doses of transfected siRNA duplexes have been observed to quicken rebound of target transcript levels in both Chinese hamster ovary (CHO) cell culture and live mice (Hong *et al.*, 2005). The rebound can also be quickened by additional injection with unrelated siRNAs, indicating that the siRNA duplexes trigger an anti-siRNA response.

As a known RNAi antagonist, ADAR1 was a putative effector of this mechanism. ADAR1 expression four days after injection of mice with 10 μg of a target-less siRNA was therefore examined by quantitative PCR and found to be up-regulated ~3-fold (Hong *et al.*, 2005). Consequently, treating mouse tumours by injecting an siRNA against c-MYC can be made more efficient by additional injection of an siRNA against ADAR1 (Hong *et al.*, 2007), which has implications for potential medical uses of siRNAs.

The ability of different parts of the ADAR1 promoter region to mediate siRNA-induced expression was tested by placing them in front of a GFP gene in a construct injected into CHO cells. Counting the base before the AUG start codon as −1, a 9-bp PU.1 motif around position −190 was the only region found to be necessary for the response (T. Wu *et al.*, 2009). Indeed, knocking down PU.1 prevented siRNA transfection from inducing ADAR1 (T. Wu *et al.*, 2009).

PU.1 is an E-twenty-six (ETS) transcription factor family member that is highly expressed in monocytic, granulocytic and B lymphoid lineages. It is expressed less in mature erythrocytes and not at all in mature T-cells. Its relative expression level determines whether cells differentiate into macrophages, granulocytes or lymphocytes (Gupta *et al.*, 2009). In line with this expression pattern, it induces transcription of many lymphoid – and myeloid-specific genes. Common to these genes is the lack of a TATA box in their promoter regions. Instead, PU.1 binds the TATA-binding protein (TBP) to recruit it to the promoter region (Gupta *et al.*, 2009).

How PU.1 responds to high siRNA duplex concentrations remains to be determined. This could be mediated by a factor that recognizes small RNA duplexes in the cytoplasm. In that case, the PU.1 response would be limited to transfected siRNA duplexes. However, it is not clear how such a response would have evolved since unsequestered siRNA duplexes are unlikely to be common in nature. Alternatively, the response could be mediated by a signal triggered by active Dicer or RISC, not dependent on the sequence of their loaded

RNA. If so, PU.1-mediated ADAR1 up-regulation could also be induced by endogenous and exogenous long dsRNAs, and perhaps even miRNAs.

Another outstanding question is how ADAR1 up-regulation could antagonize siRNA function. The quantitative PCR methods used to measure ADAR1 expression levels did not distinguish between ADAR1p110 and ADAR1p150. One possible antagonism mechanism therefore relies on sequestration of siRNA duplexes by ADAR1p150, as explained below.

Sequestration of siRNA duplexes by ADAR1p150

ADARs are unlikely to affect siRNA duplexes catalytically, since dsRNAs shorter than 29 bp are not edited by ADAR1p150 *in vitro* (Yang *et al.*, 2005). However, ADAR1p150 can still influence these duplexes by sequestering them. In fact, 19-bp siRNAs are bound by ADAR1p150 with a K_d of ~ 0.2 nM, by ADAR1p110 with a K_d of ~ 3 nM and by ADAR2 with a K_d of ~ 10 nM. These values are comparable to those for longer dsRNAs and indicate stronger binding than the ~ 500 nM K_d with which Dicer binds such 19-bp siRNAs (Lima *et al.*, 2009). By contrast, ADAR3 does not noticeably bind to < 37-bp siRNAs (Yang *et al.*, 2005).

Why ADARs can bind but not edit short dsRNAs is unclear. 19 bp may be enough to contact an ADAR's dsRBDs but not reach its catalytic site. Alternatively, the catalytic mechanism of ADAR might require a longer dsRNA even if the catalytic site is reached. Both the dsRBDs and the deaminase domain of an ADAR contact ~ 14 bp. While this is too much for two domains to bind a 19-bp dsRNA in tandem, these domains contact only one side of the dsRNA (Macbeth *et al.*, 2005; Stefl *et al.*, 2010), so it is possible that several domains could simultaneously bind different sides of the same 19-bp duplex.

Regardless of how ADARs sequester siRNA duplexes, the significance can be major. The dissociation rate constant for the ADAR1p150–siRNA complex is less than 10^{-4}/s (Yang *et al.*, 2005) and the low K_d values of other ADARs indicate comparable dissociation rate constants. This rate is slow enough to attenuate the RNAi response. Gradual release of the siRNA duplex could spread out the effect of the siRNA treatment over time, making its effect at any one moment less severe. Alternatively, binding of the siRNA duplex could cause its degradation by some unknown mechanism. Either way, sequestration inhibits siRNA function.

This has been observed via the enhancement of siRNA silencing of a GFP expressed in mouse embryonic fibroblasts upon ADAR1 knockout (Yang *et al.*, 2005). This enhancement was reversed by transfection with an ADAR1p150 expression plasmid and could not be observed upon ADAR2 knockout, consistent with the notion that only cytoplasmic ADAR1p150 can sequester cytoplasmic siRNAs.

Assessment of the amount of ADAR1p150 and siRNA in the cell indicated that sequestration is stoichiometric. It is effective as long as there is more ADAR1p150 in the cell than there are siRNA duplexes, but increasingly inefficient otherwise. Silencing efficiency would therefore be expected to be lowered if ADAR1p150 expression is induced, for example by the interferon response.

Interestingly, knocking out the interferon response protein PKR did not affect silencing (Yang *et al.*, 2005). This suggests that any ADAR1p150 observed resulted perhaps from low level of expression in the absence of the interferon response (Patterson and Samuel, 1995). siRNA sequestration may therefore be important in all mammalian cells, not just those with a triggered interferon response. Furthermore, the extent of this sequestration could be increased dramatically by up-regulation of ADAR1p150 were the interferon response triggered. This could naturally affect the conclusion drawn from any siRNA knockdown experiment.

Modulation of miRNA biogenesis and function by ADARs

Editing of a pri-miRNA can affect its maturation and function (Fig. 6.4) as well as its stability. Many biological consequences of such editing have been discovered, and more are likely to be discovered in the future, as only a few pri-miRNA editing events have so far been studied in depth.

The effect of editing is commonly assessed using 'pre-edited' RNAs that contain A-to-G

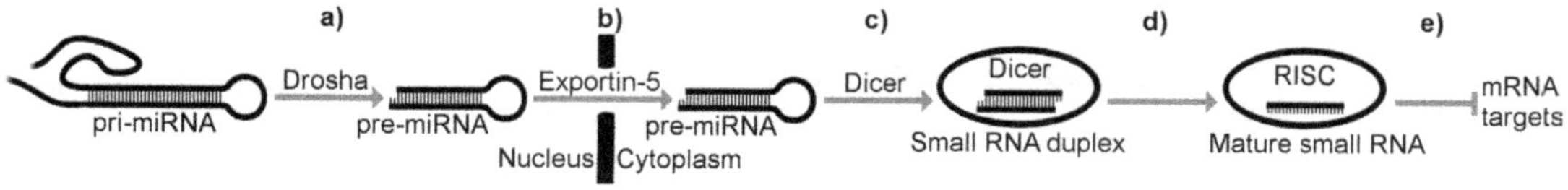

Figure 6.4 Possible effects of editing on miRNA biogenesis and function. A maturing miRNA undergoes (a) Drosha cleavage, (b) Exportin-5 export, (c) Dicer cleavage, and (d) RISC loading and strand selection before it can (e) silence its target mRNAs. Each steps has potential to be modulated by editing. Additionally, the stability of the pri-miRNA has been shown to be modulated by editing.

substitutions at the editing site. Because guanosine basepairs like inosine, the effect of this substitution is assumed to be the same as that of editing. A large set of experiments validates this assumption. However, as Tudor staphylococcal nuclease (Tudor-SN) exemplifies in one of the below examples, pre-editing might not be useful when studying effects of editing mediated by RNA–protein interactions.

Steps modulated by pri-miRNA editing

miRNAs and their precursors can be recognized by various traits. The single-stranded (ss)RNA regions surrounding a pri-miRNA, the 5′ phosphate and 3′ overhang left by Drosha and Dicer cleavage, and the hairpin loop are beyond the ability of ADARs to alter. However, sequence and bulges can be altered. Whether a step in the life of a miRNA can be modulated by pri-miRNA editing comes down to whether the relevant enzymes rely on traits in the first or latter group.

Modulation of Drosha cleavage

Drosha-DGCR8 processing is the only miRNA maturation step that does not use a 5′ phosphate and 3′ overhang for substrate recognition. This makes Drosha cleavage particularly susceptible to modulation by ADAR editing. Indeed, out of six randomly selected human pri-miRNAs edited in UAG motifs, editing modulated *in vitro* Drosha-DGCR8 processing efficiency of five (Fig. 6.5) (Kawahara *et al.*, 2008).

Drosha-DGCR8's binding activity is mediated mainly by DGCR8 (Zeng and Cullen, 2005; Han *et al.*, 2006; Yeom *et al.*, 2006), which binds the lower part of the pri-miRNA hairpin stem and the flanking ssRNA (Fig. 6.6). The length of the stem and the flexibility of the hairpin loop also matter, but less so (Han *et al.*, 2006; Yeom *et al.*, 2006; Zhang and Zeng, 2010). While these traits cannot be altered by ADAR editing, Drosha cleavage is also affected by bulges in the pri-miRNA stem.

Introducing two consecutive mismatches on either side of a Drosha cleavage site can inhibit cleavage *in vitro* (Han *et al.*, 2006). Similarly, an SNP identified in the stem of Kaposi's sarcoma-associated herpesvirus (KSHV) pri-miR-K5 creates a bulge that prevents Drosha processing (Gottwein *et al.*, 2006). It is unclear whether bulges affect binding, catalysis or both. However, this allows ADAR editing to modulate Drosha cleavage as exemplified by pri-miR-142, Epstein–Barr Virus pri-miR-BART6, and additional pri-miRNAs illustrated in Fig. 6.5. These examples demonstrate that bulges can also aid cleavage.

In addition to cleavage efficiency, editing could potentially modulate the choice of cleavage site. This choice relies on binding of the Drosha–DGCR8 complex to the ssRNA–dsRNA junction, which positions Drosha's active site (Han *et al.*, 2006). The junction-to-cleavage site distance is ~11 bp, but varies somewhat between different pri-miRNA species. This variation indicates the importance of bulges and other traits in determining the exact cleavage site. Editing could therefore shift the Drosha cleavage site, which would shift the seed sequence and alter any resulting mature miRNA's target set.

While there is no proven example of editing causing such shifts, there are hints. It has for example been pointed out that the edited pri-miRNA pri-miR-142 has an especially variable Drosha cleavage site (H. Wu *et al.*, 2009). Additionally, mature miRNA high-throughput sequencing data contains hints of this phenomenon, as some edited mature miRNAs have

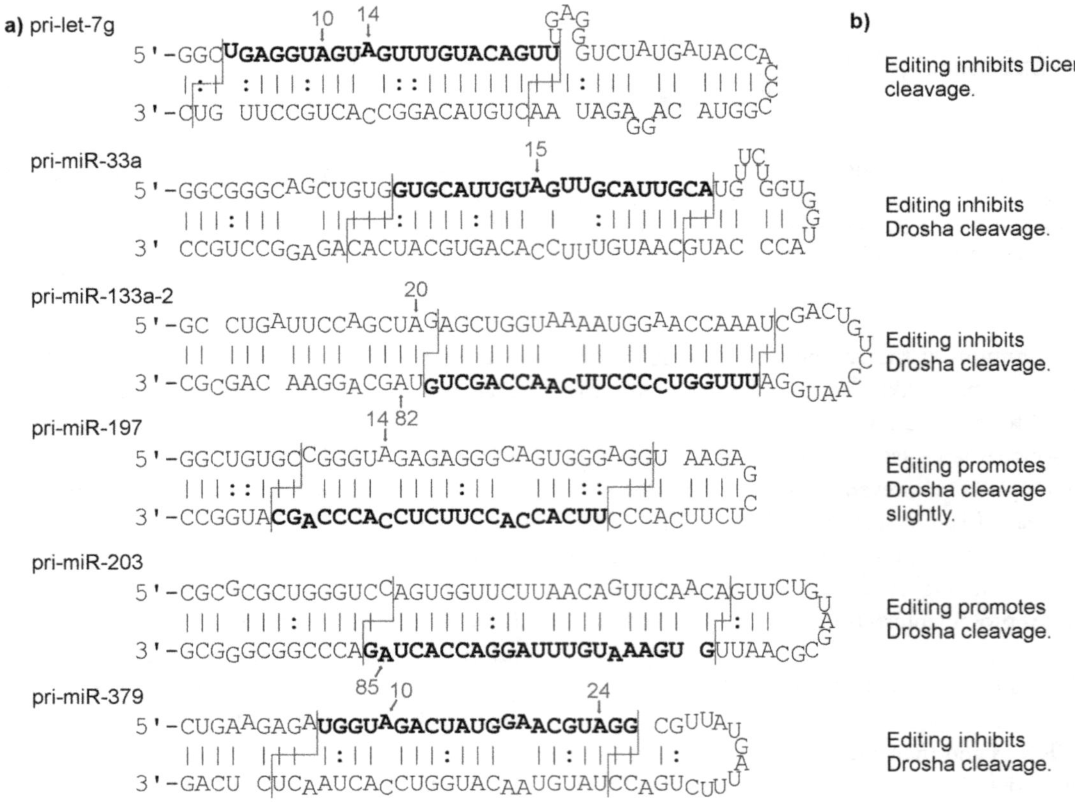

Figure 6.5 Effect of pri-miRNA editing on *in vitro* Drosha and Dicer processing. (a) Six human pri-miRNAs edited in UAG motifs were randomly selected for study (Kawahara *et al.*, 2008). Their editing sites are indicated by arrows. Conventional numeral identifiers of the editing sites are given. Common Drosha and Dicer cleavage sites are indicated by grey lines, while strands commonly loaded onto RISC are indicated in bold. (b) The effect of pre-editing on *in vitro* processing by Drosha-DGCR8 and Dicer-TRBP was determined (Kawahara *et al.*, 2008). For pri-let-7g, only site 14 was pre-edited for this experiment. For pri-miR-133a-2, only site 20 was pre-edited. For pri-miR-379, only site 10 was pre-edited.

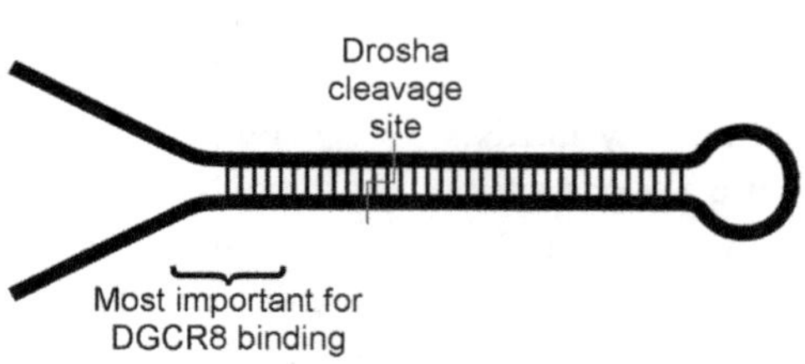

Figure 6.6 Drosha-DGCR8 binding of pri-miRNAs. DGCR8 helps Drosha bind its substrates. DGCR8 mainly binds the ssRNA flanking the pri-miRNA and the adjacent stem. The presence of the stem further removed from the ssRNA and a flexible loop are also conducive to DGCR8 binding (Zeng and Cullen, 2005; Han *et al.*, 2006; Yeom *et al.*, 2006; Zhang and Zeng, 2010). Bulges in the stem structure could affect binding in ways not well understood.

different end distributions than their unedited counterparts. One experiment sequenced mature miRNAs from mouse brain (Chiang *et al.*, 2010). Stringent screening left 16 putative edited mature miRNAs, of which seven had 5′ end distributions significantly different from their unedited counterparts, including miR-411-5p and miR-376a-5p.

Modulation of pri-miRNA stability

Tudor-SN is an RNase specific to inosine-containing dsRNA. Its activity was first observed in *Xenopus* oocyte and HeLa nuclear extracts (Scadden and Smith, 2001b). Tudor-SN was thereafter isolated from *Xenopus* oocyte extract by its ability to exclusively bind dsRNA affinity matrices containing inosine and identified using mass spectrometry (Scadden, 2005). Three other proteins that also specifically bound affinity matrices containing inosine could not be identified by mass spectrometry and remain unidentified.

The sequence requirements of Tudor-SN have been studied by *in vitro* cleavage of synthetic inosine-containing oligonucleotides in *Xenopus* oocyte extracts (Scadden and Smith, 2001b). Two, three or four consecutive I·U or U·I basepairs allow increasingly tight binding to Tudor-SN. The surrounding sequence seems to be of little influence. Which strand carries the inosine residues is also important, as Tudor-SN seemingly prefers to cleave across from an inosine. A dsRNA with four inosines on one strand is therefore cleaved only on the opposite strand. When both strands contain inosine, both can get cleaved.

Despite this emerging knowledge about substrate requirements, the only known natural Tudor-SN substrate is pri-miR-142 (Yang *et al.*, 2006). Edited pri-miR-142 is unable to mature, as discussed later. Tudor-SN therefore prevents accumulation of the non-maturing pri-miRNA. This suggests other pri-miRNAs prevented from maturing by editing as candidate Tudor-SN substrates, although all known cases except for pri-miR-142 lack the sequential editing sites required for Tudor-SN cleavage (compare Figs. 6.5 and 6.10). An edited pri-miRNA not prevented from maturing might even be processed by Drosha and exported before it can encounter Tudor-SN.

However, there are hints that substrates could be handed directly from ADAR to Tudor-SN. In particular, Tudor-SN shuttles between the nucleus and cytoplasm, where it interacts with ADAR1p150 in an RNA-independent manner (Weissbach and Scadden, 2012). It is tempting to wonder whether Tudor-SN also interacts with ADAR1p110 in the nucleus. Although such interaction remains speculative, it would be another example of co-transcriptional wrangling over pri-miRNAs (Fig. 6.7).

The ability of Tudor-SN to act specifically on inosine-containing dsRNA is a rare instance of functional distinction between inosine and guanosine. While their common basepairing to cytidine make them indistinguishable to basepairing-mediated recognition, guanosine has an additional amine group pointing into the dsRNA minor grove. This group can be recognized by proteins in two ways. Firstly, proteins may probe an open conformation of the dsRNA to make more contacts with, and better recognize, the inosine. Secondly, they may probe the dsRNA minor grove to recognize of the absence of the guanosine amine group. Despite a solved crystal structure (Li *et al.*, 2008), it is not yet clear which mechanism Tudor-SN employs.

Modulation of nuclear export

Nuclear export of pre-miRNAs is carried out by Exportin-5, with the help of the GTP-binding protein Ran. The crystal structure of this complex reveals that its cargo recognition relies largely on some degree of double-strandedness ending

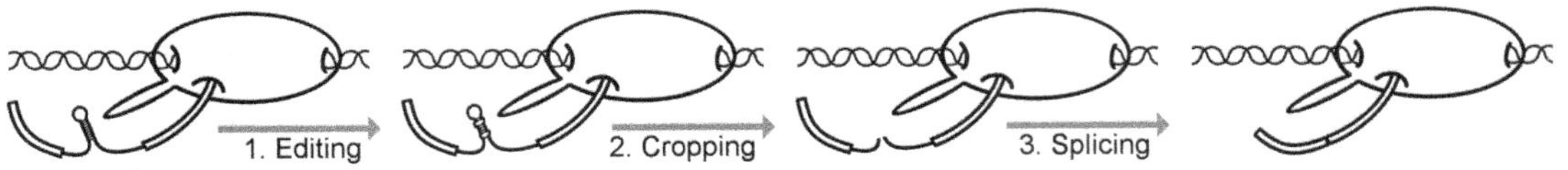

Figure 6.7 Co-transcriptional editing and Drosha cleavage. Pri-miRNA editing takes place before cropping of the pri-miRNA by Drosha. If the pri-miRNA is located in an intron, this in turn takes place before splicing. ADAR2 is likely able to achieve this early editing by localizing to the site of transcription through interaction with the CTD of RNAPII (Laurencikiene *et al.*, 2006; Ryman *et al.*, 2007). A larger complex may form around the nascent RNA to co-coordinate these events and keep the RNA's 5′ end from diffusing away following cropping by Drosha.

in a 3′ overhang (Okada *et al.*, 2009). It has very little other structural specificity, and also exports some RNAs structurally very different from pre-miRNAs. It is therefore unlikely that editing could affect pre-miRNA export, at least by altering the pre-miRNAs affinity for Exportin-5.

Modulation of Dicer cleavage

Dicer's PAZ domain (Fig. 6.1b) binds the pre-miRNA 5′ phosphate and 3′ overhang while its RNase III domain cleaves off the hairpin loop. The crystal structure of *Giardia intestinalis* Dicer indicates how the distance between these domains measures the length of pre-miRNA stem to form the miRNA duplex (Macrae *et al.*, 2006). While this distance is set, the number of basepairs that it can contain can be altered by bulges in the stem. Editing therefore has the potential to shift Dicer cleavage sites.

Hints of this phenomenon come from high-throughput sequencing of mature miRNAs, where edited miRNAs occasionally have end distributions different from those of their unedited counterparts. For example, one experiment sequenced mature miRNAs from mouse brain (Chiang *et al.*, 2010). Stringent screening left 16 putative edited mature miRNA, of which seven had 5′ end distributions significantly different from their unedited counterparts. For example, unedited miR-337–3p was about equally cleaved at two neighbouring phosphates, whereas edited miR-337–3p was cleaved only at one of these phosphates.

Editing can also prevent Dicer-TRBP processing altogether, as has been demonstrated by mouse pre-miR-151 (Kawahara *et al.*, 2007a) and human pre-let-7g (Fig. 6.5) (Kawahara *et al.*, 2008). This could occur by inhibition of either binding or catalysis. EMSA studies show that the K_d for binding of Dicer-TRBP to pre-miR-151 is ~ 1.5 nM independent of editing at sites −1, +3 or both, leading to the conclusion that it is catalysis which is inhibited by editing. This might not be surprising since at least part of Dicer's affinity for its substrate relies on the ~ 2-nt 3′ overhang and 5′-phosphate. Furthermore, both the −1 and +3 sites are very close to Dicer's cleavage site (Fig. 6.8). Perhaps editing adjacent to the base of the pre-miRNA, bound by Dicer's PAZ domain, could have a larger effect on binding.

Modulation of RISC loading

Following Dicer cleavage, one of the miRNA duplex strands is loaded onto RISC as the guide strand while the other is degraded. A few duplexes can have either strand loaded. However, one strand is generally designated for loading by having the less stably basepaired 5′ end (Schwarz *et al.*, 2003). In fact, almost half of all strands to be loaded have a mismatch at their first residue (Han *et al.*, 2006).

This relative end stability could clearly be altered by editing. An end could be stabilized by replacing an A·C mismatch by an I·C basepair or destabilized by replacing an A·U basepair with an I·U wobble. This could change which strand

a)

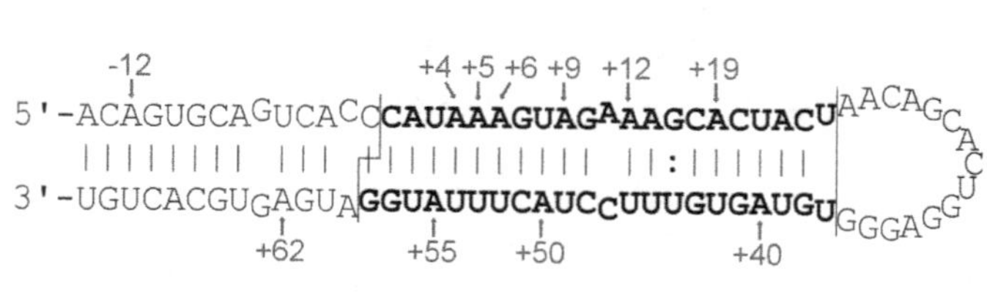

b)

Site	ADAR
-12	2
+4	2
+5	1p110
+6	2
+9	2
+12	2
+19	2
+40	1p110&2
+50	2
+55	2
+62	2

Figure 6.8 Mouse pri-miR-151. (a) Pri-miR-151 editing sites are assigned identifiers reflecting their position relative to the 5′ end of miR-151–3p (Kawahara *et al.*, 2007a). (b) Both ADAR1p110 and ADAR1p150 are able to edit pri-miR-151 at the −1 and +3 sites *in vitro*. ADAR2 knockout does not to affect editing at these sites in mouse cerebellum, cerebral cortex or brain stem. The effect of ADAR1 knockout is difficult to assess because ADAR1 knockout mice die at E12. However, the small extent of editing seen at positions −1 and +3 at E11.5 in wild-type mice is absent in ADAR1 knockout mice (Kawahara *et al.*, 2007a).

is loaded, leading to silencing of a different set of mRNAs.

No examples of this phenomenon have yet been discovered. However, examples have been hinted at. For example, high-throughput sequencing of mature miRNAs has revealed that mouse miR-540-5p is loaded particularly often in brain, where editing events are particularly frequent (Chiang *et al.*, 2010). However, the observed editing is not extensive enough to fully explain the strand switch (Chiang *et al.*, 2010).

In addition to changing the strand utilized, editing can prevent RISC loading altogether. While it is less clear what the mechanistic basis of this would be, it has been demonstrated for Epstein–Barr virus miR-BART6. Without further information on what structural constraints a miRNA duplex must conform to in order to be loaded onto RISC, it is hard to predict how widespread this phenomenon may be.

Modulation of mature miRNA target set

When editing does not prevent miRNA maturation as described above, an edited miRNA can get loaded onto RISC and silence a set of target mRNAs different from that of the unedited miRNA. This is especially the case when editing takes place in miRNA nucleotides 2–8, as this 'seed sequence' is especially important for target selection.

Such retargeting is exemplified by miRNAs of the human and mouse miR-376 clusters. Additionally, high-throughput sequencing of mature miRNAs has identified mature miRNAs that are seemingly edited, although they have still to be validated and studied further (Ruby *et al.*, 2006; Landgraf *et al.*, 2007; Babiarz *et al.*, 2008; Kuchenbauer *et al.*, 2008; Morin *et al.*, 2008; Suzuki *et al.*, 2009; Chiang *et al.*, 2010; Linsen *et al.*, 2010; Schulte *et al.*, 2010; Berezikov *et al.*, 2011). Several of these seem to be edited in their seed sequences, indicating the possibility of major effects of editing on target selection.

Co-transcriptional competition for pri-miRNAs

ADAR editing of mRNA exons generally relies on basepairing with an adjacent intron to form dsRNA. This requires mRNA editing to take place co-transcriptionally, before splicing can occur. In the case of ADAR2, this is dependent on interaction between ADAR2 and the C-terminal domain (CTD) of RNAPII. This has been shown for editing of mRNAs encoding GluR-B and ADAR2 transcripts (Laurencikiene *et al.*, 2006; Ryman *et al.*, 2007). It remains to be seen whether ADAR1 can similarly interact with the CTD.

About half of all pri-miRNAs are located within pre-mRNA introns. Even though splicing takes place co-transcriptionally, these pri-miRNAs are cleaved out of their surrounding transcript prior to splicing (Kim and Kim, 2007; Morlando *et al.*, 2008; Pawlicki and Steitz, 2008). Co-transcriptional Drosha cleavage is also the case for intergenic pri-miRNAs not located in introns (Morlando *et al.*, 2008; Ballarino *et al.*, 2009). How Drosha localizes to sites of transcription is not clear, but it could be mediated by interaction with pre-mRNA processing factors (Gregory *et al.*, 2004; Shiohama *et al.*, 2007; Morlando *et al.*, 2008).

For ADAR to act on a pri-miRNA, it would have to act before Drosha, which in turn acts before splicing (Fig. 6.7). This indicates the importance of ADAR localization to transcription sites for pri-miRNA editing. ADAR2 likely localizes by interaction with the CTD, whereas the basis for ADAR1 localization is an open question.

Determining whether the ADAR2–CTD interaction, and any mechanism behind ADAR1 localization, is constitutive or regulated could be highly elucidating. It might help explain why editing varies seemingly randomly across different tissues and pri-miRNA editing sites (Blow *et al.*, 2006), largely independent of ADAR expression levels. Furthermore, it would better delineate what pri-miRNAs may get edited. For example, if all ADAR2 localization is RNAPII CTD-dependent, we may reason that ADAR2 editing does not take place in RNA polymerase III-transcribed pri-miRNAs.

Possible editing of pre-miRNAs

ADAR1p150 can edit 36-bp dsRNAs *in vitro*, but not 29-bp dsRNAs (Yang *et al.*, 2005). While this prevents miRNA duplex editing, pre-miRNAs are at this length threshold and therefore potential ADAR substrates.

Such editing may be exemplified by pre-miR-151 (Kawahara *et al.*, 2007a). Pri-miR-151 is ~41% edited in human amygdala and ~29% edited in mouse cerebral cortex. Yet pre-miR-151 is 100% edited in both of these tissues. The explanation may be editing of pre-miR-151. Indeed, pre-miR-151 can be edited *in vitro* by both ADAR1p110 and ADAR1p150.

However, an alternative explanation for this highly edited pre-miRNA also exists (Kawahara *et al.*, 2007a). Editing inhibits the Dicer processing of pre-miR-151, which might cause accumulation of enough edited pre-miRNA to drown out the unedited pre-miRNA during sequencing. Current data cannot refute either explanation, and both could be true. Since ADAR1p110 and ADAR1p150 edit pre-miR-151 *in vitro*, it is not obvious what would prevent them from doing so in the cell.

Sequestration of pri-miRNAs *by ADAR2*

ADARs are commonly viewed as exerting their influence through their catalytic activity. Yet by tightly binding dsRNA, they can also sequester dsRNAs. As discussed earlier, ADAR1p150 sequestration of siRNAs has important consequences for siRNA experiments (Yang *et al.*, 2005). Similarly, ADAR2 sequesters pri-miRNAs (Heale *et al.*, 2009).

This has been demonstrated for human pri-miR-376a-2, which is seemingly edited by ADAR1p110 or ADAR2, depending on the site in question (see Figs. 6.12 and 6.13). Pre-editing at the +4 and +44 editing sites does not inhibit *in vitro* Drosha cleavage by HeLa nuclear extract. However, addition of purified ADAR2 does, indicating inhibition independent of editing.

This conclusion was bolstered by experiments in HEK293T cells (Heale *et al.*, 2009). Co-transfection with constructs expressing either human pri-miR-376a-1 or pri-miR-376a-2, and a luciferase transcript with the relevant miRNA target sites reported on the efficiency of pri-miRNA maturation by the extent of luciferase knockdown. Interestingly, co-transfection with a catalytically inactive mutant ADAR2 up-regulated luciferase by ~2-fold. These results could not be recreated with pri-let-7a-1, not believed to be edited, or with ADAR1p110 in place of ADAR2 (Heale *et al.*, 2009).

The conclusion is that ADAR2 can sequester some of its pri-miRNA substrates well enough to delay or prevent their Drosha processing. Because ADARs might bind more miRNAs than they edit, many miRNAs may be affected by ADARs solely in an editing-independent manner. ADAR2's superior ability to sequester pri-miRNAs of the miR-376 family might explain its previously observed ability to inhibit ADAR1 editing of mouse pri-miR-376b and pri-miR-376c at the +44 site (Kawahara *et al.*, 2007b). Whether other pri-miRNAs may be preferentially sequestered by ADAR1p110 remains to be seen. It is also not clear what the natural effect of this sequestration may be in the cell, where editing and Drosha cleavage occur in certain regulated co-transcriptional time frames (Laurencikiene *et al.*, 2006; Ryman *et al.*, 2007; Morlando *et al.*, 2008; Ballarino *et al.*, 2009).

Examples of pri-miRNA editing

The consequences of pri-miRNA editing have been examined in detail for a limited number of pri-miRNAs, providing valuable information on the possible functions of pri-miRNA editing.

Pri-miR-142: modulation of Drosha processing and stability

Both the 5p and 3p strands of miR-142 are highly expressed in haematopoietic tissues, where they regulate differentiation (Chen *et al.*, 2004). Pri-miR-142 is edited at several sites (Fig. 6.9), which prevents Drosha processing and induces degradation by Tudor-SN (Yang *et al.*, 2006).

Modulation of Drosha cleavage

The effect of editing on miR-142 biogenesis has been tested in HEK293 cells. Transfection with a plasmid expressing pri-miR-142 showed that pre-editing caused accumulation of pri-miR-142 and depletion of pre – and mature miR-142. A similar depletion of pre – and mature miRNA-142 was seen in cells overexpressing unedited pri-miR-142 upon additional overexpression of ADAR1p110

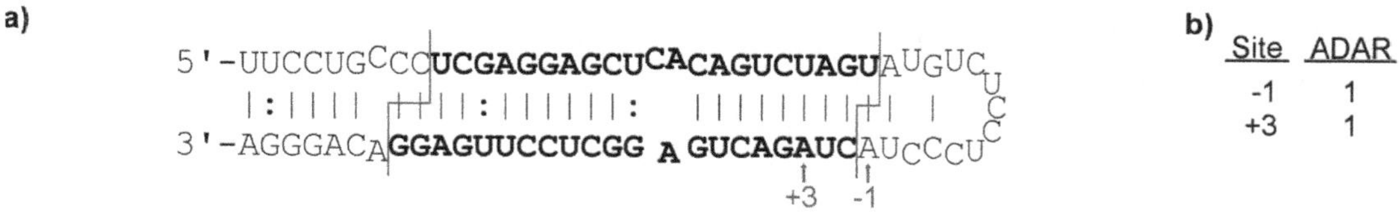

Figure 6.9 Human pri-miR-142. (a) Pri-miR-142 editing sites are assigned identifiers reflecting their position relative to the 5′ end of miR-142–5p (Yang *et al.*, 2006). (b) The enzyme most able to edit each site was determined by *in vitro* editing assays using recombinant ADAR1p110, ADAR1p150 or ADAR2 (Yang *et al.*, 2006).

or ADAR2. This shows that editing prevents processing of pri-miR-142 into pre-miR-142, presumably by preventing Drosha processing (Yang *et al.*, 2006).

The specific editing sites able to inhibit Drosha processing were investigated *in vitro*. Processing of pri-miR-142 by recombinant Drosha-DGCR8 and Dicer-TRBP showed that pre-editing at sites +4, +5, +40 and +50 or *in vitro* editing by ADAR1p110 and ADAR2 prevented Drosha cleavage. This could also be achieved by pre-editing only at sites +4 and +5, but not by pre-editing only at site +40.

Based on these results, one would expect editing to affect mature miR-142 expression. This was tested using ADAR2 knockout mice and mice that had ADAR1 knocked out specifically in B-lineage cells using Cre-Lox recombination (Yang *et al.*, 2006). Although these knockouts did not affect miR-142–3p expression, miR-142–5p expression in mouse spleen increased 2.5-fold upon ADAR1 knockout and 3.3-fold upon ADAR2 knockout. Similarly, miR-142–5p expression in mouse thymus increased 3.0-fold upon ADAR2 knockout. This confirms that miR-142 expression levels are antagonized by editing.

Modulation of stability

As discussed above, pre-editing of pri-miR-142 has been observed to prevent its Drosha processing and cause accumulation of the pri-miRNA. Overexpression of ADAR1p110 or ADAR2 likewise prevented processing of pri-miR-142, but curiously did not cause accumulation of the pri-miRNA. Something degraded edited, but not pre-edited, pri-miR-142.

This distinction between editing and pre-editing suggested processing by Tudor-SN, which exclusively cleaves inosine-containing dsRNA. Indeed, *in vitro* editing of pri-miR-142 increased susceptible to *in vitro* degradation by recombinant Tudor-SN. This degradation gave about eight bands on a Northern blot, suggesting that double-stranded cleavage took place at roughly two sites in the pri-miR-142 stem. Furthermore, treating HEK293 cells overexpressing pri-miR-142 and either ADAR1p110 or ADAR2 with the specific competitive inhibitor of Tudor-SN 2′-deoxythimidine 3′,5′-bisphosphate (pdTp) caused accumulation of edited pri-miR-142 (Yang *et al.*, 2006).

The Tudor-SN cleavage of edited pri-miR-142, which is unable to mature, prevents its accumulation. This further adds to the scope of miRNA antagomism by editing. To this time, edited pri-miR-142 remains the only known example of a natural Tudor-SN substrate.

Medical relevance

Pri-miR-142 is highly expressed in mouse haematopoietic tissues including fetal liver, bone marrow, spleen and thymus, but almost undetectable in non-haematopoietic tissues except lung (Chen *et al.*, 2004). Its overexpression in mouse bone marrow haematopoietic progenitor cells induces differentiation into T-cells without affecting differentiation into B-cells (Chen *et al.*, 2004), whereas overexpression in chronic myeloid leukaemia cell lines attenuates proliferation (Sun *et al.*, 2010). These results suggest induction of haematopoietic differentiation as a role of miR-142, which could implicate lack of miR-142 in cancers.

Indeed, a chromosomal translocation fusing a truncated MYC gene to the 3′ end of the pri-miR-142 gene is associated with B-cell prolymphocytic leukaemia (Gauwerky *et al.*, 1989;

Robbiani *et al.*, 2009). This has earned miR-142 the alternative name B-cell leukaemia/lymphoma 3 (BCL3). Abnormal miR-142 expression levels are also associated with leukaemia and lymphomas, including B-cell acute lymphoblastic leukaemia (B-ALL) (Ramkissoon *et al.*, 2006; Ju *et al.*, 2009; Lv *et al.*, 2011).

By editing miR-142, ADARs could inhibit their ability to induce differentiation. Consequently, overactive editing may be complicit in leukaemia and lymphomas. Interestingly, B-ALL patients have abnormally high ADAR1p110 expression levels, as determined by both quantitative PCR and Western analysis (Ma *et al.*, 2011). Among these patients, higher ADAR1p110 expression is associated with better prognosis and expression levels return to normal in patients achieving complete remission. It is tempting to speculate that pri-miR-142 editing could be part of this ADAR1p110-B-ALL connection.

Pri-miR-151: modulation of Dicer cleavage

Both the 5p and 3p strands miR-151 are highly expressed in retina, brain, lung and spleen (Xu *et al.*, 2007; Luedde, 2010). Pri-miR-151 is edited at two sites (Fig. 6.8) in the mouse nervous system. Human pri-miR-151 is edited at the +3 site in more than 50% of transcripts in cerebral cortex, amygdala and thymus, and in less than 15% of transcripts in cerebellum, lung and spleen (Blow *et al.*, 2006; Kawahara *et al.*, 2007a). In all of these tissues, the −1 site is also edited, though in less than 10% of transcripts. Neither site is edited in HeLa and HEK293 cells (Kawahara *et al.*, 2007a). As discussed previously, there are some indications that miR-151 may also be edited at the pre-miRNA stage. No matter the stage that editing takes place, it inhibits Dicer processing of the pre-miRNA.

Modulation of Dicer cleavage

Sequencing of > 50 clones of mature miR-151-3p derived from human amygdala, mouse cerebral cortex and mouse lung showed that none were edited. By contrast, all pre-miR-151 detected in human amygdala and mouse cerebral cortex was completely edited at the +3 site. This discrepancy suggested that editing prevented Dicer processing of pre-miR-151, which was confirmed by *in vitro* Dicer processing of synthetic pre-miR-151. Inosine at either the −1 or +3 site substantially inhibited Dicer processing, whereas inosine at both sites resulted in no processing at all.

Pri-miR-BART6: modulation of Drosha cleavage and RISC loading

Pri-miR-BART6 (Fig. 6.10) is one of 25 pri-miRNAs known to be encoded by the Epstein–Barr virus (EBV) genome. Four of these – pri-miR-BHRF1–1, pri-miR-BART6, pri-miR-BART8 and pri-miR-BART16 – are known to be edited by ADARs (Iizasa *et al.*, 2010). Pri-miR-BART6 is most highly edited and has therefore been subjected to further studies. 10% of its transcripts are edited in C666–1 nasopharyngeal carcinoma cells, 50% are edited in Daudi Burkitt's lymphoma cells and 70% are edited in GM607 lymphoblastoid cells.

The sequence of pri-miR-BART6 varies between different EBV-infected cell lines (Fig. 6.10) (Iizasa *et al.*, 2010). The pri-miR-BART6 hairpin loop in Daudi or C666-1 cells has a UUU deletion not found in GM607 cells. In the presence of the deletion, editing prevents *in vitro* processing by the Drosha–DGCR8 complex. In the absence of the deletion, editing prevents *in*

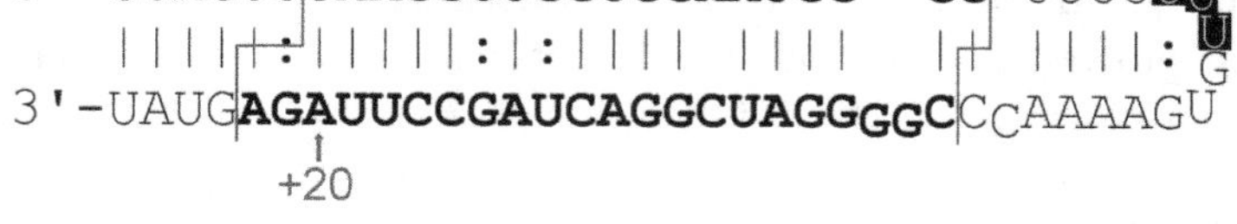

Figure 6.10 EBV pri-miR-BART6. Pri-miR-BART6's editing site is commonly given an identifier based on its position relative to the 5′ end of miR-BART6–3p, as shown. The three uracil residues deleted in Daudi and C666–1 cells are boxed in black.

vitro RISC loading of the miRNA duplex (Iizasa *et al.*, 2010).

In vitro Drosha-DGCR8 and Dicer-TRBP processing of pri-miR-BART6 with or without the UUU deletion and pre-editing has shown that neither deletion nor pre-editing affect processing on their own. However, combination of pre-editing and UUU deletion absolutely prevented Drosha-DGCR8 processing (Iizasa *et al.*, 2010).

The disruption of Drosha–DGCR8 processing could be due to editing preventing either binding or catalytic cleavage of pri-miR-BART6. Nearly all binding of pri-miRNAs by Drosha–DGCR8 seems to be mediated by DGCR8 (Han *et al.*, 2006; Yeom *et al.*, 2006). Mobility shift assays (EMSAs) indicate that pri-miR-BART6 binding by DGCR8 is about 5 nM independent of pre-editing (Iizasa *et al.*, 2010), which suggests that editing affects Drosha catalysis.

Modulation of RISC loading

Following Dicer cleavage, the miRNA duplex is loaded onto RISC, and one strand becomes the guide strand while the other is discarded. For BART6, this process has been shown to be prevented by pre-editing.

Incubation of pre-miR-BART6 with Dicer-TRBP and RISC should have led to Dicer cleavage and RISC loading. Addition of radiolabelled miR-BART6-5p targets allowed the extent of RISC loading to be assessed by target cleavage. Interestingly, pre-editing decreased the extent of target cleavage by two-thirds, indicating that editing prevented miR-BART6-5p RISC loading. This was not due to miR-BART6–3p being loaded instead of the 5p strand, as targets of unedited or edited miR-BART6-3p were not significantly cleaved whether the pre-miR-BART6 used was pre-edited or not.

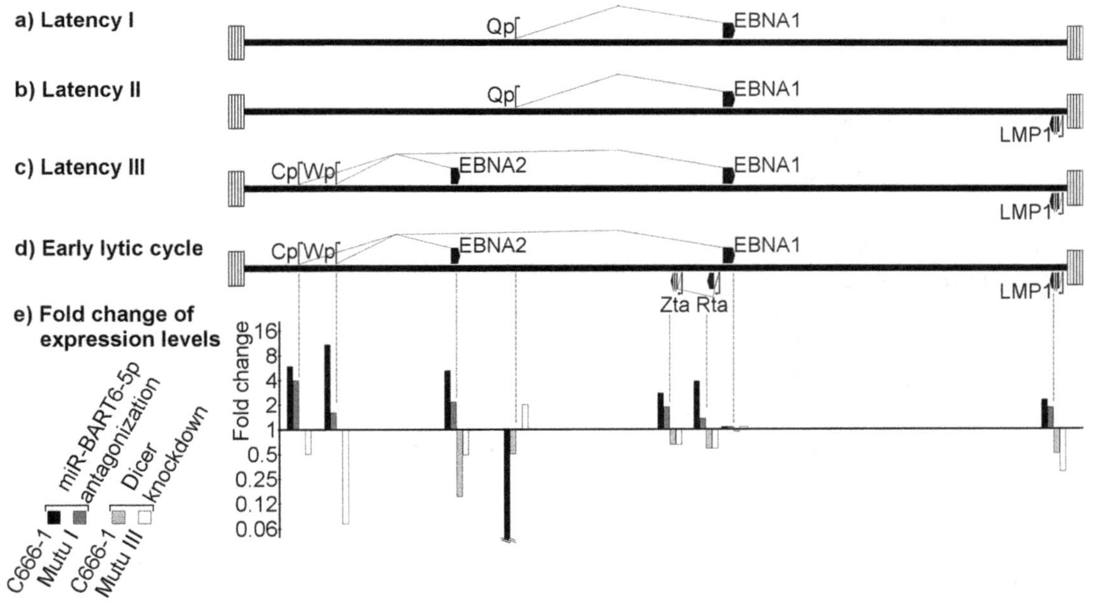

Figure 6.11 Distinctive forms of EBV infections have distinctive transcription programmes. (a) Latency I, (b) latency II, (c) latency III and (d) the early lytic cycle have unique transcription programmes. The proteins expressed and promoters active in each programme are illustrated to scale on the ~172-kb EBV genome. (e) Illustrated are fold changes in Western blot-determined protein levels and quantitative PCR-determined promoter-specific transcript levels upon the indicated treatments. The effect of Dicer knockdown on promoter activities in C666–1 cells has not been determined. Note that miR-BART6–5p antagonism consistently up-regulates factors associated with more aggressive forms of infection and down-regulates factors specific to less immunoresponse-prone forms of infection. Dicer knockdown has the opposite effect.

The mechanism by which RISC loading is prevented is not clear. It is possible that following Dicer cleavage, Dicer is unable to properly hand the miRNA duplex to RISC. Alternatively, RISC may be unable to discard the non-guide strand subsequent to the handover.

Medical Relevance

For EBV pri-miR-BART6 with UUU deletion, editing prevents Drosha cleavage. Without the deletion, editing prevents RISC loading. Either way, editing disrupts BART6 function (Iizasa *et al.*, 2010).

The transcript encoding Dicer is a target of miR-BART6-5p and is therefore up-regulated when editing suppresses miR-BART6-5p expression (Iizasa *et al.*, 2010). Transfection of HEK293 cells with a pri-miR-BART6 expression plasmid down-regulates Dicer protein by ~60%. When the plasmid encodes a pre-edited BART6, the down-regulation is only 25% (Iizasa *et al.*, 2010).

This Dicer up-regulation causes global up-regulation of mature miRNAs. Although the detailed consequences of this remain poorly understood, one important outcome is known. It shifts the virus towards increasingly immunoresponse-prone latency types and the early lytic cycle (Fig. 6.11a–d). This shift has been observed in both C666-1 cells and Mutu I cells upon antagonism of miR-BART6-5p by antagomir treatment (Fig. 6.11e). On the other hand, aiding miR-BART6-5p's function by knocking down Dicer in C666-1 cells and Mutu III cells has been observed to have the opposite effect (Fig. 6.11e).

Why inducing this shift by editing would be advantageous for either the virus or the cell is not clear, but the implications are important (Kutok and Wang, 2006). 95% of adults are infected by EBV, which sometimes causes Hodgkin's lymphoma, Burkitt's lymphoma and nasopharyngeal carcinoma. These tragic outcomes are particularly likely in people who are immunosuppressed due to immunosuppressants, HIV or chronic malaria infections. Additionally, although EBV infection is asymptomatic early in life, later infections are associated with infectious mononucleosis. It is tempting to speculate that antagonism of pri-miR-BART6 editing could aid in treating these ailments.

miR-376 family: modulation of mature miRNA target set

Pri-miRNAs of the miR-376 cluster are transcribed as one transcript from which each is excised as a separate pre-miRNA. They have high sequence similarity, and many are edited at the sites corresponding to sites +4 and +44 of human miR-376a-1 (Figs. 6.12 and 6.13). This editing does not prevent maturation and can therefore produce RISC-loaded mature miRNAs with altered mRNA target sets.

Modulation of mature miRNA target set

Unedited and edited mouse miR-376a-5p are predicted by the algorithm Diana-MicroT2 to have mRNA target sets with only two targets in common (Fig. 6.14) (Kawahara *et al.*, 2007b), and similar results hold for human miR-376a-1-5p.

The reliability of this prediction has been examined by the experimental verification of three random putative targets of the unedited miRNA (TTK, SFRS11 and SLC16A1) and three random putative targets of the edited miRNA (PRPS1, ZNF513 and SNX19). HeLa cells were co-transfected with miR-376a-5p RNAs and luciferase reporter constructs containing the relevant putative target sites in their 3′UTRs. Co-transfection with unedited miR-376a-5p repressed luciferase activity exclusively for TTK, SFRS11 and SLC16A1 3′UTRs. Co-transfection with miR-376a-5p pre-edited at the +4 site repressed luciferase activity exclusively for PRPS1, ZNF513 and SNX19 3′UTRs. Because pri-miR-376a +4 sites are ~50% edited in mouse cortex, but barely edited in mouse liver, these results mean that the target set of pri-miR-376a is altered in a spatially controlled manner to give tissue-specific silencing.

Medical relevance

One mRNA verified to be targeted exclusively by the edited form of mouse miR-376a-5p is phosphoribosylpyrophosphate synthetase 1 (PRPS1), an essential housekeeping enzyme involved in the synthesis of purines. The importance of proper control of PRPS1 expression is demonstrated by a set of X-chromosome-linked disorders (de Brouwer *et al.*, 2010).

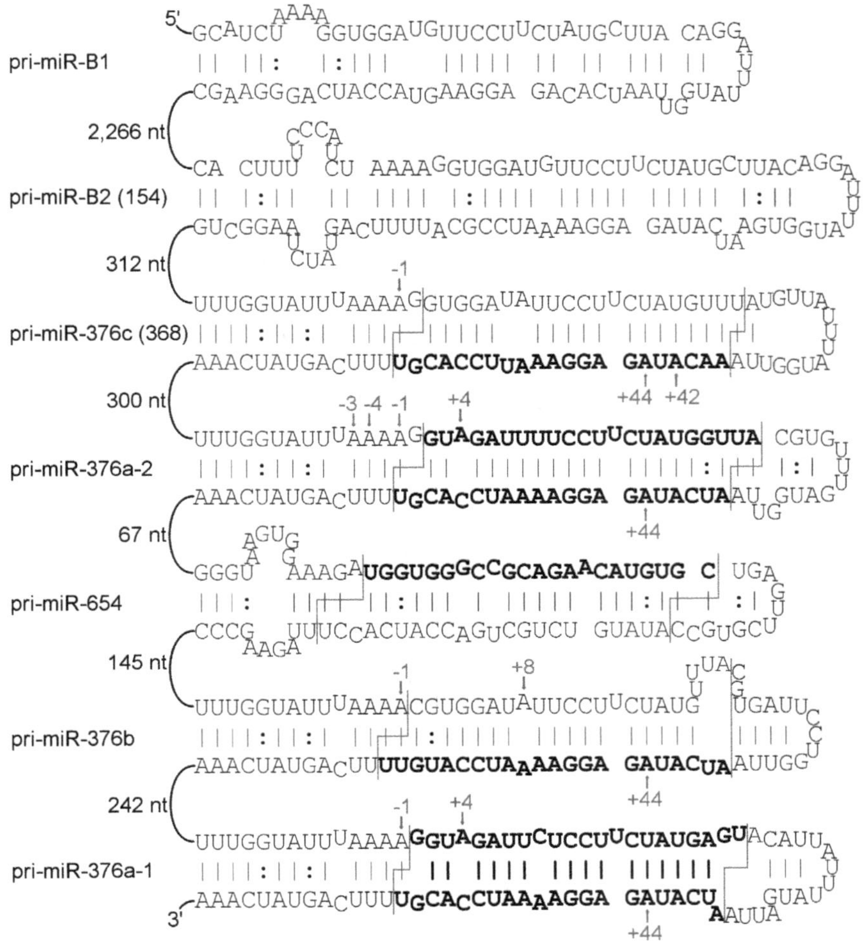

Figure 6.12 The human miR-376 cluster. Editing sites in pri-miR-376a-1 are given identifiers based on their position relative to the 5′ end of miR-376a-1–5p. Editing sites in other miR-376 family members are given designations based on the identifier of the homologous pri-miR-376a-1 site. Pri-miR-B2 is alternatively known as pri-miR-154, while pri-miR-376c is alternatively known as pri-miR-368. Pri-miR-654 does not belong to the miR-376 family, but is still part of the same transcript as the family members. The cleavage sites and resulting mature miRNAs of pri-miR-B1 and pri-miR-B2 remain poorly characterized.

Hypomorphic PRPS1 mutations cause Arts syndrome and Charcot–Marie–Tooth disease. As purines are important to energy storage and transport, this causes problems for tissues requiring large amounts of energy. This can result in neurological problems, immune dysfunction and peripheral neuropathy.

On the other hand, hypermorphic PRPS1 mutations cause phosphoribosylpyrophosphate synthetase superactivity. This usually entails a 2 – to 4-fold increase in PRPS1 levels. Excessive amounts of purines are generated and broken down into excessive amounts of uric acid, which accumulate to cause gout and neurodevelopmental impairment, among other ailments.

The importance of miR-376 family editing to PRPS1 control has been shown by a ~2-fold up-regulation of both PRPS1 and uric acid levels in the cortex of ADAR2 knockout mice. By contrast, PRPS1 and uric acid levels in the liver were unaffected, which shows the tissue-specificity of PRPS1 control by ADAR2 (Kawahara *et al.*, 2007b). Interestingly, there exists a somewhat milder version of phosphoribosylpyrophosphate synthetase superactivity that is due to PRPS1 up-regulation without genetic cause in or around

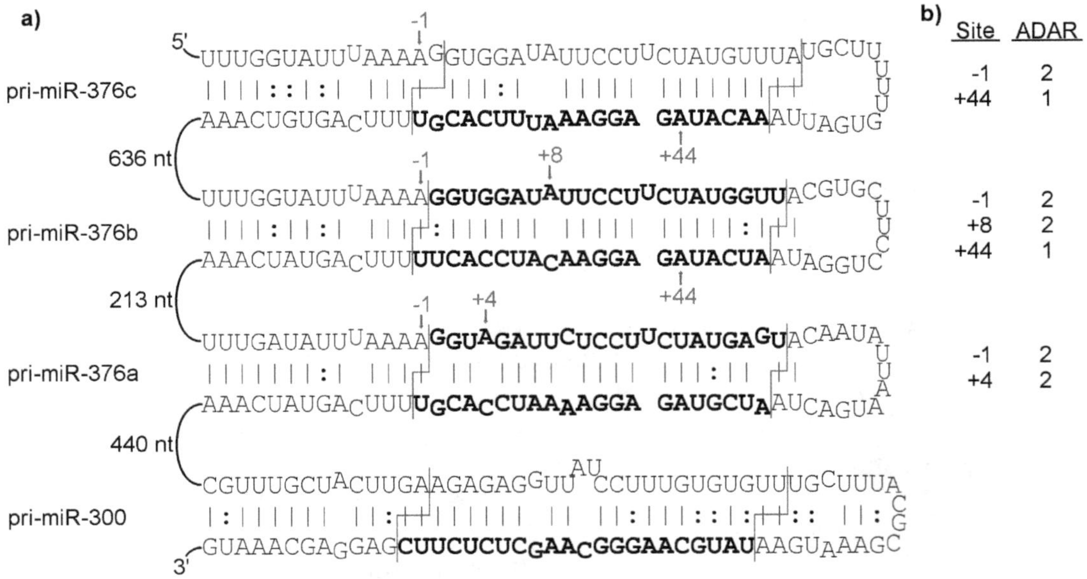

Figure 6.13 The mouse miR-376 cluster. (a) Editing sites in mouse miR-376 family members are given designations based on the designation of the homologous site in human pri-miR-376a-1. Pri-miR-300 does not belong to the miR-376 family, but is still part of the same transcript as the family members. (b) The ADAR responsible for editing each site was determined by examining editing at the site in ADAR2 knockout cortex and ADAR1 knockout E11.5 embryos.

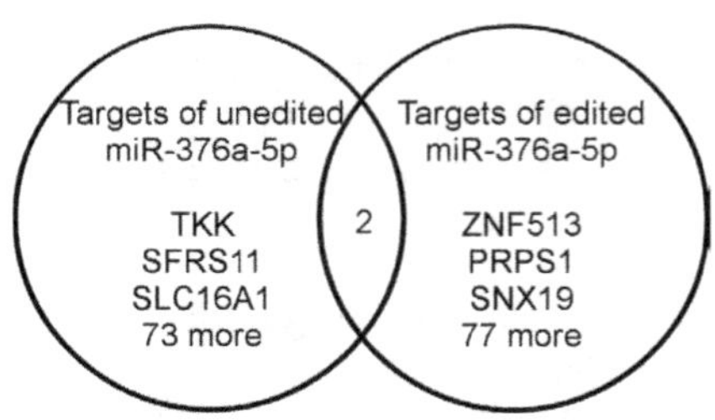

Figure 6.14 mRNA targets of unedited and edited mouse miR-376a-5p. The relationship between the targets predicted for unedited and edited mouse miR-376a-5p is illustrated with a Venn diagram. The experimentally verified targets are named explicitly. TTK is threonine and tyrosine kinase, SFRS11 is arginine/serine-rich splicing factor 11, and SLC16A1 is solute carrier family 16-A1. ZNF513 is zinc finger protein 513, PRPS1 is phosphoribosylpyrophosphate synthetase 1, and SNX19 is sorting nexin 19.

the PRPS1 gene. Anything disrupting the ability of ADAR2 to edit the +4 site of miR-376 family members could potentially be a cause of this ailment.

Identification of miRNA editing sites

miRNA A-to-I editing was first reported for pri-miR-22 from mouse brain and human brain, lung and testis (Luciano *et al.*, 2004). Reverse transcription and PCR amplification across the miRNA hairpin followed by sequencing revealed several editing sites. The editing frequencies, however, were less than 10%. Although the biological consequences of this editing remain unclear, the establishment of pri-miRNAs as possible ADAR substrates prompted more systematic surveys for pri-miRNA editing sites.

Systematic surveys for pri-miRNA editing sites

The first systematic survey for pri-miRNA editing sites considered all 231 human pri-miRNAs registered in miRBase at the time (Blow *et al.*, 2006). Each pri-miRNA's cDNA and corresponding genomic DNA was sequenced for ten different tissues: adult human brain, heart, liver, lung, ovary, placenta, skeletal muscle, small intestine, spleen and testis. This succeeded for 99 pri-miRNAs.

Six pri-miRNAs showed A-to-I editing for at least one tissue: pri-miR-99a, pri-miR-151, pri-miR-379, pri-miR-223, pri-miR-376a-1 and pri-miR-197. This indicates that about 6% of pri-miRNAs are edited. However, this might be an underestimate. The fact that editing in pri-miR-22 was not identified might indicate that low levels of editing could not be detected.

One additional editing site was found in a novel pri-miRNA-like hairpin within the region amplified to sequence pri-miR-374. This hairpin has since been annotated as pri-miR-545 in miRBase. Five additional U-to-C changes were also identified in the hairpins. As no known enzyme could catalyse these conversions, they might represent transcription and editing of the opposite strand. This is possible because the experimental procedure used could not distinguish between reverse complements.

Known editing sites indicate that ADARs prefer to edit UAG motifs. This information guided a second systematic survey (Kawahara *et al.*, 2008). Reverse transcription, PCR amplification and sequencing was attempted for the 257 pri-miRNAs in miRBase at the time that contained UAG motifs in their stems with at most one mismatch either at the U or G. 209 pri-miRNAs were successfully amplified and sequenced from human brain. 43 editing sites were found in UAG motifs and another 43 in non-UAG motifs. In total, 47 pri-miRNAs were edited.

This suggested a new estimate for the frequency of pri-miRNA editing. Using the number of non-UAG editing sites identified to infer the frequency of editing in the 217 known miRNAs at the time not containing UAG motifs indicated that ~16% of pri-miRNAs are edited.

No new systematic surveys of pri-miRNA editing have since been undertaken, even though miRBase now lists 1527 human pri-miRNAs. This means that a large number of recently discovered pri-miRNAs could harbour interesting editing events. Additionally, previous surveys may have missed editing events specific to tissues not investigated.

High-throughput sequencing of mature miRNA

As Sanger sequencing of reverse-transcribed and PCR-amplified pri-miRNA has ebbed off, high-throughput sequencing of mature miRNAs has increasingly provided new data on editing sites. Large-scale sequencing of small RNAs (< 35 nt) has been carried out for a variety of organisms, cell lines and tissues (Ruby *et al.*, 2006; Landgraf *et al.*, 2007; Babiarz *et al.*, 2008; Kuchenbauer *et al.*, 2008; Morin *et al.*, 2008; Suzuki *et al.*, 2009; Chiang *et al.*, 2010; Linsen *et al.*, 2010; Schulte *et al.*, 2010; Berezikov *et al.*, 2011; Peng *et al.*, 2012; Vesely *et al.*, 2012). Many of the sequences generated correspond to mature miRNAs. By comparing these sequences to their genomic origins, bioinformatics approaches could identify novel editing. However, unique identification of a mature miRNA's proper genomic origins remains a major challenge.

Mature miRNAs are only 19–22 nt long, may have 3′ non-templated nucleotide additions and commonly belong to families with high similarity. All of these traits make miRNAs highly susceptible to cross-mapping, whereby a miRNA originating from one genomic locus gets mapped to another (de Hoon *et al.*, 2010). This problem is even more prominent when trying to map an edited miRNA, which by definition will not have a perfect match to its genomic origin. For example, an edited miRNA may align to its proper origin with the only mismatch at the editing site. However, if another member of the same miRNA family encodes a guanosine at this site, proper alignment is unlikely to be achieved and the editing site will not be discovered.

Very stringent mapping of miRNA sequences can to some extent overcome this problem (Chiang *et al.*, 2010; Linsen *et al.*, 2010; Berezikov *et al.*, 2011). The effectiveness of this approach is indicated by overrepresentation of UAG motifs around the resulting putative editing sites and identification of previously validated editing sites. However, this stringency discards all but a few mature miRNA sequences, causing the number of putative editing sites to dwindle.

One algorithm has been developed specifically to overcome this problem (de Hoon *et al.*, 2010). First, sequences are mapped by any conventional method. From the alignments to each genomic origin, the origin's expression level and error profile is estimated. The mapping procedure is repeated with sequences preferentially divided to highly expressed origins with matching error profiles. This procedure is repeated until convergence is achieved. Application of this algorithm to the 236 mature miRNAs sequenced in the FANTOM4 small RNA sequence library indicated that editing took place at only one miRNA, miR-376c from a human macrophage cell line (Suzuki *et al.*, 2009). However, this algorithm may favour alignments that give mismatches near miRNA ends over mismatches that indicate editing, leading to under-reporting of editing. Ultimately, further experiments to distinguish real and false editing sites are therefore needed. This could for example rely on confirming that the editing event is no longer observed following chemical manipulations that prevent amplification of inosine-containing RNA (Sakurai and Suzuki, 2011) or following ADAR knockout.

Effect of ADAR knockout on mature miRNAs

One way to identify edited miRNA is to compare miRNA sequences from ADAR knockout and wild-type animals. Editing sites are more likely real if they disappear upon ADAR knockout. If a mature miRNA is not detectably edited but has its expression increased upon ADAR knockout, it may be edited in a way that prevents its maturation. This relies on the reasonable assumption that effects due to direct editing of miRNAs are large and therefore distinguishable from those due to changes in transcription programmes upon knockout.

One study of this kind has used high-throughput sequencing to compare mature miRNA expression in ADAR2 knockout and ADAR1, ADAR2 double-knockout mice to that in wild-type mice (Vesely *et al.*, 2012). Sequencing of small RNAs from three female E11.5 mouse embryos of each genotype gave 418 mature miRNAs with > 300 reads. The expression of each was normalized by the total number of miRNA reads from that mouse.

Thirty-five per cent of the miRNAs were up-regulated > 1.3-fold upon ADAR2 knockout, whereas only 4% were down-regulated > 1.3-fold. This matches the notion that editing generally antagonizes miRNA maturation. Several of the up-regulated miRNA were known to be edited. For example, miR-133a was up-regulated 1.34-fold and miR-142-5p 1.48-fold. Both are known to have their Drosha cleave inhibited by editing (Yang *et al.*, 2006; Kawahara *et al.*, 2008).

In other cases, the effects of knockout did not match previously determined consequences of editing. For example, even though pri-miR-let-7g ADAR2 editing can hinder its Dicer cleavage (Kawahara *et al.*, 2008), miR-let-7g was down-regulated 1.50-fold by ADAR2 knockout.

An additional hard-to-explain characteristic of the data is that additional knockout of ADAR1 had nearly no overall effect. In some specific cases, the effect was contrary to previous data. For example, mouse miR-411* was up-regulated 1.38-fold by ADAR2 knockout yet only 1.12-fold by ADAR1, ADAR2 knockout, even though pri-miR-411 has previously been shown to be edited by ADAR1 and not by ADAR2 (Kawahara *et al.*, 2008).

One reason for these curious results might be lingering issues with properly aligning miRNAs. In the wild-type mice, A-to-G transitions were not more common than transitions of other types, and they were only slightly more common than A-to-G transitions in knockout mice. This indicates a very large number of transitions not due to editing, which can be both causes for and consequences of misalignment.

Modulation of miRNA target site function by ADARs

As explained previously, editing of a miRNA seed sequence can change its target set by altering its basepairing properties. Editing of target sites might similarly prevent or enable their targeting by a miRNA. This seems especially plausible since both editing and target sites are common in 3′UTRs.

This was first investigated by bioinformatics approaches (Liang and Landweber, 2007). Aligning ~28,000 putative 3′UTR editing sites with predicted miRNA target sites found ~300 seed sequence matches disrupted, and ~200 perfected, by editing. However, only two disrupted matches were edited in the same tissue where the relevant miRNA was expressed.

There are several possible explanations for this low number. Firstly, target sites may be under evolutionary pressure to avoid being edited for some reason not understood. Secondly, the dsRNA structures required for editing may sterically hinder miRNA targeting (Grimson *et al.*, 2007). Thirdly, information on the tissue specificity of miRNA expression and editing is far from complete.

Later approaches have had more success and been experimentally verified (Borchert *et al.*, 2009). One examination of ~12,000 putative human editing sites revealed ~3000 able to perfect a seed sequence match. 258 were in the motif 5′-CCUGUAA-3′ and perfected a match to the miR-513-5p seed sequence. Two hundred and fifty-two were in the motif 5′-AAUCCCA-3′ and perfected a match to a seed sequence common to miR-769-3p and miR-450b-3p. Surprisingly, approximately 190 were in the 12-nt motif 5′-CCUGUAAUCCCA-3′ and perfected a match to both seed sequences. Co-transfecting HEK293 cells with vectors expressing pri-miR-513 or pri-miR-796 and a luciferase transcript with the 12-nt motif allowed the function of this target site to be tested. Indeed, an approximately 50% reduction in luciferase activity was observed when the 12-nt motif was pre-edited.

One gene harbouring this motif in its 3′UTR is DNA fragmentation factor alpha (DFFA). Cloning of DFFA revealed that these motifs are edited in NB7 cells, but not in HEK293 cells. Co-transfection of HEK293 cells with vectors expressing pri-miR-513 or pri-miR-796 and a luciferase transcript with the 3′UTR of DFFA as cloned from NB7 or HEK293 cells repressed luciferase activity only when the 3′UTR originated from NB7 cells. Finally, overexpression of miR-796 in NB7, but not HEK293, cells caused ~60% reduction in endogenous DFFA levels.

Together, these data indicate that the miRNA–mRNA interaction can be regulated by ADARs not only from the miRNA side, but from the mRNA side as well. Recently, high-throughput sequencing identified 2474 editing sites in human 3′UTRs (Peng *et al.*, 2012). Three hundred and ninety-eight of these resided in candidate miRNA target sites whereas an additional 411 resided in sites that would become candidate miRNA target sites once edited.

Conclusions and future trends

ADARs exert important influence over post-transcriptional gene silencing as mediated by small regulatory RNAs.

esiRNA function is antagonized by the editing of their long dsRNAs precursors in the nucleus (Wu *et al.*, 2011). Some of these would otherwise go on to silence genes overlapping their genomic origin. It is not yet known whether this antagonism is constitutive or spatiotemporally regulated. In the prior case, it would be a defence against unwanted, potentially dangerous esiRNAs. In the latter case, it would be a form of control over genes overlapping the long dsRNA's genomic origin. It is also possible that either scenario could be true depending on the long dsRNA in question.

siRNA duplexes, meanwhile, are antagonized by being sequestered by ADAR1p150 in the cytoplasm, with important consequences for siRNA experiments (Yang *et al.*, 2005). This function of ADAR1p150 might explain why ADAR1 is up-regulated by PU.1 upon transfection with siRNA duplexes (T. Wu *et al.*, 2009b). However, it is currently not clear how antagonism of free siRNA duplexes evolved, since cells are unlikely to encounter free-floating siRNA duplexes in nature. Rather, this ADAR1 up-regulation might be triggered by RNAi in general. It could for example be induced by generic Dicer or RISC activity. The details of exactly what RNA species can up-regulate which ADAR1 isoforms therefore remain to be determined.

Many miRNA editing sites have been discovered thanks to systematic surveys for editing sites in pri-miRNAs (Blow *et al.*, 2006; Kawahara *et al.*,

2008). More recently, identification of miRNA editing sites has relied largely on high-throughput sequencing of mature miRNAs (Ruby *et al.*, 2006; Landgraf *et al.*, 2007; Babiarz *et al.*, 2008; Kuchenbauer *et al.*, 2008; Morin *et al.*, 2008; Suzuki *et al.*, 2009; Chiang *et al.*, 2010; Linsen *et al.*, 2010; Schulte *et al.*, 2010; Berezikov *et al.*, 2011; Peng *et al.*, 2012; Vesely *et al.*, 2012). A problem with this approach is that each editing site is reported with low confidence and needs verification. A possible future for mature miRNA editing site detection therefore entails scaling up the verification procedures as well. This could for example rely on confirming that inosine-specific chemistry prevents amplification of edited RNAs (Sakurai and Suzuki, 2011) or ADAR knockout (Vesely *et al.*, 2012) prevents detection of the putative editing event in a second round of high-throughput sequencing.

The editing events studied so far have yielded fascinating insight into how ADARs modulate miRNA function. By altering the sequence of, and bulges in, the pri-miRNA's stem, ADARs can affect Drosha processing (Yang *et al.*, 2006), Dicer processing (Kawahara *et al.*, 2007a), RISC loading (Iizasa *et al.*, 2010) and pri-miRNA stability (Yang *et al.*, 2006). If a mature miRNA is still produced despite all of this, it can target a set of mRNAs different from that of its unedited counterpart (Kawahara *et al.*, 2007b). Several additional possible consequences of pri-miRNA editing seem realistic, but have not yet been discovered. For example, editing could shift Drosha or Dicer cleavage sites, or change what miRNA duplex strand gets loaded onto RISC. This gives additional reason for further studies of miRNA editing events. There is much room for such studies, as only a few pri-miRNA editing sites have been examined in detail, and more are continuously being reported.

The pri-miRNA editing events followed up so far already provide tantalizing implications for human disease. Pri-miR-142 editing could be connected to lymphomas, editing of miR-376 family members affects purine metabolism and editing of pri-miR-BART6 affects the latency of EBV infections. Examining editing in these pri-miRNAs could have potential diagnostic value. Additionally, one would hope that remedying incorrect editing patterns could one day be a medical tool.

However, any such hopes will rely on a better understanding how pri-miRNA editing by ADARs is controlled. ADAR activity can depend on substrate suitability, ADAR expression levels, post-translational ADAR modifications, and subcellular localization of ADARs and substrates. Yet these parameters are still not sufficient to explain observed editing patterns, which vary seemingly randomly across tissues and editing sites (Blow *et al.*, 2006). Based on the need of ADARs to act before co-transcriptional Drosha processing and binding of the RNAPII CTD by ADAR2 (Laurencikiene *et al.*, 2006; Ryman *et al.*, 2007), regulated recruitment of ADARs to transcription sites is a tantalizing possibility.

Acknowledgements

B.-E.W. is supported by a Robert and Marvel Kirby Stanford Graduate Fellowship. K.N. is supported by grants from the US National Institutes of Health, the Ellison Medical Foundation and the Commonwealth Universal Research Enhancement Program, Pennsylvania Department of Health.

References

Ando, Y., Tomaru, Y., Morinaga, A., Burroughs, A.M., Kawaji, H., Kubosaki, A., Kimura, R., Tagata, M., Ino, Y., Hirano, H., *et al.* (2011). Nuclear pore complex protein mediated nuclear localization of dicer protein in human cells. PLoS One *6*, e23385.

Babiarz, J.E., Ruby, J.G., Wang, Y., Bartel, D.P., and Blelloch, R. (2008). Mouse ES cells express endogenous shRNAs, siRNAs, and other Microprocessor-independent, Dicer-dependent small RNAs. Genes Dev. *22*, 2773–2785.

Ballarino, M., Pagano, F., Girardi, E., Morlando, M., Cacchiarelli, D., Marchioni, M., Proudfoot, N.J., and Bozzoni, I. (2009). Coupled RNA processing and transcription of intergenic primary microRNAs. Mol. Cell. Biol. *29*, 5632–5638.

Berezikov, E., Robine, N., Samsonova, A., Westholm, J.O., Naqvi, A., Hung, J.H., Okamura, K., Dai, Q., Bortolamiol-Becet, D., Martin, R., *et al.* (2011). Deep annotation of *Drosophila melanogaster* microRNAs yields insights into their processing, modification, and emergence. Genome Res. *21*, 203–215.

Bhogal, B., Jepson, J.E., Savva, Y.A., Pepper, A.S., Reenan, R.A., and Jongens, T.A. (2011). Modulation of dADAR-dependent RNA editing by the *Drosophila* fragile X mental retardation protein. Nat. Neurosci. *14*, 1517–1524.

Blow, M.J., Grocock, R.J., van Dongen, S., Enright, A.J., Dicks, E., Futreal, P.A., Wooster, R., and Stratton, M.R. (2006). RNA editing of human microRNAs. Genome Biol. *7*, R27.

Borchert, G.M., Gilmore, B.L., Spengler, R.M., Xing, Y., Lanier, W., Bhattacharya, D., and Davidson, B.L. (2009). Adenosine deamination in human transcripts generates novel microRNA binding sites. Hum. Mol. Genet. *18*, 4801–4807.

de Brouwer, A.P., van Bokhoven, H., Nabuurs, S.B., Arts, W.F., Christodoulou, J., and Duley, J. (2010). PRPS1 mutations: four distinct syndromes and potential treatment. Am. J. Hum. Genet. *86*, 506–518.

Chen, C.Z., Li, L., Lodish, H.F., and Bartel, D.P. (2004). MicroRNAs modulate hematopoietic lineage differentiation. Science *303*, 83–86.

Chiang, H.R., Schoenfeld, L.W., Ruby, J.G., Auyeung, V.C., Spies, N., Baek, D., Johnston, W.K., Russ, C., Luo, S., Babiarz, J.E., *et al.* (2010). Mammalian microRNAs: experimental evaluation of novel and previously annotated genes. Genes Dev. *24*, 992–1009.

Fritz, J., Strehblow, A., Taschner, A., Schopoff, S., Pasierbek, P., and Jantsch, M.F. (2009). RNA-regulated interaction of transportin-1 and exportin-5 with the double-stranded RNA-binding domain regulates nucleocytoplasmic shuttling of ADAR1. Mol. Cell. Biol. *29*, 1487–1497.

Gauwerky, C.E., Huebner, K., Isobe, M., Nowell, P.C., and Croce, C.M. (1989). Activation of MYC in a masked t(8;17) translocation results in an aggressive B-cell leukemia. Proc. Natl. Acad. Sci. U.S.A. *86*, 8867–8871.

Gottwein, E., Cai, X., and Cullen, B.R. (2006). A novel assay for viral microRNA function identifies a single nucleotide polymorphism that affects Drosha processing. J. Virol. *80*, 5321–5326.

Gregory, R.I., Yan, K.P., Amuthan, G., Chendrimada, T., Doratotaj, B., Cooch, N., and Shiekhattar, R. (2004). The Microprocessor complex mediates the genesis of microRNAs. Nature *432*, 235–240.

Grimson, A., Farh, K.K., Johnston, W.K., Garrett-Engele, P., Lim, L.P., and Bartel, D.P. (2007). MicroRNA targeting specificity in mammals: determinants beyond seed pairing. Mol. Cell *27*, 91–105.

Gupta, P., Gurudutta, G.U., Saluja, D., and Tripathi, R.P. (2009). PU.1 and partners: regulation of haematopoietic stem cell fate in normal and malignant haematopoiesis. J. Cell. Mol. Med. *13*, 4349–4363.

Han, J., Lee, Y., Yeom, K.H., Nam, J.W., Heo, I., Rhee, J.K., Sohn, S.Y., Cho, Y., Zhang, B.T., and Kim, V.N. (2006). Molecular basis for the recognition of primary microRNAs by the Drosha–DGCR8 complex. Cell *125*, 887–901.

Hartner, J.C., Schmittwolf, C., Kispert, A., Muller, A.M., Higuchi, M., and Seeburg, P.H. (2004). Liver disintegration in the mouse embryo caused by deficiency in the RNA editing enzyme ADAR1. J. Biol. Chem. *279*, 4894–4902.

Hartner, J.C., Walkley, C.R., Lu, J., and Orkin, S.H. (2009). ADAR1 is essential for the maintenance of hematopoiesis and suppression of interferon signaling. Nat. Immunol. *10*, 109–115.

Heale, B.S., Keegan, L.P., McGurk, L., Michlewski, G., Brindle, J., Stanton, C.M., Caceres, J.F., and O'Connell, M.A. (2009). Editing independent effects of ADARs on the miRNA/siRNA pathways. EMBO J. *28*, 3145–3156.

Higuchi, M., Single, F.N., Kohler, M., Sommer, B., Sprengel, R., and Seeburg, P.H. (1993). RNA editing of AMPA receptor subunit GluR-B: a basepaired intron–exon structure determines position and efficiency. Cell *75*, 1361–1370.

Higuchi, M., Maas, S., Single, F.N., Hartner, J., Rozov, A., Burnashev, N., Feldmeyer, D., Sprengel, R., and Seeburg, P.H. (2000). Point mutation in an AMPA receptor gene rescues lethality in mice deficient in the RNA editing enzyme ADAR2. Nature *406*, 78–81.

Hong, J., Qian, Z., Shen, S., Min, T., Tan, C., Xu, J., Zhao, Y., and Huang, W. (2005). High doses of siRNAs induce eri-1 and adar-1 gene expression and reduce the efficiency of RNA interference in the mouse. Biochem. J. *390*, 675–679.

Hong, J., Zhao, Y., Li, Z., and Huang, W. (2007). esiRNA to eri-1 and adar-1 genes improving high doses of c-myc-directed esiRNA effect on mouse melanoma growth inhibition. Biochem. Biophys. Res. Commun. *361*, 373–378.

de Hoon, M.J., Taft, R.J., Hashimoto, T., Kanamori-Katayama, M., Kawaji, H., Kawano, M., Kishima, M., Lassmann, T., Faulkner, G.J., Mattick, J.S., *et al.* (2010). Cross-mapping and the identification of editing sites in mature microRNAs in high-throughput sequencing libraries. Genome Res. *20*, 257–264.

Iizasa, H., Wulff, B.E., Alla, N.R., Maragkakis, M., Megraw, M., Hatzigeorgiou, A., Iwakiri, D., Takada, K., Wiedmer, A., Showe, L., *et al.* (2010). Editing of Epstein–Barr virus-encoded BART6 microRNAs controls their dicer targeting and consequently affects viral latency. J. Biol. Chem. *285*, 33358–33370.

Jepson, J.E., and Reenan, R.A. (2009). Adenosine-to-inosine genetic recoding is required in the adult stage nervous system for coordinated behavior in *Drosophila*. J. Biol. Chem. *284*, 31391–31400.

Jepson, J.E., and Reenan, R.A. (2010). Unraveling pleiotropic functions of A-to-I RNA editing in *Drosophila*. Fly (Austin) *4*, 154–158.

Jepson, J.E., Savva, Y.A., Yokose, C., Sugden, A.U., Sahin, A., and Reenan, R.A. (2011). Engineered alterations in RNA editing modulate complex behavior in Drosophila: regulatory diversity of adenosine deaminase acting on RNA (ADAR) targets. J. Biol. Chem. *286*, 8325–8337.

Jin, Y., Zhang, W., and Li, Q. (2009). Origins and evolution of ADAR-mediated RNA editing. IUBMB Life *61*, 572–578.

Ju, X., Li, D., Shi, Q., Hou, H., Sun, N., and Shen, B. (2009). Differential microRNA expression in childhood B-cell precursor acute lymphoblastic leukemia. Pediatr. Hematol. Oncol. *26*, 1–10.

Kawahara, Y., Zinshteyn, B., Chendrimada, T.P., Shiekhattar, R., and Nishikura, K. (2007a). RNA editing of the microRNA-151 precursor blocks cleavage by the Dicer–TRBP complex. EMBO Rep *8*, 763–769.

Kawahara, Y., Zinshteyn, B., Sethupathy, P., Iizasa, H., Hatzigeorgiou, A.G., and Nishikura, K. (2007b). Redirection of silencing targets by adenosine-to-inosine editing of miRNAs. Science *315*, 1137–1140.

Kawahara, Y., Megraw, M., Kreider, E., Iizasa, H., Valente, L., Hatzigeorgiou, A.G., and Nishikura, K. (2008). Frequency and fate of microRNA editing in human brain. Nucleic Acids Res. *36*, 5270–5280.

Kawamura, Y., Saito, K., Kin, T., Ono, Y., Asai, K., Sunohara, T., Okada, T.N., Siomi, M.C., and Siomi, H. (2008). *Drosophila* endogenous small RNAs bind to Argonaute 2 in somatic cells. Nature *453*, 793–797.

Kim, Y.K., and Kim, V.N. (2007). Processing of intronic microRNAs. EMBO J. *26*, 775–783.

Knight, S.W., and Bass, B.L. (2002). The role of RNA editing by ADARs in RNAi. Mol. Cell *10*, 809–817.

Kuchenbauer, F., Morin, R.D., Argiropoulos, B., Petriv, O.I., Griffith, M., Heuser, M., Yung, E., Piper, J., Delaney, A., Prabhu, A.L., *et al.* (2008). In-depth characterization of the microRNA transcriptome in a leukemia progression model. Genome Res. *18*, 1787–1797.

Kutok, J.L., and Wang, F. (2006). Spectrum of Epstein–Barr virus-associated diseases. Annu. Rev. Pathol. *1*, 375–404.

Landgraf, P., Rusu, M., Sheridan, R., Sewer, A., Iovino, N., Aravin, A., Pfeffer, S., Rice, A., Kamphorst, A.O., Landthaler, M., *et al.* (2007). A mammalian microRNA expression atlas based on small RNA library sequencing. Cell *129*, 1401–1414.

Laurencikiene, J., Kallman, A.M., Fong, N., Bentley, D.L., and Ohman, M. (2006). RNA editing and alternative splicing: the importance of co-transcriptional coordination. EMBO Rep. *7*, 303–307.

Li, C.L., Yang, W.Z., Chen, Y.P., and Yuan, H.S. (2008). Structural and functional insights into human Tudor-SN, a key component linking RNA interference and editing. Nucleic Acids Res. *36*, 3579–3589.

Liang, H., and Landweber, L.F. (2007). Hypothesis: RNA editing of microRNA target sites in humans? RNA *13*, 463–467.

Lima, W.F., Murray, H., Nichols, J.G., Wu, H., Sun, H., Prakash, T.P., Berdeja, A.R., Gaus, H.J., and Crooke, S.T. (2009). Human Dicer binds short single-strand and double-strand RNA with high affinity and interacts with different regions of the nucleic acids. J. Biol. Chem. *284*, 2535–2548.

Linsen, S.E., de Wit, E., de Bruijn, E., and Cuppen, E. (2010). Small RNA expression and strain specificity in the rat. BMC Genomics *11*, 249.

Luciano, D.J., Mirsky, H., Vendetti, N.J., and Maas, S. (2004). RNA editing of a miRNA precursor. RNA *10*, 1174–1177.

Luedde, T. (2010). MicroRNA-151 and its hosting gene FAK (focal adhesion kinase) regulate tumor cell migration and spreading of hepatocellular carcinoma. Hepatology *52*, 1164–1166.

Lv, M., Zhang, X., Jia, H., Li, D., Zhang, B., Zhang, H., Hong, M., Jiang, T., Jiang, Q., Lu, J., *et al.* (2011). An oncogenic role of miR-142-3p in human T-cell acute lymphoblastic leukemia (T-ALL) by targeting glucocorticoid receptor-alpha and cAMP/PKA pathways. Leukemia. *26*, 769–777.

Ma, C.H., Chong, J.H., Guo, Y., Zeng, H.M., Liu, S.Y., Xu, L.L., Wei, J., Lin, Y.M., Zhu, X.F., and Zheng, G.G. (2011). Abnormal expression of ADAR1 isoforms in Chinese pediatric acute leukemias. Biochem. Biophys. Res. Commun. *406*, 245–251.

Macbeth, M.R., Schubert, H.L., Vandemark, A.P., Lingam, A.T., Hill, C.P., and Bass, B.L. (2005). Inositol hexakisphosphate is bound in the ADAR2 core and required for RNA editing. Science *309*, 1534–1539.

Macrae, I.J., Zhou, K., Li, F., Repic, A., Brooks, A.N., Cande, W.Z., Adams, P.D., and Doudna, J.A. (2006). Structural basis for double-stranded RNA processing by Dicer. Science *311*, 195–198.

Morin, R.D., O'Connor, M.D., Griffith, M., Kuchenbauer, F., Delaney, A., Prabhu, A.L., Zhao, Y., McDonald, H., Zeng, T., Hirst, M., *et al.* (2008). Application of massively parallel sequencing to microRNA profiling and discovery in human embryonic stem cells. Genome Res. *18*, 610–621.

Morlando, M., Ballarino, M., Gromak, N., Pagano, F., Bozzoni, I., and Proudfoot, N.J. (2008). Primary microRNA transcripts are processed co-transcriptionally. Nat. Struct. Mol. Biol. *15*, 902–909.

Morse, D.P., Aruscavage, P.J., and Bass, B.L. (2002). RNA hairpins in noncoding regions of human brain and *Caenorhabditis elegans* mRNA are edited by adenosine deaminases that act on RNA. Proc. Natl. Acad. Sci. U.S.A. *99*, 7906–7911.

Okada, C., Yamashita, E., Lee, S.J., Shibata, S., Katahira, J., Nakagawa, A., Yoneda, Y., and Tsukihara, T. (2009). A high-resolution structure of the pre-microRNA nuclear export machinery. Science *326*, 1275–1279.

Patterson, J.B., and Samuel, C.E. (1995). Expression and regulation by interferon of a double-stranded-RNA-specific adenosine deaminase from human cells: evidence for two forms of the deaminase. Mol. Cell. Biol. *15*, 5376–5388.

Pawlicki, J.M., and Steitz, J.A. (2008). Primary microRNA transcript retention at sites of transcription leads to enhanced microRNA production. J. Cell Biol. *182*, 61–76.

Peng, Z., Cheng, Y., Tan, B.C., Kang, L., Tian, Z., Zhu, Y., Zhang, W., Liang, Y., Hu, X., Tan, X., *et al.* (2012). Comprehensive analysis of RNA-Seq data reveals extensive RNA editing in a human transcriptome. Nat Biotechnol. *30*, 253–260.

Provost, P., Dishart, D., Doucet, J., Frendewey, D., Samuelsson, B., and Radmark, O. (2002). Ribonuclease activity and RNA binding of recombinant human Dicer. EMBO J. *21*, 5864–5874.

Ramkissoon, S.H., Mainwaring, L.A., Ogasawara, Y., Keyvanfar, K., McCoy, J.P., Jr., Sloand, E.M., Kajigaya, S., and Young, N.S. (2006). Hematopoietic-specific microRNA expression in human cells. Leuk. Res. *30*, 643–647.

Robbiani, D.F., Bunting, S., Feldhahn, N., Bothmer, A., Camps, J., Deroubaix, S., McBride, K.M., Klein, I.A., Stone, G., Eisenreich, T.R., *et al.* (2009). AID produces DNA double-strand breaks in non-Ig genes and mature B cell lymphomas with reciprocal chromosome translocations. Mol. Cell *36*, 631–641.

Ruby, J.G., Jan, C., Player, C., Axtell, M.J., Lee, W., Nusbaum, C., Ge, H., and Bartel, D.P. (2006). Large-scale sequencing reveals 21U-RNAs and additional microRNAs and endogenous siRNAs in *C. elegans*. Cell *127*, 1193–1207.

Rueter, S.M., Dawson, T.R., and Emeson, R.B. (1999). Regulation of alternative splicing by RNA editing. Nature *399*, 75–80.

Ryman, K., Fong, N., Bratt, E., Bentley, D.L., and Ohman, M. (2007). The C-terminal domain of RNA Pol II helps ensure that editing precedes splicing of the GluR-B transcript. RNA *13*, 1071–1078.

Ryter, J.M., and Schultz, S.C. (1998). Molecular basis of double-stranded RNA–protein interactions: structure of a dsRNA-binding domain complexed with dsRNA. EMBO J. *17*, 7505–7513.

Sakurai, M., and Suzuki, T. (2011). Biochemical identification of A-to-I RNA editing sites by the inosine chemical erasing (ICE) method. Methods Mol. Biol. *718*, 89–99.

Scadden, A.D. (2005). The RISC subunit Tudor-SN binds to hyperedited double-stranded RNA and promotes its cleavage. Nat. Struct. Mol. Biol. *12*, 489–496.

Scadden, A.D., and Smith, C.W. (2001a). RNAi is antagonized by A→I hyperediting. EMBO Rep. *2*, 1107–1111.

Scadden, A.D., and Smith, C.W. (2001b). Specific cleavage of hyperedited dsRNAs. EMBO J. *20*, 4243–4252.

Schulte, J.H., Marschall, T., Martin, M., Rosenstiel, P., Mestdagh, P., Schlierf, S., Thor, T., Vandesompele, J., Eggert, A., Schreiber, S., *et al.* (2010). Deep sequencing reveals differential expression of microRNAs in favorable versus unfavorable neuroblastoma. Nucleic Acids Res. *38*, 5919–5928.

Schwarz, D.S., Hutvagner, G., Du, T., Xu, Z., Aronin, N., and Zamore, P.D. (2003). Asymmetry in the assembly of the RNAi enzyme complex. Cell *115*, 199–208.

Shiohama, A., Sasaki, T., Noda, S., Minoshima, S., and Shimizu, N. (2007). Nucleolar localization of DGCR8 and identification of eleven DGCR8-associated proteins. Exp. Cell Res. *313*, 4196–4207.

Stefl, R., Oberstrass, F.C., Hood, J.L., Jourdan, M., Zimmermann, M., Skrisovska, L., Maris, C., Peng, L., Hofr, C., Emeson, R.B., *et al.* (2010). The solution structure of the ADAR2 dsRBM–RNA complex reveals a sequence-specific readout of the minor groove. Cell *143*, 225–237.

Sun, W., Shen, W., Yang, S., Hu, F., Li, H., and Zhu, T.H. (2010). miR-223 and miR-142 attenuate hematopoietic cell proliferation, and miR-223 positively regulates miR-142 through LMO2 isoforms and CEBP-beta. Cell Res. *20*, 1158–1169.

Suzuki, H., Forrest, A.R., van Nimwegen, E., Daub, C.O., Balwierz, P.J., Irvine, K.M., Lassmann, T., Ravasi, T., Hasegawa, Y., de Hoon, M.J., *et al.* (2009). The transcriptional network that controls growth arrest and differentiation in a human myeloid leukemia cell line. Nat. Genet. *41*, 553–562.

Tonkin, L.A., and Bass, B.L. (2003). Mutations in RNAi rescue aberrant chemotaxis of ADAR mutants. Science *302*, 1725.

Tonkin, L.A., Saccomanno, L., Morse, D.P., Brodigan, T., Krause, M., and Bass, B.L. (2002). RNA editing by ADARs is important for normal behavior in Caenorhabditis elegans. EMBO J. *21*, 6025–6035.

Vesely, C., Tauber, S., Sedlazeck, F.J., von Haeseler, A., and Jantsch, M.F. (2012). Adenosine deaminases that act on RNA induce reproducible changes in abundance and sequence of embryonic miRNAs. Genome Res. *22*, 1468–1476.

Wang, Q., Khillan, J., Gadue, P., and Nishikura, K. (2000). Requirement of the RNA editing deaminase ADAR1 gene for embryonic erythropoiesis. Science *290*, 1765–1768.

Wang, Q., Miyakoda, M., Yang, W., Khillan, J., Stachura, D.L., Weiss, M.J., and Nishikura, K. (2004). Stress-induced apoptosis associated with null mutation of ADAR1 RNA editing deaminase gene. J. Biol. Chem. *279*, 4952–4961.

Weissbach, R., and Scadden, A.D. (2012). Tudor-SN and ADAR1 are components of cytoplasmic stress granules. RNA *18*, 462–471.

Wu, D., Lamm, A.T., and Fire, A.Z. (2011). Competition between ADAR and RNAi pathways for an extensive class of RNA targets. Nat. Struct. Mol. Biol. *18*, 1094–1101.

Wu, H., Ye, C., Ramirez, D., and Manjunath, N. (2009). Alternative processing of primary microRNA transcripts by Drosha generates 5′ end variation of mature microRNA. PLoS One *4*, e7566.

Wu, T., Zhao, Y., Hao, Z., Zhao, H., and Wang, W. (2009). Involvement of PU.1 in mouse adar-1 gene transcription induced by high-dose esiRNA. Int. J. Biol. Macromol. *45*, 157–162.

Xu, S., Witmer, P.D., Lumayag, S., Kovacs, B., and Valle, D. (2007). MicroRNA (miRNA) transcriptome of mouse retina and identification of a sensory organ-specific miRNA cluster. J. Biol. Chem. *282*, 25053–25066.

Yang, W., Wang, Q., Howell, K.L., Lee, J.T., Cho, D.S., Murray, J.M., and Nishikura, K. (2005). ADAR1 RNA deaminase limits short interfering RNA efficacy in mammalian cells. J. Biol. Chem. *280*, 3946–3953.

Yang, W., Chendrimada, T.P., Wang, Q., Higuchi, M., Seeburg, P.H., Shiekhattar, R., and Nishikura, K. (2006). Modulation of microRNA processing and expression through RNA editing by ADAR deaminases. Nat. Struct. Mol. Biol. *13*, 13–21.

Yeom, K.H., Lee, Y., Han, J., Suh, M.R., and Kim, V.N. (2006). Characterization of DGCR8/Pasha, the essential cofactor for Drosha in primary miRNA processing. Nucleic Acids Res. *34*, 4622–4629.

Zamore, P.D., Tuschl, T., Sharp, P.A., and Bartel, D.P. (2000). RNAi: double-stranded RNA directs the

ATP-dependent cleavage of mRNA at 21 to 23 nucleotide intervals. Cell *101*, 25–33.

Zeng, Y., and Cullen, B.R. (2005). Efficient processing of primary microRNA hairpins by Drosha requires flanking nonstructured RNA sequences. J. Biol. Chem. *280*, 27595–27603.

Zhang, X., and Zeng, Y. (2010). The terminal loop region controls microRNA processing by Drosha and Dicer. Nucleic Acids Res. *38*, 7689–7697.

Deaminase-dependent and Deaminase-independent Functions of APOBEC1 and APOBEC1 Complementation Factor in the Context of the APOBEC Family

7

Harold C. Smith

Abstract

Two decades of research revealed the mechanism for site-specific, apolipoprotein B (*apoB*) mRNA C to U editing and its developmental and metabolic regulation. The field began to lose momentum although many open questions remained. This was due to perceived impasses in translational research endpoints:

1. Liver is the most significant organ in the metabolism of cholesterol – and triglyceride-rich lipoproteins and despite active and regulated hepatic editing in rodent models, human liver does not express the cytidine deaminase APOBEC1 required for *apoB* mRNA editing.
2. Mammals express APOBEC1 in their small intestines, where 100% of the *apoB* mRNA is edited in adults, but this activity is constitutive.
3. Expression of APOBEC1 is not essential for life in mice. In the past few years there has been a resurgence in interest because (a) APOBEC1 edits the 3′ UTRs of multiple mRNAs and either alone or together with its RNA-binding cofactor, A1CF, may regulate mRNA stability and translation in diverse tissues, (b) A1CF is required for embryological development, acting through a mechanism that may be unrelated to APOBEC1 and (d) discovery of dC to dU DNA mutational activity by APOBEC1 raises new questions of its oncogenic potential.

This review will consider past and current discoveries relative to the exciting new research opportunities in the field.

Discoveries that point to the next frontier

APOBEC1 (A1) and A1 complementation factor (A1CF) primarily have been studied in the context of the system in which they were first discovered, *apoB* mRNA C to U editing (reviewed in Smith, 1998; Blanc and Davidson, 2011; Smith *et al.*, 2012) (Fig. 7.1). Mouse knockout studies demonstrated that A1CF is an essential gene, affecting embryo viability at the pre-implantation stage (Blanc *et al.*, 2010). In the same study, RNAi knockdown of A1CF in McArdle rat hepatoma cells to 70% of control cell levels induced apoptosis. These findings were unanticipated as A1, the catalytic subunit and deaminase required for *apoB* mRNA editing (Teng *et al.*, 1993), is not an essential gene product (Hirano *et al.*, 1996; Nakamuta *et al.*, 1996; Fujino *et al.*, 1998; Xie *et al.*, 2003). ApoB however is an essential gene product required for lipoprotein assembly and transport within the embryo yolk sack endoderm as well as in adult tissues (Farese *et al.*, 1995; Veniant *et al.*, 1999). From the perspective of human liver, which does not express A1 or *apoB* mRNA editing activity but has maintained expression of ApoB and A1CF (Lellek *et al.*, 2000; Mehta *et al.*, 2000), an alternative function for A1CF as an RNA-binding protein has been suspected. These

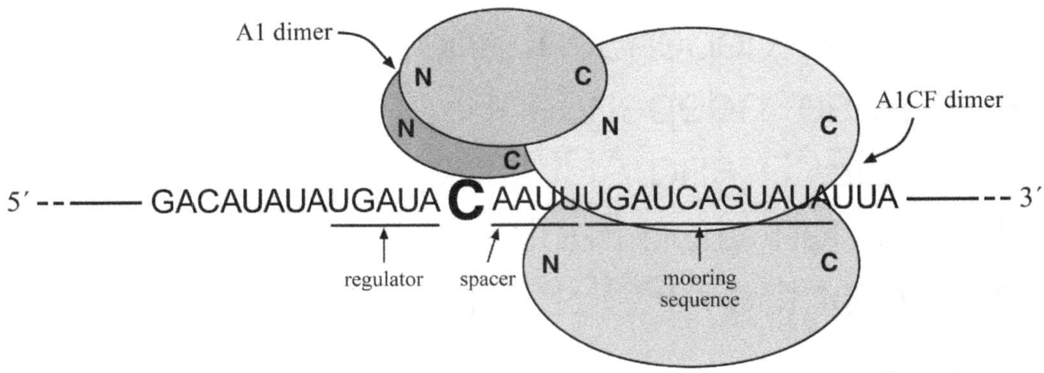

Figure 7.1 Model of the C to U editosome. Shown is a model of what may be the minimal composition of a catalytically active C to U editosome. The sequence of *apoB* mRNA tripartite editing motif is shown with the 5′ proximal regulator, the edited C6666 (in larger bold font), 3′ spacer sequence and 3′ mooring sequence. A1 and A1CF monomers are 27kDa and 64kDa as represented. A head-to-head monomer alignment of an RNA-bridged A1CF dimer is shown bound to the mooring sequence. A C-terminal dimer of A1 bound to RRMs of an A1CF monomer positions A1 for site-specific C to U editing where the spacer sequence provides an appropriate positioning of C6666 relative to an A1 catalytic site. Although shown as a dimer of side-by-side ovals, A1 is predicted to have an elongated, dimeric structure in the absence of A1CF interactions that are mediated by protein–protein interactions through the C-termini of each A1 monomer. The model does not rule out that A1 binds to RNA sequence or that the catalytic site in contact with RNA at C6666 may require structural contributions from each monomer in the dimer. The model does not rule out the possibility that each A1CF monomer individually binds to an A1 monomer or that each A1CF monomer may bind to an A1 dimer (i.e. predicted stoichiometry of A1:A1CF of 1:1 or 2:1, respectively).

functions may include deaminase-independent modulation of mRNA stability and translation by binding to 3′ UTRs (Blanc *et al.*, 2010), regulating nuclear export of mRNA (Sowden *et al.*, 2002; Galloway *et al.*, 2010a) and what may be a deaminase-dependent suppression of nonsense codon mediated mRNA decay (NMD) (Chester *et al.*, 2003). Therefore, while A1CF's role in *apoB* mRNA editing is likely to be non-essential, its requirement for development implies a yet-to-be described essential role in other cellular functions.

A1 has experienced a renaissance with the discovery that it can function autonomously *in vitro* and in both bacterial and eukaryotic cells as a single-stranded DNA (ssDNA) cytidine deaminase (Harris *et al.*, 2002) and reviewed in (Smith *et al.*, 2012). A1 has lax sequence preference when binding to ssDNA and does not require A1CF for ssDNA deaminase activity (Harris *et al.*, 2002; Beale *et al.*, 2004). As a member of the APOBEC family of 11 cytidine deaminases, many of which have activity on ssDNA and ability to bind RNA, A1 appears to be uniquely able to edit RNA (Wedekind *et al.*, 2003; Smith *et al.*, 2012). The APOBEC family of enzymes is characterized by deaminase-dependent and deaminase-independent functions including diversification of immunoglobulin genes during an immune response (Bransteitter *et al.*, 2003; Conticello *et al.*, 2007; Hamilton *et al.*, 2010) and diversification of the cancer cell genome (Skuse *et al.*, 1996; Yamanaka *et al.*, 1997; Harris *et al.*, 2002; Blanc *et al.*, 2007; Robbiani *et al.*, 2008; Chiarle *et al.*, 2011; Klein *et al.*, 2011) and anti-retroviral and anti-endogenous retroviral element host defence activities that are dependent on ssDNA and RNA binding (Harris and Liddament, 2004; Navarro *et al.*, 2005; Huthoff and Malim, 2007; Khan *et al.*, 2007; Goila-Gaur *et al.*, 2008; Huthoff *et al.*, 2009) and reviewed in Hamilton *et al.* (2010), Smith (2011) and Smith *et al.* (2012). Added to this list should be A1 binding to AU-rich elements within mRNA 3′ UTRs (Anant and Davidson, 2000; Anant *et al.*, 2004) that may also include editing of 3′ UTRs (Rosenberg *et al.*, 2011).

Gene and protein structure of A1 and A1CF

A1: a cytidine deaminase and nucleic acid-binding protein

The catalytic domain of A1 (NM_001644.3) and the mechanism of zinc-dependent hydrolytic cytidine deamination have been reviewed (Wedekind *et al.*, 2003; Smith *et al.*, 2012). The cytidine deaminase domain of A1 has not been structurally determined but models have been proposed based on amino acid sequence homology and predicted super secondary structure using crystal structures of the bacterial cytidine deaminase that is active on nucleotides/nucleosides (Navaratnam *et al.*, 1998), yeast Cdd1 cytidine deaminase that is active on free nucleotides as well as ssDNA and RNA (Smith, 1998; Xie *et al.*, 2004), APOBEC2, which has no known editing substrate (Prochnow *et al.*, 2007; Krzysiak *et al.*, 2012) but may be involved in other forms of cellular and tissue regulation (Etard *et al.*, 2010; Sato *et al.*, 2010; Lada *et al.*, 2011; Vonica *et al.*, 2011) and the catalytic domain of APOBEC3G that is a cytidine deaminase on ssDNA (Shandilya *et al.*, 2010) and reviewed in (Zhang *et al.*, 2007; Chelico *et al.*, 2008; Bransteitter *et al.*, 2009; Harjes *et al.*, 2009; Autore *et al.*, 2010; Chelico *et al.*, 2010). The oligomeric state of A1 family members in solution remains controversial with data suggesting a catalytically active monomer or elongated homomultimers with subunits interacting through protein-protein and/or protein-nucleic acid contacts (Kreisberg *et al.*, 2006; Opi *et al.*, 2006; Wedekind *et al.*, 2006; Chen *et al.*, 2007; Prochnow *et al.*, 2007; Soros *et al.*, 2007; Zhang *et al.*, 2007; Bennett *et al.*, 2008; Chelico *et al.*, 2008, 2010; Goila-Gaur *et al.*, 2008; Nowarski *et al.*, 2008; Bransteitter *et al.*, 2009; Friew *et al.*, 2009; Harjes *et al.*, 2009; Huthoff *et al.*, 2009; Salter *et al.*, 2009; Shandilya *et al.*, 2010; McDougall *et al.*, 2011; Shlyakhtenko *et al.*, 2011; Krzysiak *et al.*, 2012; Smith *et al.*, 2012).

A1CF is a member of the Elav/HelN1/HuR family of RNA-binding proteins

A1CF (NM_014576.3) is encoded on mouse chromosome 19 within 80 kb predicted to include 15 exons (Dur *et al.*, 2004; Blanc *et al.*, 2010). Transcription is initiated from a TATA-less promoter from multiple sites within a cluster of Sp1 sites (Dur *et al.*, 2004). Translation of full-length A1CF is initiated within exon 2 and exons 5, 6 and 11 are spliced out. Alternatively spliced *a1cf* mRNA variants have been detected in all tissues and are differentially expressed throughout development, with 75–89% of the spliced variants encoding functional proteins (Farese *et al.*, 1995; Dance *et al.*, 2002; Dur *et al.*, 2004; Sowden *et al.*, 2004). In liver there is one dominant variant (80% of the total *a1cf* mRNA) encoding a 64 kDa protein (ACF64) and eight minor expressed variants (Lellek *et al.*, 2000; Dance *et al.*, 2002; Dur *et al.*, 2004; Sowden *et al.*, 2004; Blanc *et al.*, 2010). There is 94% amino acid identity (and 98% similarity) between human and mouse ACF64. RNA recognition motifs (RRM) are encoded by exons 3–4, 7–8 and 8–10 and these are all essential for A1CF complementation of A1 in *apoB* mRNA editing (Farese *et al.*, 1995; Lellek *et al.*, 2000; Mehta *et al.*, 2000; Blanc *et al.*, 2001a; Dance *et al.*, 2002; Mehta and Driscoll, 2002; Sowden *et al.*, 2004). The predominant A1CF proteins detected in intestinal cells are smaller (49 to 41 kDa) but include all three RRMs (Teng and Davidson, 1992; Harris *et al.*, 1993; Henderson *et al.*, 2001; Sowden *et al.*, 2004).

A1CF high- and low-molecular-weight protein variants were first characterized through ultraviolet light cross-linking assays for its ability to selectively bind to an 11-nt RNA sequence motif required for *apoB* mRNA editing (the mooring sequence) within *apoB* mRNA or heterologous RNA constructs containing the mooring sequence (Shah *et al.*, 1991; Driscoll *et al.*, 1993; Harris *et al.*, 1993; Navaratnam *et al.*, 1993; Mehta *et al.*, 1996). Molecular cloning of *a1cf* cDNA (Lellek *et al.*, 2000; Mehta *et al.*, 2000) enabled proof that A1CF bound to A1 and *apoB* mRNA, and was necessary and sufficient to complement A1 in *apoB* mRNA editing *in vitro* and in cells (Farese *et al.*, 1995; Lellek *et al.*, 2000; Mehta *et al.*, 2000; Sowden *et al.*, 2004). A1CF bound to the mooring sequence with a high-affinity (K_D of 8 nM) (Mehta *et al.*, 1996; Mehta and Driscoll, 2002) and recruited A1 C to U deaminase activity to *apoB* mRNA by the formation of an editing complex for site-specific C to U editing, know as the C to U editosome (Backus and Smith, 1991; Smith *et al.*, 1991; Sowden *et al.*, 2002).

The domains in ACF64 required for RNA and A1 binding/complementing activity are as follows: (i) amino acids 1–129, A1 binding; (ii) RRM1 (58–123) and RRM2 (138–208), the majority of surface for binding to *apoB* mRNA; (iii) RRM3 (208–315) enhancer for *apoB* RNA-binding; and (iv) sequences following RRM3 contain a non-canonical nuclear import motif required for nuclear localization and editing activity (360–420) (Blanc *et al.*, 2003), an RG-rich region (380–442) and a weak double-stranded RNA-binding domain (446–523) that further enhanced *apoB* RNA binding (Blanc *et al.*, 2001a; Mehta and Driscoll, 2002).

AU-rich sequences within UTRs are common locations for binding sites for the Elav/HelN1/HuR family of RNA-binding proteins (Burd and Dreyfuss, 1994; Brennan and Steitz, 2001; Kielkopf *et al.*, 2004). Alignment of ACF64 RRM1 and RRM2 with the 1.8 Å crystal structure of HuD RRM1-RRM2 bound to an 11 nt segment of AU-rich RNA (Wang and Tanaka Hall, 2001) revealed that 23% of the aligned ACF64 residues were identical to HuD and 51% were similar (Lehmann *et al.*, 2007). The HuD RRM structure was typical in that it consisted of a four-stranded anti-parallel β-sheet supported by two α helices. RNA binding has been attributed to the β sheets and protein–protein interactions through the α helices or *vice versa* (Burd and Dreyfuss, 1994; Brennan and Steitz, 2001; Kielkopf *et al.*, 2004). The β sheets of HuD RRM1 and RRM2 bound single-stranded RNA. NMR analysis also predicted that the RRMs in A1CF have a structural fold consisting of four antiparallel β-strands packed by two alpha-helices in a $\beta_1\alpha_1\beta_2\beta_3\alpha_2\beta_4$ topology (Maris *et al.*, 2005a,b). A1CF RRM interaction with RNA were predicted to require conserved aromatic residues in β-strands 1 and 3 that contained the canonical RNA-binding motifs RNP2 and RNP1, respectively. Although one RRM may be sufficient to bind to a minimum of two nucleotides, as exemplified by AlkB8 (Pastore *et al.*, 2012), CBP20 (Calero *et al.*, 2002; Mazza *et al.*, 2002) and Nucleolin RRM2 (Johansson *et al.*, 2004), it is a more common occurrence that there are multiple copies of RRMs. This may enable an increased specificity and/or affinity for RNA substrates.

Various studies have sought to understand the role of three RRMs in the known function of A1CF, editosome assembly. A1CF truncated to amino acid 304 still bound to *apoB* RNA (Maris *et al.*, 2005a). A1CF binding to A1 only could be demonstrated for A1CF truncations 1–380, 1–384 and 1–391 (Blanc *et al.*, 2001a; Mehta and Driscoll, 2002; Galloway *et al.*, 2010b). ACF truncated beyond 377 failed to bind A1 (Mehta and Driscoll, 2002). Correspondingly, *in vitro apoB* mRNA editing activity was reduced by truncating the C-terminus of A1CF beyond amino acid 380 and was lost when any one of the three RRMs had been deleted (Mehta and Driscoll, 2002; Maris *et al.*, 2005a). Portions of RRMs 2 and 3 have been implicated in A1CF interaction with APOBEC1 include amino acids 144–257 (Blanc *et al.*, 2001a; Mehta and Driscoll, 2002). Insulin-dependent phosphorylation of serine 154 within RRM2 enhanced A1CF binding to A1 and A1CF complementation of RNA editing (Lehmann *et al.*, 2006, 2007). The multiple RRMs within A1CF are also involved in A1CF dimerization as recently shown in live cell FRET studies (Galloway *et al.*, 2010b) (Fig. 7.1). Given these studies, a reasonable supposition is that all three RRMs are required for complementation of editing activity in living cells (Blanc *et al.*, 2001a, 2003; Mehta and Driscoll, 2002; Sowden *et al.*, 2004; Galloway *et al.*, 2010). However the C-terminal portion of ACF may modulate RRM function and/or A1CF folding as all A1CF truncation variants do not support A1 complementation as well as ACF64 (Blanc *et al.*, 2001a, 2003; Mehta and Driscoll, 2002; Sowden *et al.*, 2004).

RNAs bound by A1 and A1CF and functional implications

Nucleic acids as substrates and/or binding partners

It is a common held belief that A1 cannot edit RNA in the absence of A1CF. However, recombinant A1 actually can carry out low efficiency editing on reporter RNAs containing the *apoB* mRNA editing site when the reactions were heated to 45°C (Chester *et al.*, 2004). It has been suggested that A1CF is required for robust editing in cells because A1 prefers to bind single-stranded AU-rich RNA (Driscoll and Zhang, 1994; Anant *et al.*, 1995, 2004; Fujino *et al.*, 1998; Navaratnam *et al.*, 1998; Teng *et al.*, 1999) and A1CF facilitates A1 access to these sequences by binding to the mooring sequence and melting duplex RNA (Chester *et al.*, 2004; Maris *et al.*, 2005b).

Amino acids in and proximal to the zinc-dependent cytidine deaminase domain located centrally within A1 were required for low affinity binding to RNA (K_D > 400 nM) (MacGinnitie *et al.*, 1995; Navaratnam *et al.*, 1995; Navaratnam *et al.*, 1998; Anant and Davidson, 2000). Non-conservative mutations of H61, E63, F66, F87 and C96 severely impaired A1 binding to RNA (Fujino *et al.*, 1998; Navaratnam *et al.*, 1998; Teng *et al.*, 1999; Anant *et al.*, 2004). A1 also required the zinc-dependent deaminase domain for dC to dU editing of ssDNA and it is likely that RNA and ssDNA binding to A1 and other APOBEC family members have similar amino acid sequence requirements (Harris *et al.*, 2002; Beale *et al.*, 2004; Xie *et al.*, 2004; Chen *et al.*, 2007, 2008; Huthoff and Malim, 2007; Prochnow *et al.*, 2007; Bransteitter *et al.*, 2009; Rausch *et al.*, 2009; Autore *et al.*, 2010; Carpenter *et al.*, 2010; Chelico *et al.*, 2010; Shandilya *et al.*, 2010; Bulliard *et al.*, 2011). This is not to say that we know whether RNA and ssDNA bind in an similar manner to APOBEC proteins, or in the case of APOBEC proteins with two deaminase domains, whether RNA and ssDNA bind to the same or different domains within these proteins (Hache *et al.*, 2005; Navarro *et al.*, 2005; Kreisberg *et al.*, 2006; Huthoff and Malim, 2007; Khan *et al.*, 2007; Zhang *et al.*, 2007; Goila-Gaur *et al.*, 2008; Huthoff *et al.*, 2009; McDougall and Smith, 2011; Smith *et al.*, 2012). Given the apparent autonomous nucleic acid binding and editing capability of A1, the question of what biological role this may play is highly significant.

Whether A1 autonomously can bind to and edit RNA or requires A1CF to bind RNA has become a very important question because of some interesting biology that has been associated with A1 and A1CF. A1 expression increased the half live of both *c-myc* (Anant and Davidson, 2000) and *cox-2* (Anant *et al.*, 2004) mRNAs by a mechanism that is currently only known for its dependency on A1 binding to AU-rich sequences within the 3′ UTRs of these mRNAs. Whether A1CF and a mooring sequence are required for A1-dependent mRNA stabilization has not been determined. A1CF increased the half-life of *IL-6* mRNA during liver regeneration through a mechanism that required RRM2 and RRM3 binding to AU-rich elements in the 3′ UTR of *IL-6* mRNA (Blanc *et al.*, 2010). RNA editing is not required for *IL6* mRNA stabilization. Elav/HelN1/HuR family of proteins are known to bind to AU-rich elements (Burd and Dreyfuss, 1994; Brennan and Steitz, 2001; Kielkopf *et al.*, 2004) and stabilize or destabilize various mRNAs depending on the tissue type, thereby modulating cell type specific protein expression (Burd and Dreyfuss, 1994; Brennan and Steitz, 2001).

Transcriptome-wide comparative RNA sequencing (RNA-Seq) of non-genomic C to U transitions revealed novel editing events within mRNAs expressed in the small intestine in wild-type mice compared with $a1^{-/-}$ mice (Rosenberg *et al.*, 2011). The analysis did not identify C to U disparities within coding sequences; however, nucleotide transitions were predicted within the 3′ UTR of several mRNAs, and 32 C to U sites were validated as being edited in intestinal cells (Table 7.1). These non-genomically encoded C to U transitions were identified within AU-rich sequences, and editing of these sites was A1 and mooring sequence dependent, with transcript editing frequency varying from 20% to >90%. Analysis of mRNA editing in other cell types that express A1 will be necessary before the prevalence of 3′ UTR editing and the mechanism whereby mRNAs are selected for A1 interaction can be determined. In addition, the role of A1CF in selecting which cytidines within 3′UTRs were edited remains to be determined.

In the early days of the field, before A1CF was discovered, multiple proteins were identified that bind to A1, A1CF or RNA and either stimulated or inhibited editing activity (reviewed in Smith *et al.*, 2005; Blanc and Davidson, 2010) (Table 7.2). It is not known whether these proteins can direct A1 editing activity to subsets of mRNAs. Such RNA recognition interactions may or may not require an *apoB*-like mooring sequence (Yamanaka *et al.*,

Table 7.1 Examples of mRNAs that are C to U edited in their 3′ UTRs

Gene[a]	Protein	Consensus function
2010106E10Rik	Uncharacterized expressed transcript	?
Aldh6a1	Mitochondrial aldehyde dehydrogenase, nuclear gene	Valine and pyrimidine oxidative catabolic pathways
Ank3	Ankyrin 3	Links integral membrane proteins to spectrin/actin cytoskeleton, binds to neurofascin and voltage-gated sodium channels
App[b]	Amyloid precursor protein	Membrane protein, synaptic formation and repair
Atf2	Activating transcription factor 2	Leucine zipper DNA-binding transcription factor and histone H2B, H4 acetyltransferase
B2m	Beta-2-microglobulin	Peripheral membrane MHC protein
BC003331	Uncharacterized expressed transcript	?
BC013529	Uncharacterized expressed transcript	?
Bche	Butyrylcholinesterase	Non-specific cholinesterase that hydrolyses many different choline esters in the ER
Casp6	Capase 6	Cysteine-aspartic acid protease
Clic5	Mitochondrial chloride ion channel 5, nuclear gene	Chloride ion channel
Cyp4v3	Cytochrome P450 family 4 subfamily v, polpeptide 3	Endoplasmic reticulum monooxygenase
Dpyd[b]	Dihydropyrimidine dehydrogenase	Rate-limiting enzyme in the pathway of uracil and thymidine catabolism
Gramd1c	GRAM domain-containing 1C	Integral membrane protein
Hprt1	Hypoxanthine guanine phosphoribosyl transferase	Conversion of hypoxanthine to inosine and guanine to monophosphates
Iqgap2[b]	IQ motif-containing GTPase-activating protein 2	Regulator of cell morphology and motility
Lrrc19	Leucine-rich repeat containing 19	Activates NF-kappaB and induces production of proinflammatory cytokines
Mfsd7b	Major facilitator superfamily domain-containing 7B	Membrane protein, haem transport
Ptpn3	Protein tyrosine phosphatase, non-receptor type 3	Phosphatase in signal transduction
Rb1	Retinoblastoma 1	Tumour suppressor, cell cycle regulation
Rnf128	Ring zinc finger protein 28	E3 ubiquitin ligase activity
Rrbp1	Ribosome-binding protein variant 1	Ribosome-associated protein
Sep15	Selenoprotein 15	Redox and protein folding regulator
Serine1	Serine incorporator 1 membrane protein	Membrane protein, L-serine transport, phospholipid metabolism
Sh3bgrl	SH3 domain-binding, glutamic acid-rich-like	SH3/SH2 adapter activity, SH3 domain building
Sult1d1	Sulfotransferase 1D	Sulfur conjugating enzyme
Tmbim6	Transmembrane BAX inhibitor motif-containing 6	Membrane protein, negative regulator of apoptosis
Tmem30a	Transmembrane protein 30A	Endoplasmic reticulum membrane protein regulator of ER protein export
Usp25	Ubiquitin-specific peptidase 25	Deubiquitinating enzyme

[a]Data compiled from (Rosenberg *et al.*, 2011).
[b]An mRNA with more than one C to U editing event in its 3′ UTR.

Table 7.2 Proteins shown to modulate apoB mRNA editing

Protein	Role
Essential for editing	
A1CF[a]	Binds A1/RNA; site specific *apoB* mRNA editing
Stimulates editing	
GRY-RBP	Binds A1/A1CF/RNA
HnRNP A/B (ABBP-1)	Binds A1/RNA
Hsp70 (ABBP-2)	Protein chaperone
KSRP	Alternative splicing factor; binds RNA
Inhibits editing	
αl2 serum protease inhibitor (p240) BAG-4	Protein sequesterants
ARCD1	Binds A1/A1CF
CUGBP-2	Binds A1/RNA
DnaJ	Binds A1
GRY-RBP	Binds A1/A1CF/RNA
HnRNP C and D	Binds A1/RNA

[a]A1CF refers specifically to ACF65 and ACF64. Other spliced variants of A1CF do not have equivalent ability to bind the mooring sequence or A1 and therefore may not all be essential for editing (Dur *et al.*, 2004; Sowden *et al.*, 2004).

1996). Therefore selection of mRNAs for editing may or may not follow known paradigms.

Clearly 3′ UTRs are not the only sites for A1 or A1CF binding and editing activity (Table 7.3). Of note, mooring motifs that are predictive for editing in 3′UTRs are also widespread in coding regions (Rosenberg *et al*, 2011); however, the prevalence of C to U editing of protein coding sequence or non-messenger RNA editing may be low. Site-directed mutagenesis studies of the cis-acting elements required for *apoB* mRNA editing provided a detailed characterization of the cis-acting sequence requirements for *apoB* mRNA editing (Chen *et al.*, 1990; Backus and Smith, 1991, 1992, 1994; Shah *et al.*, 1991; Driscoll *et al.*, 1993; Backus *et al.*, 1994; Sowden *et al.*, 1996b, 1998; Hersberger and Innerarity, 1998; Hersberger *et al.*, 1999; Nakamuta *et al.*, 1999). The most important element for editing and A1CF RNA binding is the mooring sequence (UGAUCAGUAUA: nt 6671–6681) that is the 3′ most element of a 21-nt tripartite motif that also comprises an enhancer element (immediately 5′ of C6666) and a spacer element between C6666 and the mooring sequence. RNAs that otherwise did not support editing, could be made to do so when the mooring sequence was inserted immediately 3′ of a cytidine (Backus and Smith, 1991, 1994; Driscoll *et al.*, 1993; Sowden *et al.*, 1998). The significance of the tripartite motif for editing site utilization has been demonstrated by *in vitro* mutagenesis of apoB mRNA, transient transfection, in transgenic mice and the finding that chicken *apoB* mRNA is not edited in any species because of a degenerate mooring sequence (Backus and Smith, 1991, 1992; Shah *et al.*, 1991; Driscoll *et al.*, 1993; Backus *et al.*, 1994; Sowden *et al.*, 1996b, 1998; Hersberger and Innerarity, 1998; Hersberger *et al.*, 1999; Sowden and Smith, 2001). Another naturally occurring mooring sequence-dependent editing site was identified in *apoB* mRNA at nt 6802. Unlike C6666 (where a Gln codon is edited to a translation stop codon), editing of 6802 created a sense change but C6802 editing was associated with C6666 editing therefore unlikely to be expressed in a protein (Navaratnam *et al.*, 1991). These findings prompted a search for mooring sequence homologues and led to a serendipitous discovery of a novel mooring sequence-dependent, edited mRNA encoding the tumour suppressor involved in neurofibromatosis type I mRNA (Skuse *et al.*, 1996; Mukhopadhyay *et al.*, 2002). NF1 mRNA editing is predicted to disrupt the function of

Table 7.3 Examples of mRNA coding sequences with mooring sequence homologues

Mouse
ApoB
BMI-1 proto-oncogene protein
Cell adhesion molecule CD44
Coagulation factor XI
Collagen type XI
Cyclo-oxygenase
Cytokeratin
Deoxycytidine kinase
Oestrogen receptor
ETVI transcription factor
Fatty acid synthase
Gamma-aminobutyraldehyde dehydrogenase
Giantin
Glycosylphosphatidylinositol-anchored protein
Histidase
HLA-DMB
Lysosomal acid lipase
Myosin regulator light chain
Methylthioadenosine phosphorylase
NAT-1
NF-AT4c transcription factor
Non-clathrin-coated vesicular coat protein
P1 protein
Prostaglandin synthase homolog
RAG-1 transcription factor
Rb tumour suppressor
Retinoic acid-inducible E3
Serum albumin
Serine/threonine kinase receptor 2
tITIN
Transcription factor ISGF-3
Tyrosine kinase TEC
YAP tumour suppressor
Zfx transcription factor
Human
ApoB
MTAP
NF1
RAG-1
Rb (C1744/C1745)
Rb (C2103/C2104)
Rb (C2221)

Mouse
Rat[a]
ApoB
ApoB alternate site (6802)
NF-1
Chicken
ApoB homologue
Guinea pig
ApoB
Pig
ApoB

[a]Additional comparative studies of species-specific apoB mRNA mooring sequences can be found in (Greeve *et al.*, 1993; Hersberger *et al.*, 1999; Nakamuta *et al.*, 1999).

the encoded protein (neurofibromin) as a Ras regulatory protein (GAP) due to the conversion of an Arg codon to a STOP codon within the GTPase activating domain. Cells derived from glioblastoma, neurofibrosarcoma, neuroblastoma, chronic myelogenous leukaemia and lymphoblastoma edited NF1 and apoB mRNA upon transfection. McArdle rat hepatoma cells also edited NF1 mRNA upon transfection. Whether NF1 mRNA editing in non-hepatic cell lines requires A1 and A1CF is unresolved (Skuse *et al.*, 1996; Mukhopadhyay *et al.*, 2002).

Liver-specific overexpression of A1 led to promiscuous editing of *apoB* mRNA (Sowden *et al.*, 1996a, 1998) and hyper editing of other mRNAs (Yamanaka *et al.*, 1995, 1996, 1997). Even with this deregulation, editing was mooring sequence and A1CF dependent (Sowden *et al.*, 1998). Among the mRNAs affected were those encoding the kinase Tec (Yamanaka *et al.*, 1996) and the translation factor eIF4G (Yamanaka *et al.*, 1995) with multiple sites of editing causing several sense changes. In these studies, hyper editing due to the high levels of A1 overexpression induced hyperplastic and neoplastic disease. In a different context, overexpression of A1 in HIV-infected cells resulted in C to U editing of the RNA viral genome and this appeared to be independent of A1CF and the mooring sequence (Bishop *et al.*, 2004).

Given these occurrences of exon editing, another approach to predicting the prevalence of C to U editing is to search for mooring sequence homologues. In studies carried out in 1998, the editing efficiency of naturally occurring editing sites, promiscuous and hyper editing sites and the functional outcome from site-directed mutagenesis of the *apoB* mRNA editing site, were compiled as a weighted matrix (Smith *et al.*, 2005). Each nucleotide position of the 11 nucleotide mooring sequence was given a score from 0 to 10 based on the efficiency of editing observed on RNAs containing that nucleotide variant within the mooring sequence (nucleotides within the *apoB* mooring sequence that supported the highest level of editing were scored as 10). This matrix suggested a consensus mooring sequence of UGAUpy(A/T)NN(A/T)pyN. Several studies have demonstrated that the efficiency of editing not only depended on the mooring sequence, but increased with the AT-richness of the 5′ and 3′ flanking sequences (Backus and Smith, 1991, 1994; Smith, 1993; Hersberger and Innerarity, 1998; Hersberger *et al.*, 1999) and decreased with increasing proximity of GC-rich secondary structure (Backus and Smith, 1994) and editing site proximity to RNA splice junctions (Sowden *et al.*, 1996b; Yang *et al.*, 2000; Sowden and Smith, 2001). This information was used in conjunction with the annotated, non-redundant cDNA database of human, mouse and rat available at that time for the computational prediction of high homology mooring sequences in the transcriptome.

Approximately 100 ‘hits’ were obtained in each species consisting of largely known mRNAs (<20% were unknown ORFs) that either contained a perfect match to the *apoB* mRNA mooring sequence or deviated by one or two nucleotides. Seventy per cent of the mooring sequence homologues were located within the 5′ or 3′ UTRs of mRNAs, with the remainder within coding regions (representative results are shown in Table 7.3). Although several coding sequences from this list were evaluated in transfected, editing-competent, McArdle rat hepatoma cells, none of the transcripts expressed from cDNAs supported mooring sequence-dependent C to U editing (unpublished results). Some of these tripartite editing motifs would have been split across exon junctions and according to the gating hypothesis (Sowden *et al.*, 1996b), the efficiency of editosome assembly and editing at these sites would be impaired by the assembly and function of pre-mRNA splicing complexes (Sowden *et al.*, 1996b; Sowden and Smith, 2001). This of course does not explain why they could not support editing when expressed in transfected cells as fully processed transcripts (cDNAs). Together with the recent RNA-seq analysis (Rosenberg *et al.*, 2011), the data suggest that only a few exon sequences support C to U editing (Table 7.3).

Cell and tissue biology

A1 and A1CF are mobile proteins within the cell

A1 and A1CF shuttle between inactive 60S cytoplasmic complexes and active 27S nuclear complexes (Sowden *et al.*, 1996b) that are assembled on nascent RNAs (Lau *et al.*, 1991; Sowden *et al.*, 1996b). Shuttling and *apoB* mRNA editing are regulated hormonally and metabolically (Harris *et al.*, 1993; Yang and Smith, 1997; Yang *et al.*, 2001; Sowden *et al.*, 2002; Blanc *et al.*, 2003; Chester *et al.*, 2003; Lehmann *et al.*, 2006; Galloway *et al.*, 2010a). Although both A1 (Lau *et al.*, 1994; Fujino *et al.*, 1998; Teng *et al.*, 1999) and A1CF (Galloway *et al.*, 2010b) form homomultimers, their interactions and stoichiometry within 60S and 27S complexes has not been determined fully (Fig. 7.1). Nuclear localization of A1 is dependent on a bipartite nuclear localization in the N-terminus and a unique domain ('M') located in the middle of A1 (Yang *et al.*, 2001). Together these function as a weak nuclear localization signal (NLS) (Yang and Smith, 1997; Yang *et al.*, 2001). It remains unclear whether A1 can enter the nucleus in the absence of A1CF due the apparent limited molecular mass that the NLS of A1 will import (Yang and Smith, 1997; Yang *et al.*, 2001) and the presence of a powerful C-terminal, CRM1-independent cytoplasmic localization signal (Yang *et al.*, 2001; Blanc *et al.*, 2003). It remains to be determined whether A1 uses its own NLS for nuclear import or binds to A1CF and co-imports using the NLS in A1CF (Yang and Smith, 1997; Chester *et al.*, 2003).

From the observation that human liver A1CF shuttles in the absence of A1 (Galloway *et al.*, 2010a) it is reasonable to propose that A1CF does not require A1 for nuclear import. A1CF enters the nucleus in a hypophosphorylated state but nuclear retention is metabolically regulated by protein kinase C-dependent serine phosphorylation (Lehmann *et al.*, 2007). Nuclear retention of A1CF enhanced A1 binding to A1CF and editing activity but also resulted in nuclear retention of edited and unedited *apoB* mRNA, thereby reducing ApoB translation and lipoprotein secretion from hepatocytes (Galloway *et al.*, 2010a). Also, dephosphorylation of A1CF by protein phosphatase 1 was required for nuclear export (Lehmann *et al.*, 2006, 2007). A1 is predicted to be phosphorylated at two different sites that may differentially inhibit or stimulate *apoB* mRNA editing (Chen *et al.*, 2001). In support of these predictions, the activities of other APOBEC family deaminases also were modulated by phosphorylation (Basu *et al.*, 2005; Douaisi *et al.*, 2005; Chatterji *et al.*, 2007; Shirakawa *et al.*, 2008; Demorest *et al.*, 2011).

A1CF may remain bound to the AU-rich mooring sequences throughout the lifespan of *apoB* mRNA. This is suggested by the finding that: (i) nuclear retention of A1CF resulted in retention of *apoB* mRNA (Galloway *et al.*, 2010a), (ii) A1CF 60S complexes are localized on the cytoplasmic surface of the rough endoplasmic reticulum (Sowden *et al.*, 2002), (iii) A1CF stabilizes edited *apoB* mRNA from NMD (Chester *et al.*, 2003; Blanc *et al.*, 2010) and (iv) AICF promoted mRNA translation (Chester *et al.*, 2003; Blanc *et al.*, 2010; Galloway *et al.*, 2010a).

Cytoplasmic A1 may be localized with cytoplasmic A1CF as biochemical isolated inactive cytoplasmic 60 complexes can be reactivated *in vitro* to form editing-competent 27S editosomes (Harris *et al.*, 1993; Yang *et al.*, 1997; Sowden *et al.*, 2002). The macromolecular composition of 60S complexes is not known nor is it apparent why normally they lack editing activity on cytoplasmic mRNA. Nuclear import and post-translational modification of editing factors alone may not be sufficient to explain the lack of mooring sequence-dependent *apoB* mRNA editing in the cytoplasm because ectopic overexpression of A1 in editing competent cells induced cytoplasmic mRNA editing (Yang *et al.*, 2000). These findings suggest that the abundance and/or the stoichiometry of A1 to A1CF may be a critical determinant for editing activity. However there have been reports where hepatic editing activity can be induced by ethanol without *de novo* protein synthesis (Giangreco *et al.*, 2001) (presumably due to A1CF hyperphosphorylation; Lehmann *et al.*, 2006) and thyroid hormone-dependent conditions wherein the expression of A1 and A1CF became rate-limiting for editing in intestine but not in liver (Inui *et al.*, 1994). It also is unclear why overexpression of A1 was not sufficient to induce editing in cells that natively did not edit *apoB* mRNA (Yang and Smith, 1997; Yang *et al.*, 2001; Galloway and Smith, 2010) even when A1CF was co-expressed (Galloway and Smith, 2010). Given that A1, A1CF and *apoB* mRNA can be co-expressed with little or no editing in certain cells types (see the next section), an interesting possibility is that the regulation of A1 and A1CF may require coordinate regulation of other cellular activator or suppressor proteins that have been shown to bind to A1, A1CF or both (Table 7.1). Although these proteins have other functions in the cell, they can affect editing activity on *apoB* mRNA when overexpressed (Schock *et al.*, 1996; Lau *et al.*, 1997, 2001; Greeve *et al.*, 1998; Lellek *et al.*, 2000; Anant, 2001a,b; Blanc *et al.*, 2001b; Lau, 2001; Lau and Chan, 2003).

A1CF and A1 expression in adult tissues and during development

Analysis of adult tissues by Northern blotting and/or Western blotting with A1CF peptide-specific antibodies demonstrated broad expression (Demmer *et al.*, 1986; Wu *et al.*, 1990; Greeve *et al.*, 1993; Farese *et al.*, 1995; Funahashi *et al.*, 1995; Lellek *et al.*, 2000; Mehta *et al.*, 2000; Dur *et al.*, 2004; Blanc *et al.*, 2010) that was not ubiquitous (most notably lacking expression was placenta). A1CF is an essential gene product as complete *a1cf* gene ablation resulted in the failure of 3.5 day blastocysts to develop normally and implant (Blanc *et al.*, 2010). This finding supported a very early requirement for A1CF during development that was not linked to A1 editing or *apoB* mRNA. A1CF expression was haplosufficient for development but adult *a1cf* +/− mice may be deficient in their capacity for tissue repair as suggested by their inability to regenerate liver following partial hepatectomy (Blanc *et al.*, 2010).

A1 also has a broad tissue expression that is determined by multiple TATA-less promoters and tissue-specific expression of alternatively spliced mRNAs (Wu *et al.*, 1990; Driscoll and Zhang, 1994; Funahashi *et al.*, 1995). The expression of A1CF in adult tissues was not always accompanied by expression of A1 or *apoB* mRNAs, suggesting that in the adult there need not be coordinate regulation. The relative expression levels of A1CF and A1 and *apoB* mRNA in adult tissues also was not a predictor for whether a tissue supported high, low or no editing activity (Teng and Davidson, 1992; Greeve *et al.*, 1993; Driscoll and Zhang, 1994; Funahashi *et al.*, 1995; Siddiqui *et al.*, 1999; Steinburg *et al.*, 1999; Galloway and Smith, 2010). Hormonal and metabolic regulation of the editing factors at the level of their expression, alternative splicing, post-translational modification and shuttling to the nucleus frequently determined assembly of functional editosomes on *apoB* mRNA and the level of mRNA editing activity (Baum *et al.*, 1990; Lau *et al.*, 1991; Inui *et al.*, 1992; Funahashi *et al.*, 1995; Phung *et al.*, 1996; Sowden *et al.*, 1996b; Yang and Smith, 1997; von Wronski *et al.*, 1998; Chen *et al.*, 2001; Giangreco *et al.*, 2001; Dance *et al.*, 2002; Sowden *et al.*, 2002, 2004; Blanc *et al.*, 2003; Lehmann *et al.*, 2006, 2007; Galloway *et al.*, 2010a).

During development the expression of *a1* and *a1cf* mRNAs in mice is regulated. (Huguchi *et al.*, 1992). Editing of *apoB* mRNA remained at low levels but showed a sharp increase in intestine and liver on day 18 of gestation and 15 days post partum respectively (Farese *et al.*, 1995; Funahashi *et al.*, 1995). It is logical to postulate that the molecular mechanism for the expression of editing activity during development involves the timing of *a1* and *a1cf* mRNA expression. However there was no correlation with the timing of their expression and the onset of *apoB* mRNA editing activity nor did all tissue express the editing factors at the same time during development (Demmer *et al.*, 1986; Huguchi *et al.*, 1992; Inui *et al.*, 1992; Farese *et al.*, 1995; Funahashi *et al.*, 1995). A possible exception to this apparently discordant regulation is in liver where the induction of hepatic *apoB* mRNA editing postpartum more closely paralleled the expression of *a1 and a1cf* mRNAs that rose after the first day postpartum (Farese *et al.*, 1995; Blanc *et al.*, 2010). Taken together, the data suggested that the expression levels of A1 and A1CF in a given tissue may not be predictive of the level of editing activity. The regulation of the interaction of editing factors with each other and with RNA substrates, and the circumstances leading to the induction of editing activity remain very much an open question.

ApoB mRNA expression is essential for development beyond midgestation (day 10.5) (Xie *et al.*, 2003). A1 however is not an essential gene product. Homogenous *a1* knockout mice developed normally with only subtle perturbations in lipoprotein profiles in response to their diets (Hirano *et al.*, 1996; Xie *et al.*, 2003). The induction of editing activity in intestine could not be explained by *a1* mRNA expression that remained low in fetal intestine until day 21 (Funahashi *et al.*, 1995). During development, intestinal (and hepatic) expression of *a1cf* mRNA was relatively constant (Farese *et al.*, 1995; Blanc *et al.*, 2010). In fact, *in vitro* differentiation of the human intestinal Caco2 cell line also showed that although expression of A1 and A1CF could be induced by transfection, this was not sufficient to activate *apoB* mRNA editing (Galloway and Smith, 2010) until cells became differentiated (Farese *et al.*, 1995; Galloway and Smith, 2010). Therefore, the induction of *apoB* mRNA editing during development could not be explained by the developmental timing of the expression of the intestinal editing factors.

Conclusions

In conclusion, a few exon sequences are known to support C to U editing activity and several mRNAs are edited in their 3′ UTRs. As with many forms of RNA modification, A1 and A1-dependent C to U RNA editing are non-essential. The mechanism and regulation of C to U mRNA editing is known in the context of *apoB* mRNA. Recognition of C6666 in *apoB* mRNA for site-specific editing requires the interaction of A1CF and A1 and their binding to a 3′ mooring sequence to assemble a 27S editosome. A1, A1CF and *apoB* mRNA are expressed in several tissues but the reason why some tissues support mooring sequence-dependent editing and others do not is not known. Additional control mechanisms may determine the onset of editing activity because expression of apoB mRNA, A1 and A1CF proteins was clearly not sufficient for robust editing activity during the early stages of intestine and liver development. Editing may be important qualitatively in the adult organism and in this regard numerous metabolic regulatory mechanisms have been described that modulate *apoB* mRNA editing activity, mRNA stability and mRNA translation. In this regard, the numerous spliced variants for A1CF and A1 and the regulation of their expression hormonally and during development suggest a purpose in tissue function. Traditionally, the field has been asking what this means in terms of the induction and regulation of C to U *apoB* mRNA editing. However what the author hopes will be brought to light in this review is the importance of categorizing A1 and A1CF functionality in terms of deaminase-dependent (involving C to U editing) and deaminase-independent (A1 and A1CF binding but not editing) activities. This terminology has been helpful in focusing research approaches and delineating data interpretation for the activities of other APOBEC family members (Smith, 2009; Smith *et al.*, 2012). In addition, the lack of developmental coordination in the expression of A1, A1CF and *apoB* mRNA and RNA editing activity and the requirement for A1CF expression during embryological development suggest that there is a bigger picture, perhaps one in which the expression of A1 and A1CF serves a function(s) in different cell types or at different stages of development that do not fit within our established mechanistic models for these proteins.

Future trends

A major limitation in the field is the lack of structural information. A1 is difficult to express as a soluble, mono-dispersed protein and in sufficient abundance for structural studies. This is perhaps the explanation for why there are crystal structures for a 40 amino acid N-terminal truncation of APOBEC2 (Prochnow *et al.*, 2007), an NMR structure for full-length A2 (Krzysiak *et al.*, 2012), the C-terminus of APOBEC3G (Shandilya *et al.*, 2010) and the yeast A1 homologue Cdd1 (Xie *et al.*, 2004) but not for A1. Based on these structures for A1 homologues, A1 could either form an elongated dimeric structure interacting through protein-protein contacts through either its N-terminus or C-terminus (i.e. like N-terminally truncated A2 or native full-length A3G), a monomer (i.e. full-length A2) or compact tetramer (i.e. Cdd1). A greater effort towards structural studies with a goal of crystallizing A1 or A1CF alone, or A1 or A1CF bound to RNA/ssDNA, or A1/A1CF/RNA complexes are essential for understanding the structural constraints governing site-specific nucleic acids interaction, not only for these proteins but for other members of the APOBEC family as well.

Conditional *a1cf* knockout mice need to be developed and used to determine exactly when and why A1CF is essential for life beyond the blastula. If the answer involves RNAs that A1CF binds to and affects their stability, subcellular localization and/or translation, then the regulation of these RNAs could become the next frontier of discovery in developmental biology and genetic diseases.

We need to understand how A1 and A1CF regulate mRNA stability and translation by binding to the 3′ UTR and how A1CF inhibits NMD. Are these mechanisms deaminase dependent, and do they involve changes in targeting of mRNAs by RNAi? Alternatively, if the mechanism is deaminase independent, what is that mechanism? It is easy to envisage a deaminase-independent function where A1 or A1CF might bind to AU-rich RNA sequences and thereby physically block some interactions while recruiting others. Likewise, it can be envisaged how A1 deaminase–dependent interactions with RNAs might be site-specific if they are directed by A1CF and the mooring sequence or another RNA binding protein that can tether A1 to an editing site. Data show that A1 can edit cytidines that are as far as 50 nt 5′ of a mooring sequence given an interaction with A1CF (Sowden *et al.*, 1996a). It is less obvious how A1 alone might select mRNAs for binding and deamination other than perhaps a random scanning mechanism (Pham *et al.*, 2011). Given that mRNA is coated with protein in a living cell, it is likely that A1 uses proteins other than A1CF to target mRNAs. Studies should focus on how the lax RNA and ssDNA sequence requirements for A1 binding enable A1 to select its substrates and whether A1 always requires an AU-rich, mooring sequence-like motif for editing mRNAs (Rosenberg *et al.*, 2011) and hyperediting (Yamanaka *et al.*, 1996; Chen *et al.*, 2010) or promiscuous editing (Sowden *et al.*, 1996a, 1998) of mRNAs.

Conservation in the mooring sequence across species for any given mRNA only was observed for 25% of the predicted sequences. It is anticipated therefore that mooring sequence-dependent binding of A1 or A1CF will be species-specific if not also tissue-specific. Therefore, editing site utilization should be evaluated in cell types and tissues were transcripts are natively expressed. If the mooring sequence is not the only means by which A1 and A1CF bind to mRNAs, then it may be more difficult to predict mRNA binding partners. This would be particularly true if A1 binding or C to U editing is not for all circumstances dependent on the mooring sequence and A1CF. It may be possible for distal mooring sequences to drive editing of cytidines that apparently are not adjacent to a mooring sequence if A1CF RNA binding and dimerization bridged the two by folding RNA. At this point, other than *apoB, nat1, tec* and *nf1* mRNA coding sequences, the occurrence of additional exon sequence C to U editing and A1 or A1CF binding to these sequences remain open questions.

The determinants for A1 and A1CF intracellular shuttling and their regulation remain to be determined. What is the macromolecular composition and organization of inactive cytoplasmic 60S complexes and active nuclear 27S complexes? Do A1 and A1CF import to the nucleus alone or as complexes? Answers to these questions are important for understanding not only how editing activity and editosome assembly are regulated, but also how A1 and A1CF interact with RNAs and regulate their abundance and expression. Subcellular localization of other APOBEC family members also serves to regulate their interactions and activities (reviewed in Smith *et al.*, 2012).

A1 is non-essential for mouse development and other members of the APOBEC family [e.g. AID (Muramatsu *et al.*, 1999, 2000), A2 and A3 (Mikl *et al.*, 2005; Sato *et al.*, 2010)] also are not required for mouse development. Whether APOBEC proteins also are non-essential in other model organisms or primates remains to be determined.

Why all members of the APOBEC family bind RNA as well as ssDNA but only A1 edits RNA, is an open question in terms of the functional utility of binding RNA. AID, APOBEC3F and APOBEC3G ssDNA deaminase activities are inactivated by RNA binding (Bransteitter *et al.*, 2003; Soros *et al.*, 2007; Stopak *et al.*, 2007; Hamilton *et al.*, 2010) and reviewed in (Smith, 2009) and in the example of APOBEC3G (A3G), RNA and ssDNA binding to A3G are mutually exclusive events (McDougall and Smith, 2011).

Addressing the above-mentioned open questions may reveal novel therapeutic targets. Low or moderate A1 overexpression in mice has been shown to increase hepatic editing and decrease circulating ApoB100-containing very low-density lipoprotein (VLDL) and low-density lipoprotein (LDL) (the bad forms of cholesterol) without inducing cancer (Teng *et al.*, 1997; Qian *et al.*, 1998; Yang *et al.*, 2002). As additional A1 and A1CF mechanisms become apparent, endogenously expressed proteins might be targeted by small molecules for research purposes or as chemical scaffolds for drug discovery in obesity-induced diabetes, atherosclerosis or disorders in development. A1 and A1CF are members of protein families with family themes for structure and function. Therefore, the open questions described here concern topics that will be of broad impact and should be of general interest.

Acknowledgements

The author thanks Nina Papavasiliou and Ryan Bennett for their helpful suggestions and Jenny M.L. Smith for the preparation of the figure and tables. The work described from the author's laboratory was supported by Public Health Services Grants NIDDK DK43739, NIAID AI54369, AI095007, AI58789, NINDS NS067671, an Air Force Office of Scientific Research Grant F49620 and a Bill and Melinda Gates foundation Grand Challenges in Exploration grant awarded to H.C.S. Additional support was from Developmental Center for AIDS Research grant (NIAID P30 078498) awarded to Steve Dewhurst, a Public Health Services Grant NIAID AI076085 awarded to Joseph E. Wedekind.

References

Anant, S., MacGinnitie, A.J., and Davidson, N.O. (1995). apobec-1, the catalytic subunit of the mammalian apolipoprotein B mRNA editing enzyme, is a novel RNA-binding protein. J. Biol. Chem. *270*, 14762–14767.

Anant, S., and Davidson, N.O. (2000). An AU-rich sequence element (UUUN[A/U]U) downstream of the edited C in apolipoprotein B mRNA is a high-affinity binding site for Apobec-1: binding of Apobec-1 to this motif in the 3′ untranslated region of c – myc increases mRNA stability. Mol. Cell. Biol. *20*, 1982–1992.

Anant, S., Henderson, J.O., Mukhopadhyay, D., Navaratnam, N., Kennedy, S., Min, J., and Davidson, N.O. (2001a). Novel role for RNA-binding protein CUGBP2 in mammalian RNA editing. J. Biol. Chem. *276*, 47338–47351.

Anant, S., Mukhopadhyay, D., Sankaranand, V., Kennedy, S., Henerson, J.O., Davidson,N.O. (2001b). ARCD-1, an apobec-1-related cytidine deaminase, exerts a dominant negative effect on C to U mRNA editing. Am. J. Physiol. Cell Physiol. *281*, 1904–1916.

Anant, S., Murmu, N., Houchen, C.W., Mukhopadhyay, D., Riehl, T.E., Young, S.G., Morrison, A.R., Stenson, W.F., and Davidson, N.O. (2004). Apobec-1 protects intestine from radiation injury through posttranscriptional regulation of cyclooxygenase-2 expression. Gastroenterology *127*, 1139–1149.

Autore, F., Bergeron, J.R., Malim, M.H. Fraternail, F., and Hufhoff, H. (2010). Rationalization of the differences between APOBEC3G structures from crystallography and NMR studies by molecular dynamics simulations. PLoS One 5, e11515.

Backus, J.W., and Smith, H.C. (1991). Apolipoprotein B mRNA sequences 3′ of the editing site are necessary and sufficient for editing and editosome assembly. Nucleic Acids Res. *19*, 6781–6786.

Backus, J.W., and Smith, H.C. (1992). Three distinct RNA sequence elements are required for efficient apolipoprotein B (apoB) RNA editing *in vitro*. Nucleic Acids Res. *20*, 6007–6014.

Backus, J.W., and Smith, H.C. (1994). Specific 3′ sequences flanking a minimal apolipoprotein B (apoB) mRNA editing 'cassette' are critical for efficient editing *in vitro*. Biochim. Biophys. Acta *1217*, 65–73.

Backus, J.W., Schock, D., and Smith, H.C. (1994). Only cytidines 5′ of the apolipoprotein B mRNA mooring sequence are edited. Biochim. Biophys. Acta *1219*, 1–14.

Basu, U., Chaudhuri, J., Alpert, C., Dutt, S., Ranganath, S., Li, G., Schrum, J.P., Manis, J.P., and Alt, F.W. (2005). The AID antibody diversification enzyme is regulated by protein kinase A phosphorylation. Nature *438*, 508–511.

Baum, C.L., Teng, B.B., and Davidson, N.O. (1990). Apolipoprotein B messenger RNA editing in the rat liver. Modulation by fasting and refeeding a high carbohydrate diet. J. Biol. Chem. *265*, 19263–19270.

Beale, R.C., Petersen-Mahrt, S.K., Watt, I.N., Harris, R.S., Rada, C., and Neuberger, M.S. (2004). Comparison of the differential context-dependence of DNA deamination by APOBEC enzymes: correlation with mutation spectra *in vivo*. J. Mol. Biol. *337*, 585–596.

Bennett, R.P., Salter, J.D., Liu, X., Wedekind, J.E., and Smith, H.C. (2008). APOBEC3G subunits self-associate via the C-terminal deaminase domain. J. Biol. Chem. *283*, 33329–33336.

Bishop, K.N., Holmes, R.K., Sheehy, A.M., and Malim, M.H. (2004). APOBEC-mediated editing of viral RNA. Science *305*, 645.

Blanc, V., and Davidson, N.O. (2010). APOBEC-1-mediated RNA editing. Wiley Interdiscip. Rev. Syst. Biol. Med. *2*, 594–602.

Blanc, V., and Davidson, N.O. (2011). Mouse and other rodent models of C to U RNA editing. Methods Mol. Biol. *718*, 121–135.

Blanc, V., Henderson, J.O., Kennedy, S., and Davidson, N.O. (2001a). Mutagenesis of apobec-1 complementation factor reveals distinct domains that modulate RNA binding, protein–protein interaction with apobec-1, and complementation of C to U RNA editing activity. J. Biol. Chem. *276*, 46386–46393.

Blanc, V., Navaratnam, N., Henderson, J.O., Anant, S., Kennedy, S., Jarmuz, A., Scott, J., and Davidson, N.O. (2001b). Identification of GRY-RBP as an apolipoprotein B RNA-binding protein that interacts with both apobec-1 and apobec-1 complementation factor to modulate C to U editing. J. Biol. Chem. *276*, 10272–10283.

Blanc, V., Kennedy, S., and Davidson, N.O. (2003). A novel nuclear localization signal in the auxiliary domain of apobec-1 complementation factor regulates nucleocytoplasmic import and shuttling. J. Biol. Chem. *278*, 41198–41204.

Blanc, V., Sessa, K.J., Kennedy, S., Luo, J., and Davidson, N.O. (2010). Apobec-1 complementation factor modulates liver regeneration by post-transcriptional regulation of interleukin-6 mRNA stability. J. Biol. Chem. *285*, 19184–19192.

Blanc, V., Henderson, J.O., Newberry, R.D., Xie, Y., Cho, S.J., Newberry, E.P., Kennedy, S., Rubin, D.C., Wang, H.L., Luo, J., *et al.* (2007). Deletion of the AU-rich RNA binding protein Apobec-1 reduces intestinal tumor burden in Apc(min) mice. Cancer Res. *67*, 8565–8573.

Bransteitter, R., Pham, P., Scharff, M.D., and Goodman, M.F. (2003). Activation-induced cytidine deaminase deaminates deoxycytidine on single-stranded DNA but requires the action of RNase. Proc. Natl. Acad. Sci. U.S.A. *100*, 4102–4107.

Bransteitter, R., Prochnow, C., and Chen, X.S. (2009). The current structural and functional understanding of APOBEC deaminases. Cell. Mol. Life Sci. *66*, 3137–3147.

Brennan, C.M., and Steitz, J.A. (2001). HuR and mRNA stability. Cell. Mol. Life Sci. *58*, 266–277.

Bulliard, Y., Narvaiza, I., Bertero, A., Peddi, S., Rohrig, U.F., Ortiz, M., Zoete, V., Castro-Diaz, N., Turelli, P., Telenti, A., *et al.* (2011). Structure-function analyses point to a polynucleotide-accommodating groove essential for APOBEC3A restriction activities. J. Virol. *85*, 1765–1776.

Burd, C.G., and Dreyfuss, G. (1994). Conserved structures and diversity of functions of RNA-binding proteins. Science *265*, 615–621.

Calero, G., Wilson, K.F., Ly, T., Rios-Steiner, J.L., Clardy, J.C., and Cerione, R.A. (2002). Structural basis of m7GpppG binding to the nuclear cap-binding protein complex. Nat. Struct. Biol. *9*, 912–917.

Carpenter, M.A., Rajagurubandara, E., Wijesinghe, P., and Bhagwat, A.S. (2010). Determinants of sequence-specificity within human AID and APOBEC3G. DNA Repair (Amst) *9*, 579–587.

Chatterji, M., Unniraman, S., McBride, K.M., and Schatz, D.G. (2007). Role of activation-induced deaminase protein kinase A phosphorylation sites in Ig gene conversion and somatic hypermutation. J. Immunol. *179*, 5274–5280.

Chelico, L., Sacho, E.J., Erie, D.A., and Goodman, M.F. (2008). A model for oligomeric regulation of APOBEC3G cytosine deaminase-dependent restriction of HIV. J. Biol. Chem. *283*, 13780–13791.

Chelico, L., Prochnow, C., Erie, D.A., Chen, X.S., and Goodman, M.F. (2010). A structural model for deoxycytidine deamination mechanisms of the HIV-1 inactivation enzyme APOBEC3G. J. Biol. Chem. *285*, 16195–16205.

Chen, K.M., Martemyanova, N., Lu, Y., Shindo, K., Matsuo, H., and Harris, R.S. (2007). Extensive mutagenesis experiments corroborate a structural model for the DNA deaminase domain of APOBEC3G. FEBS Lett. *581*, 4761–4766.

Chen, K.M., Harjes, E., Gross, P.J., Fahmy, A., Lu, Y., Shindo, K., Harris, R.S., and Matsuo, H. (2008).

Structure of the DNA deaminase domain of the HIV-1 restriction factor APOBEC3G. Nature *452*, 116–119.

Chen, S.H., Li, X.X., Liao, W.S., Wu, J.H., and Chan, L. (1990). RNA editing of apolipoprotein B mRNA. Sequence specificity determined by *in vitro* coupled transcription editing. J. Biol. Chem. *265*, 6811–6816.

Chen, Z., Eggerman, T.L., and Patterson, A.P. (2001). Phosphorylation is a regulatory mechanism in apolipoprotein B mRNA editing. Biochem. J. *357*, 661–672.

Chen, Z., Eggerman, T.L., Bocharov, A.V., Baranova, I.N., Vishnyakova, T.G., Csako, G., and Patterson, A.P. (2010). Hypermutation induced by APOBEC-1 overexpression can be eliminated. RNA *16*, 1040–1052.

Chester, A., Somasekaram, A., Tzimina, M., Jarmuz, A., Gisbourne, J., O'Keefe, R., Scott, J., and Navaratnam, N. (2003). The apolipoprotein B mRNA editing complex performs a multifunctional cycle and suppresses nonsense-mediated decay. EMBO J. *22*, 3971–3982.

Chester, A., Weinreb, V., Carter, C.W., Jr., and Navaratnam, N. (2004). Optimization of apolipoprotein B mRNA editing by APOBEC1 apoenzyme and the role of its auxiliary factor, ACF. RNA *10*, 1399–1411.

Chiarle, R., Zhang, Y., Frock, R.L., Lewis, S.M., Molinie, B., Ho, Y.J., Myers, D.R., Choi, V.W., Compagno, M., Malkin, D.J., *et al.* (2011). Genome-wide translocation sequencing reveals mechanisms of chromosome breaks and rearrangements in B cells. Cell *147*, 107–119.

Conticello, S.G., Langlois, M.A., Yang, Z., and Neuberger, M.S. (2007). DNA deamination in immunity: AID in the context of its APOBEC relatives. Adv. Immunol. *94*, 37–73.

Dance, G.S.C., Sowden, M.P., Cartegni, L., Cooper, E., Krainer, A.R., and Smith, H.C. (2002). Two proteins essential for apolipoprotein B mRNA editing are expressed from a single gene through alternative splicing. J. Biol. Chem. *277*, 12703–12709.

Demmer, L.A., Levin, M.S., Elovson, J., Reuben, M.A., Lusis, A.J., and Gordon, J.I. (1986). Tissue-specific expression and developmental regulation of the rat apolipoprotein B gene. Proc. Natl. Acad. Sci. U.S.A. *83*, 8102–8106.

Demorest, Z.L., Li, M., and Harris, R.S. (2011). Phosphorylation directly regulates the intrinsic DNA cytidine deaminase activity of activation-induced deaminase and APOBEC3G protein. J. Biol. Chem. *286*, 26568–26575.

Douaisi, M., Dussart, S., Courcoul, M., Bessou, G., Lerner, E.C., Decroly, E., and Vigne, R. (2005). The tyrosine kinases Fyn and Hck favor the recruitment of tyrosine-phosphorylated APOBEC3G into vif-defective HIV-1 particles. Biochem. Biophys. Res. Commun. *329*, 917–924.

Driscoll, D.M., and Zhang, Q. (1994). Expression and characterization of p27, the catalytic subunit of the apolipoprotein B mRNA editing enzyme. J. Biol. Chem. *269*, 19843–19847.

Driscoll, D.M., Lakhe-Reddy, S., Oleksa, L.M., and Martinez, D. (1993). Induction of RNA editing at heterologous sites by sequences in apolipoprotein B mRNA. Mol. Cell. Biol. *13*, 7288–7294.

Dur, S., Krause, K., Pluntke, N., and Greeve, J. (2004). Gene structure and expression of the mouse APOBEC-1 complementation factor: multiple transcriptional initiation sites and a spliced variant with a premature stop translation codon. Biochim. Biophys. Acta *1680*, 11–23.

Etard, C., Roostalu, U., and Strahle, U. (2010). Lack of Apobec2-related proteins causes a dystrophic muscle phenotype in zebrafish embryos. J. Cell Biol. *189*, 527–539.

Farese, R.V., Jr., Ruland, S.L., Flynn, L.M., Stokowski, R.P., and Young, S.G. (1995). Knockout of the mouse apolipoprotein B gene results in embryonic lethality in homozygotes and protection against diet-induced hypercholesterolemia in heterozygotes. Proc. Natl. Acad. Sci. U.S.A. *92*, 1774–1778.

Friew, Y.N., Boyko, V., Hu, W.S., and Pathak, V.K. (2009). Intracellular interactions between APOBEC3G, RNA, and HIV-1 Gag: APOBEC3G multimerization is dependent on its association with RNA. Retrovirology *6*, 56.

Fujino, T., Navaratnam, N., and Scott, J. (1998). Human apolipoprotein B RNA editing deaminase gene (APOBEC1). Genomics *47*, 266–275.

Funahashi, T., Giannoni, F., DePaoli, A.M., Skarosi, S.F., and Davidson, N.O. (1995). Tissue-specific, developmental and nutritional regulation of the gene encoding the catalytic subunit of the rat apolipoprotein B mRNA editing enzyme: functional role in the modulation of apoB mRNA editing. J. Lipid Res. *36*, 414–428.

Galloway, C.A., and Smith, H.C. (2010). The expression of apoB mRNA editing factors is not the sole determinant for the induction of editing in differentiating Caco-2 cells. Biochem. Biophys. Res. Commun. *391*, 659–663.

Galloway, C.A., Ashton, J., Sparks, J.D., Mooney, R.A., and Smith, H.C. (2010a). Metabolic regulation of APOBEC-1 complementation factor trafficking in mouse models of obesity and its positive correlation with the expression of ApoB protein in hepatocytes. Biochim. Biophys. Acta *1802*, 976–985.

Galloway, C.A., Kumar, A., Krucinska, J., and Smith, H.C. (2010b). APOBEC-1 complementation factor (ACF) forms RNA-dependent multimers. Biochem. Biophys. Res. Commun. *398*, 38–43.

Giangreco, A., Sowden, M.P., Mikityansky, I., and Smith, H.C. (2001). Ethanol stimulates apolipoprotein B mRNA editing in the absence of de novo RNA or protein synthesis. Biochem. Biophys. Res. Commun. *289*, 1162–1167.

Goila-Gaur, R., Khan, M.A., Miyagi, E., Kao, S., Opi, S., Takeuchi, H., and Strebel, K. (2008). HIV-1 Vif promotes the formation of high molecular mass APOBEC3G complexes. Virology *372*, 136–146.

Greeve, J., Altkemper, I., Dieterich, J.H., Greten, H., and Windler, E. (1993). Apolipoprotein B mRNA editing in 12 different mammalian species: hepatic expression is reflected in low concentrations of apoB-containing plasma lipoproteins. J. Lipid Res. *34*, 1367–1383.

Greeve, J., Lellek, H., Rautenberg, P., and Greten, H. (1998). Inhibition of the apolipoprotein B mRNA editing enzyme-complex by hnRNP C1 protein and 40S hnRNP complexes. Biol. Chem. *379*, 1063–1073.

Hache, G., Liddament, M.T., and Harris, R.S. (2005). The retroviral hypermutation specificity of APOBEC3F and APOBEC3G is governed by the C-terminal DNA cytosine deaminase domain. J. Biol. Chem. *280*, 10920–10924.

Hamilton, C.E., Papavasiliou, F.N., and Rosenberg, B.R. (2010). Diverse functions for DNA and RNA editing in the immune system. RNA Biol. *7*, 220–228.

Harjes, E., Gross, P.J., Chen, K.M., Lu, Y., Shindo, K., Nowarski, R., Gross, J.D., Kotler, M., Harris, R.S., and Matsuo, H. (2009). An extended structure of the APOBEC3G catalytic domain suggests a unique holoenzyme model. J. Mol. Biol. *389*, 819–832.

Harris, R.S., and Liddament, M.T. (2004). Retroviral restriction by APOBEC proteins. Nat. Rev. Immunol. *4*, 868–877.

Harris, R.S., Petersen-Mahrt, S.K., and Neuberger, M.S. (2002). RNA editing enzyme APOBEC1 and some of its homologs can act as DNA mutators. Molecular Cell *10*, 1247–1253.

Harris, S.G., Sabio, I., Mayer, E., Steinberg, M.F., Backus, J.W., Sparks, J.D., Sparks, C.E., and Smith, H.C. (1993). Extract-specific heterogeneity in high-order complexes containing apolipoprotein B mRNA editing activity and RNA-binding proteins. J. Biol. Chem. *268*, 7382–7392.

Henderson, J.O., Blanc, V., and Davidson, N.O. (2001). Isolation, characterization and developmental regulation of the human apobec-1 complementation factor (ACF) gene. Biochim. Biophys. Acta *1522*, 22–30.

Hersberger, M., and Innerarity, T.L. (1998). Two efficiency elements flanking the editing site of cytidine 6666 in the apolipoprotein B mRNA support mooring-dependent editing. J. Biol. Chem. 273, 9435–9442.

Hersberger, M., Patarroyo-White, S., Arnold, K.S., and Innerarity, T.L. (1999). Phylogenetic analysis of the apolipoprotein B mRNA editing region. Evidence for a secondary structure between the mooring sequence and the 3′ efficiency element. J. Biol. Chem. *274*, 34590–34597.

Hirano, K.I., Young, S.G., Farese, R.V., Ng, J., Sande, E., Warburton, C., Powell-Braxton, L.M., and Davidson, N.O. (1996). Targeted disruption of the mouse apobec-1 gene abolishes apolipoprotein B mRNA editing and eliminates apolipoprotein B48. J. Biol. Chem. *271*, 9887–9890.

Huguchi, K., Kitagawa, K., Kogishi, K., and Takeda, T. (1992). Developmental and age-related changes in apoB mRNA editing in mice. J. Lipid Res. *33*, 1753–1764.

Huthoff, H., and Malim, M.H. (2007). Identification of amino acid residues in APOBEC3G required for regulation by human immunodeficiency virus type 1 Vif and Virion encapsidation. J. Virol. *81*, 3807–3815.

Huthoff, H., Autore, F., Gallois-Montbrun, S., Fraternali, F., and Malim, M.H. (2009). RNA-dependent oligomerization of APOBEC3G is required for restriction of HIV-1. PLoS Pathog. *5*, e1000330.

Inui, Y., Giannoni, F., Funahashi, T., and Davidson, N.O. (1994). REPR and complementation factor(s) interact to modulate rat apolipoprotein B mRNA editing in response to alterations in cellular cholesterol flux. J. Lipid Res. *35*, 1477–1489.

Inui, Y., Hausman, A.M., Nanthakumar, N., Henning, S.J., and Davidson, N.O. (1992). Apolipoprotein B messenger RNA editing in rat liver: developmental and hormonal modulation is divergent from apolipoprotein A-IV gene expression despite increased hepatic lipogenesis. J. Lipid Res. *33*, 1843–1856.

Johansson, C., Finger, L.D., Trantirek, L., Mueller, T.D., Kim, S., Laird-Offringa, I.A., and Feigon, J. (2004). Solution structure of the complex formed by the two N-terminal RNA-binding domains of nucleolin and a pre-rRNA target. J. Mol. Biol. *337*, 799–816.

Khan, M.A., Goila-Gaur, R., Opi, S., Miyagi, E., Takeuchi, H., Kao, S., and Strebel, K. (2007). Analysis of the contribution of cellular and viral RNA to the packaging of APOBEC3G into HIV-1 virions. Retrovirology *4*, 48.

Kielkopf, C.L., Lucke, S., and Green, M.R. (2004). U2AF homology motifs: protein recognition in the RRM world. Genes Dev. *18*, 1513–1526.

Klein, I.A., Resch, W., Jankovic, M., Oliveira, T., Yamane, A., Nakahashi, H., Di Virgilio, M., Bothmer, A., Nussenzweig, A., Robbiani, D.F., *et al.* (2011). Translocation-capture sequencing reveals the extent and nature of chromosomal rearrangements in B lymphocytes. Cell *147*, 95–106.

Kreisberg, J.F., Yonemoto, W., and Greene, W.C. (2006). Endogenous factors enhance HIV infection of tissue naive CD4 T cells by stimulating high molecular mass APOBEC3G complex formation. J. Exp. Med. *203*, 865–870.

Krzysiak, T.C., Jung, J., Thompson, J., Baker, D., and Gronenborn, A.M. (2012). APOBEC2 is a monomer in solution: implications for APOBEC3G models. Biochemistry *51*, 2008–2017 .

Lada, A.G., Krick, C.F., Kozmin, S.G., Mayorov, V.I., Karpova, T.S., Rogozin, I.B., and Pavlov, Y.I. (2011). Mutator effects and mutation signatures of editing deaminases produced in bacteria and yeast. Biochemistry (Mosc) *76*, 131–146.

Lau, P.P., and Chan, L. (2003). Involvement of a chaperone regulator, Bcl2-associated Athanogene-4 (BAG-4), in apolipoprotein B mRNA editing. J. Biol. Chem. *298*, 52988–52996.

Lau, P.P., Chang, B.H., and Chan, L. (2001). Two-hybrid cloning identifies an RNA-binding protein, GRY-RBP, as a component of apobec-1 editosome. Biochem. Biophys. Res. Commun. *282*, 977–983.

Lau, P.P., Villanueva, H., Kobayashi, K., Nakamuta, M., Chang, H.J., and Chan, L. (2001). A DnaJ protein, Apobec-1-binding protein-2, modulates apolipoprotein B mRNA editing. J. Biol. Chem. *276*, 46445–46452.

Lau, P.P., Xiong, W.J., Zhu, H.J., Chen, S.H., and Chan, L. (1991). Apolipoprotein B mRNA editing is an intranuclear event that occurs posttranscriptionally coincident with splicing and polyadenylation. J. Biol. Chem. *266*, 20550–20554.

Lau, P.P., Zhu, H.-J., Baldini, H.A., Charnsangavej, C., and Chan, L. (1994). Dimeric structure of a human apo B mRNA editing protein and cloning and chromosomal localization of its gene. Proc. Natl. Acad. Sci. U.S.A. *91*, 8522–8526.

Lau, P.P., Zhu, H.J., Nakamuta, M., and Chan, L. (1997). Cloning of an Apobec-1-binding protein that also interacts with apolipoprotein B mRNA and evidence for its involvement in RNA editing. J. Biol. Chem. *272*, 1452–1455.

Lehmann, D.M., Galloway, C.A., Sowden, M.P., and Smith, H.C. (2006). Metabolic regulation of apoB mRNA editing is associated with phosphorylation of APOBEC-1 complementation factor. Nucleic Acids Res. *34*, 3299–3208.

Lehmann, D.M., Galloway, C.A., Macelrevey, C., Sowden, M.P., Wedekind, J.E., and Smith, H.C. (2007). Functional characterization of APOBEC-1 complementation factor phosphorylation sites. Biochim. Biophys. Acta *1773*, 408–418.

Lellek, H., Kirsten, R., Diehl, I., Apostel, F., Buck, F., and Greeve, J. (2000). Purification and molecular cloning of a novel essential component of the apolipoprotein B mRNA editing enzyme-complex. J. Biol. Chem. *275*, 19848–19856.

MacGinnitie, A.J., Anant, S., and Davidson, N.O. (1995). Mutagenesis of APOBEC-1, the catalytic subunit of the mammalian apolipoprotein B mRNA editing enzyme, reveals distinct domains that mediate cytosine nucleoside deaminase, RNA-binding, and RNA editing activity. J. Biol. Chem. *270*, 14768–14775.

Maris, C., Dominguez, C., and Allain, F.H. (2005a). The RNA recognition motif, a plastic RNA-binding platform to regulate post-transcriptional gene expression. FEBS J. *272*, 2118–2131.

Maris, C., Masse, J., Chester, A., Navaratnam, N., and Allain, F.H. (2005b). NMR structure of the apoB mRNA stem–loop and its interaction with the C to U editing APOBEC1 complementary factor. RNA *11*, 173–186.

Mazza, C., Segref, A., Mattaj, I.W., and Cusack, S. (2002). Large-scale induced fit recognition of an m(7)GpppG cap analogue by the human nuclear cap-binding complex. EMBO J. *21*, 5548–5557.

McDougall, W.M., and Smith, H.C. (2011). Direct evidence that RNA inhibits APOBEC3G ssDNA cytidine deaminase activity. Biophys. Biochem. Res. Commun. *412*, 612–617.

McDougall, W.M., Okany, C., and Smith, H.C. (2011). Deaminase activity on single-stranded DNA (ssDNA) occurs *in vitro* when APOBEC3G cytidine deaminase forms homotetramers and higher-order complexes. J. Biol. Chem. *286*, 30655–30661.

Mehta, A., and Driscoll, D.M. (2002). Identification of domains in APOBEC-1 complementation factor required for RNA binding and Apolipoprotein B mRNA editing. RNA *8*, 69–82.

Mehta, A., Banerjee, S., and Driscoll, D.M. (1996). Apobec-1 interacts with a 65-kDa complementing protein to edit apolipoprotein-B mRNA *in vitro*. J. Biol. Chem. *271*, 28294–28299.

Mehta, A., Kinter, M.T., Sherman, N.E., and Driscoll, D.M. (2000). Molecular cloning of apobec-1 complementation factor, a novel RNA – binding protein involved in the editing of apolipoprotein B mRNA. Mol. Cell. Biol. *20*, 1846–1854.

Mikl, M.C., Watt, I.N., Lu, M., Reik, W., Davies, S.L., Neuberger, M.S., and Rada, C. (2005). Mice deficient in APOBEC2 and APOBEC3. Mol. Cell. Biol. *25*, 7270–7277.

Mukhopadhyay, D., Anant, S., Lee, R.M., Kennedy, S., Viskochil, D., and Davidson, N.O. (2002). C→U editing of neurofibromatosis 1 mRNA occurs in tumors that express both the type II transcript and apobec-1, the catalytic subunit of the apolipoprotein B mRNA editing enzyme. Am. J. Human Genet. *70*, 38–50.

Muramatsu, M., Sankaranand, V., Anant, S., Sugai, M., Kinoshita, K., Davidson, N., and Honjo, T. (1999). Specific expression of activation-induced cytidine deaminase (AID), a novel member of the RNA editing deaminase family in germinal center B cells. J. Biol. Chem. *274*, 18470–18476.

Muramatsu, M., Kinoshita, K., Fagarasan, S., Yamada, S., Shinkai, Y., and Honjo, T. (2000). Class switch recombination and hypermutation require activation-induced cytidine deaminase (AID), a potential RNA editing enzyme. Cell *102*, 553–563.

Navaratnam, N., Patel, D., Shah, R.R., Greeve, J.C., Powell, L.M., Knott, T.J., and Scott, J. (1991). An additional editing site is present in apolipoprotein B mRNA. Nucleic Acids Res. *19*, 1741–1744.

Navaratnam, N., Shah, R., Patel, D., Fay, V., and Scott, J. (1993). Apolipoprotein B mRNA editing is associated with UV crosslinking of proteins to the editing site. Proc. Natl. Acad. Sci. U.S.A. *90*, 222–226.

Navaratnam, N., Bhattacharya, S., Fujino, T., Patel, D., Jarmuz, A.L., and Scott, J. (1995). Evolutionary origins of apoB mRNA editing: catalysis by a cytidine deaminase that has acquired a novel RNA-binding motif at its active site. Cell *81*, 187–195.

Navaratnam, N., Fujino, T., Bayliss, J., Jarmuz, A., How, A., Richardson, N., Somasekaram, A., Bhattacharya, S., Carter, C., and Scott, J. (1998). *Escherichia coli* cytidine deaminase provides a molecular model for ApoB RNA editing and a mechanism for RNA substrate recognition. J. Mol. Biol. *275*, 695–714.

Nakamuta, M., Chang, B.H.J., Zsigmond, E., Kobayashi, K., Lei, H., Ishida, B.Y., Oka, K., Li, E., and Chan, L. (1996). Complete phenotypic characterization of the apobec-1 knockout mice with a wild-type genetic background and a human apolipoprotein B transgenic background, and restoration of apolipoprotein B mRNA editing by somatic gene transfer of Apobec-1. J. Biol. Chem. *271*, 25981–25988.

Nakamuta, M., Tsai, A., Chan, L., Davidson, N.O., and Teng, B.B. (1999). Sequence elements required for apolipoprotein B mRNA editing enhancement activity from chicken enterocytes. Biochem. Biophys. Res. Commun. *254*, 744–750.

Navarro, F., Bollman, B., Chen, H., Konig, R., Yu, Q., Chiles, K., and Landau, N.R. (2005). Complementary function of the two catalytic domains of APOBEC3G. Virology *333*, 374–386.

Nowarski, R., Britan-Rosich, E., Shiloach, T., and Kotler, M. (2008). Hypermutation by intersegmental transfer of APOBEC3G cytidine deaminase. Nat. Struct. Mol. Biol. *15*, 1059–1066.

Opi, S., Takeuchi, H., Kao, S., Khan, M.A., Miyagi, E., Goila-Gaur, R., Iwatani, Y., Levin, J.G., and Strebel, K. (2006). Monomeric APOBEC3G is catalytically active and has antiviral activity. J. Virol. *80*, 4673–4682.

Pastore, C., Topalidou, I., Forouhar, F., Yan, A.C., Levy, M., and Hunt, J.F. (2012). Crystal structure and RNA binding properties of the RNA recognition motif (RRM) and AlkB domains in human AlkB homolog 8 (ABH8), an enzyme catalyzing tRNA hypermodification. J. Biol. Chem. *287*, 2130–2143.

Pham, P., Calabrese, P., Park, S.J., and Goodman, M.F. (2011). Analysis of a single-stranded DNA-scanning process in which activation-induced deoxycytidine deaminase (AID) deaminates C to U haphazardly and inefficiently to ensure mutational diversity. J. Biol. Chem. *286*, 24931–24942.

Phung, T.L., Sowden, M.P., Sparks, J.D., Sparks, C.E., and Smith, H.C. (1996). Regulation of hepatic apolipoprotein B RNA editing in the genetically obese Zucker rat. Metabolism *45*, 1056–1058.

Prochnow, C., Bransteitter, R., Klein, M.G., Goodman, M.F., and Chen, X.S. (2007). The APOBEC-2 crystal structure and functional implications for the deaminase AID. Nature *445*, 447–451.

Qian, X., Balestra, M.E., Yamanaka, S., Boren, J., Lee, I., and Innerarity, T.L. (1998). Low expression of the apolipoprotein B mRNA editing transgene in mice reduces LDL but does not cause liver dysplasia or tumors. Arteriosc. Thromb. Vasc. Biol. *18*, 1013–1020.

Rausch, J.W., Chelico, L., Goodman, M.F., and Le Grice, S.F. (2009). Dissecting APOBEC3G substrate specificity by nucleoside analog interference. J. Biol. Chem. *284*, 7047–7058.

Robbiani, D.F., Bothmer, A., Callen, E., Reina-San-Martin, B., Dorsett, Y., Difilippantonio, S., Bolland, D.J., Chen, H.T., Corcoran, A.E., Nussenzweig, A., *et al.* (2008). AID is required for the chromosomal breaks in c-myc that lead to c-myc/IgH translocations. Cell *135*, 1028–1038.

Rosenberg, B.R., Dewell, S., and Papavasiliou, F.N. (2011). Identifying mRNA editing deaminase targets by RNA-Seq. Methods Mol. Biol. *718*, 103–119.

Salter, J.D., Krucinska, J., Raina, J., Smith, H.C., and Wedekind, J.E. (2009). A hydrodynamic analysis of APOBEC3G reveals a monomer-dimer-tetramer self-association that has implications for anti-HIV function. Biochemistry *48*, 10685–10687.

Sato, Y., Probst, H.C., Tatsumi, R., Ikeuchi, Y., Neuberger, M.S., and Rada, C. (2010). Deficiency in APOBEC2 leads to a shift in muscle fiber type, diminished body mass, and myopathy. J. Biol. Chem. *285*, 7111–7118.

Schock, D., Kuo, S.R., Steinburg, M.F., Bolognino, M., Sparks, J.D., Sparks, C.E., and Smith, H.C. (1996). An auxiliary factor containing a 240-kDa protein complex is involved in apolipoprotein B RNA editing. Proc. Natl. Acad. Sci. U.S.A. *93*, 1097–1102.

Shah, R.R., Knott, T.J., Legros, J.E., Navaratnam, N., Greeve, J.C., and Scott, J. (1991). Sequence requirements for the editing of apolipoprotein B mRNA. J. Biol. Chem. *266*, 16301–16304.

Shandilya, S.M., Nalam, M.N., Nalivaika, E.A., Gross, P.J., Valesano, J.C., Shindo, K., Li, M., Munson, M., Royer, W.E., Harjes, E., *et al.* (2010). Crystal structure of the APOBEC3G catalytic domain reveals potential oligomerization interfaces. Structure *18*, 28–38.

Shirakawa, K., Takaori-Kondo, A., Yokoyama, M., Izumi, T., Matsui, M., Io, K., Sato, T., Sato, H., and Uchiyama, T. (2008). Phosphorylation of APOBEC3G by protein kinase A regulates its interaction with HIV-1 Vif. Nat. Struct. Mol. Biol. *15*, 1184–1191.

Shlyakhtenko, L.S., Lushnikov, A.Y., Li, M., Lackey, L., Harris, R.S., and Lyubchenko, Y.L. (2011). Atomic force microscopy studies provide direct evidence for dimerization of the HIV restriction factor APOBEC3G. J. Biol. Chem. *286*, 3387–3395.

Siddiqui, J.F., Van Mater, D., Sowden, M.P., and Smith, H.C. (1999). Disproportionate relationship between APOBEC-1 expression and apolipoprotein B mRNA editing activity. Exp. Cell Res. *252*, 154–164.

Skuse, G.R., Cappione, A.J., Sowden, M., Metheny, L.J., and Smith, H.C. (1996). The neurofibromatosis type I messenger RNA undergoes base-modification RNA editing. Nucleic Acids Res. *24*, 478–485.

Smith, H.C. (1993). Apolipoprotein B mRNA editing: the sequence to the event. Semin. Cell Biol. *4*, 267–278.

Smith, H.C. (1998). Analysis of protein complexes assembled on apolipoprotein B mRNA for mooring sequence-dependent RNA editing. Methods *15*, 27–39.

Smith, H.C. (2009). The APOBEC1 paradigm for mammalian cytidine deaminases that edit DNA and RNA (Landes bioScience, Austin, TX).

Smith, H.C. (2011). APOBEC3G: a double agent in defense. Trends Biochem. Sci. *36*, 239–244.

Smith, H.C., Kuo, S.R., Backus, J.W., Harris, S.G., Sparks, C.E., and Sparks, J.D. (1991). *In vitro* apolipoprotein B mRNA editing: identification of a 27S editing complex. Proc. Natl. Acad. Sci. U.S.A. *88*, 1489–1493.

Smith, H.C., Wedekind, J.E., Xie, K., and Sowden, M.P., eds. (2005). Mammalian C to U Editing (Springer, Gothenburg).

Smith, H.C., Bennett, R.P., Kizilyer, A., McDougall, W.M., and Prohaska, K.M. (2012). Functions and regulation of the APOBEC family of proteins. Semin. Cell. Dev. Biol. *23*, 258–268.

Soros, V.B., Yonemoto, W., and Greene, W.C. (2007). Newly synthesized APOBEC3G is incorporated into HIV virions, inhibited by HIV RNA, and subsequently activated by RNase H. PLoS Pathog. *3*, e15.

Sowden, M.P., and Smith, H.C. (2001). Commitment of apolipoprotein B RNA to the splicing pathway regulates cytidine-to-uridine editing-site utilization. Biochem. J. *359*, 697–705.

Sowden, M., Hamm, J.K., and Smith, H.C. (1996a). Overexpression of APOBEC-1 results in mooring sequence-dependent promiscuous RNA editing. J. Biol. Chem. *271*, 3011–3017.

Sowden, M., Hamm, J.K., Spinelli, S., and Smith, H.C. (1996b). Determinants involved in regulating the proportion of edited apolipoprotein B RNAs. RNA *2*, 274–288.

Sowden, M.P., Ballatori, N., de Mesy Jensen, K.L., Hamilton Reed, L., and Smith, H.C. (2002). The editosome for cytidine to uridine mRNA editing has a native complexity of 27S: identification of intracellular domains containing active and inactive editing factors. J. Cell. Sci. *115*, 1027–1039.

Sowden, M.P., Eagleton, M.J., and Smith, H.C. (1998). Apolipoprotein B RNA sequence 3′ of the mooring sequence and cellular sources of auxiliary factors determine the location and extent of promiscuous editing. Nucleic Acids Res. *26*, 1644–1652.

Sowden, M.P., Lehmann, D.M., Lin, X., Smith, C.O., and Smith, H.C. (2004). Identification of novel alternative splice variants of APOBEC-1 complementation factor with different capacities to support ApoB mRNA editing. J. Biol. Chem. *278*, 197–206.

Steinburg, M.F., Schock, D., Backus, J.W., and Smith, H.C. (1999). Tissue-specific differences in the role of RNA 3′ of the apo B mRNA mooring sequence in editosome assembly. Biochem. Biophys. Res. Commun. *263*, 81–86.

Stopak, K.S., Chiu, Y.L., Kropp, J., Grant, R.M., and Greene, W.C. (2007). Distinct patterns of cytokine regulation of APOBEC3G expression and activity in primary lymphocytes, macrophages, and dendritic cells. J. Biol. Chem. *282*, 3539–3546.

Teng, B., Burant, C.F., and Davidson, N.O. (1993). Molecular cloning of an apolipoprotein B messenger RNA editing protein. Science *260*, 1816–1819.

Teng, B., and Davidson, N.O. (1992). Evolution of intestinal apolipoprotein B mRNA editing. Chicken apolipoprotein B mRNA is not edited, but chicken enterocytes contain *in vitro* editing enhancement factor(s). J. Biol. Chem. *267*, 21265–21272.

Teng, B., Ishida, B., Forte, T.M., Blumenthal, S., Song, L.Z., Gotto, A.M., Jr., and Chan, L. (1997). Effective lowering of plasma, LDL, and esterified cholesterol in LDL receptor-knockout mice by adenovirus-mediated gene delivery of ApoB mRNA editing enzyme (Apobec1). Arterioscler. Thromb Vasc. Biol. *17*, 889–897.

Teng, B.B., Ochsner, S., Zhang, Q., Soman, K.V., Lau, P.P., and Chan, L. (1999). Mutational analysis of apolipoprotein B mRNA editing enzyme (APOBEC1). Structure–function relationships of RNA editing and dimerization. J. Lipid Res. *40*, 623–635.

Veniant, M.M., Kim, E., McCormick, S., Boren, J., Nielsen, L.B., Raabe, M., and Young, S.G. (1999). Insights into apolipoprotein B biology from transgenic and gene-targeted mice. J. Nutr. *129*, 451S–455S.

Vonica, A., Rosa, A., Arduini, B.L., and Brivanlou, A.H. (2011). APOBEC2, a selective inhibitor of TGFbeta signaling, regulates left–right axis specification during early embryogenesis. Dev. Biol. *350*, 13–23.

Wang, X., and Tanaka Hall, T.M. (2001). Structural basis for recognition of AU-rich element RNA by the HuD protein. Nat. Struct. Biol. *8*, 141–145.

Wedekind, J.E., Dance, G.S., Sowden, M.P., and Smith, H.C. (2003). Messenger RNA editing in mammals: new members of the APOBEC family seeking roles in the family business. Trends Genet. *19*, 207–216.

Wedekind, J.E., Gillilan, R., Janda, A., Krucinska, J., Salter, J.D., Bennett, R.P., Raina, J., and Smith, H.C. (2006). Nanostructures of APOBEC3G support a hierarchical assembly model of high molecular mass ribonucleoprotein particles from dimeric subunits. J. Biol. Chem. *281*, 38122–38126.

von Wronski, M.A., Hirano, K.I., Cagen, L.M., Wilcox, H.G., Raghow, R., Thorngate, F.E., Heimberg, M., Davidson, N.O., and Elam, M.B. (1998). Insulin increases expression of apobec-1, the catalytic subunit of the apolipoprotein B mRNA editing complex in rat hepatocytes. Metabolism *47*, 869–873.

Wu, J.H., Semenkovich, C.F., Chen, S.H., Li, W.H., and Chan, L. (1990). Apolipoprotein B mRNA editing. Validation of a sensitive assay and developmental biology of RNA editing in the rat. J. Biol. Chem. *265*, 12312–12316.

Xie, K., Sowden, M.P., Dance, G.S., Torelli, A.T., Smith, H.C., and Wedekind, J.E. (2004). The structure of a yeast RNA editing deaminase provides insight into the fold and function of activation-induced deaminase and APOBEC-1. Proc. Natl. Acad. Sci. U.S.A. *101*, 8114–8119.

Xie, Y., Nassir, F., Luo, J., Buhman, K., and Davidson, N.O. (2003). Intestinal lipoprotein assembly in apobec-1-/- mice reveals subtle alterations in triglyceride secretion coupled with a shift to larger lipoproteins. Am J. Physiol. Gastrointest. Liver Physiol. *285*, G735–746.

Yamanaka, S., Balestra, M., Ferrell, L., Fan, J., Arnold, K.S., Taylor, S., Taylor, J., M., and Innerarity, T.L. (1995). Apolipoprotein B mRNA editing protein induces hepatocellular carcinoma and dysplasia in transgenic animals. Proc. Natl. Acad. Sci. U.S.A. *92*, 8483–8487.

Yamanaka, S., Poksay, K.S., Driscoll, D.M., and Innerarity, T.L. (1996). Hyperediting of multiple cytidines of apolipoprotein B mRNA by APOBEC-1 requires auxiliary protein(s) but not a mooring sequence motif. J. Biol. Chem. *271*, 11506–11510.

Yamanaka, S., Poksay, K.S., Arnold, K.S., and Innerarity, T.L. (1997). A novel translational repressor mRNA is edited extensively in livers containing tumors caused by the transgene expression of the apoB mRNA – editing enzyme. Genes Dev. *11*, 321–333.

Yang, Y., and Smith, H.C. (1997). Multiple protein domains determine the cell type-specific nuclear distribution of the catalytic subunit required for apolipoprotein B mRNA editing. Proc. Natl. Acad. Sci. U.S.A. *94*, 13075–13080.

Yang, Y., Kovalski, K., and Smith, H.C. (1997). Partial characterization of the auxiliary factors involved in apolipoprotein B mRNA editing through APOBEC-1 affinity chromatography. J. Biol. Chem. *272*, 27700–27706.

Yang, Y., Sowden, M.P., and Smith, H.C. (2000). Induction of cytidine to uridine editing on cytoplasmic apolipoprotein B mRNA by overexpressing APOBEC-1. J. Biol. Chem. *275*, 22663–22669.

Yang, Y., Sowden, M.P., Yang, Y., and Smith, H.C. (2001). Intracellular trafficking determinants in APOBEC-1, the catalytic subunit for cytidine to uridine editing of apolipoprotein B mRNA. Exp. Cell. Res. *267*, 153–164.

Yang, Y., Ballatori, N., and Smith, H.C. (2002). Apolipoprotein B mRNA editing and the reduction in synthesis and secretion of the atherogenic risk factor, apolipoprotein B100 can be effectively targeted through TAT-mediated protein transduction. Mol. Pharmacol. *61*, 269–276.

Zhang, K.L., Mangeat, B., Ortiz, M., Zoete, V., Trono, D., Telenti, A., and Michielin, O. (2007). Model structure of human APOBEC3G. PLoS One *2*, e378.

Identification of RNA Editing Sites: a Survey of the Past, Present and Future

8

Meng How Tan and Jin Billy Li

Abstract

RNA editing is a post-transcriptional mechanism whereby genomically encoded information is altered at the level of the transcript. We describe in this chapter how RNA editing sites can be identified. The pace of discovery in the past few decades was dependent on the sequencing technologies available at a particular time. At the beginning when DNA sequencing had just been developed and automated, the identification of RNA editing sites was slow and often occurred by chance. Over time, as more and more sequences were deposited in databases, it became possible for scientists to computationally mine the databases for more editing sites. In recent years, with the development of ultra-high-throughput sequencing technologies whereby millions to billions of DNA molecules are sequenced simultaneously, scientists can now uncover RNA editing sites in a genome-wide manner. However, extra care has to be taken during the analysis process to remove artefacts and to ensure that true editing sites are identified.

RNA in the limelight

A myriad of regulatory mechanisms can operate at multiple steps from a gene encoded within the chromosomes of a cell to the final activity of the gene product. For example, expression of a gene may be activated or repressed by specific transcription factors or histone marks, while the function of a protein may be altered by post-translational modifications such as phosphorylation or sumoylation. Even though numerous studies have focused on DNA and protein, the RNA can also be regulated in various ways. Recently, a process known as RNA editing, whereby genomically encoded information is changed after transcription such that nucleotides in the RNA molecule do not match perfectly the sequence of the DNA from which the RNA is derived, has been thrown into the spotlight. Multiple research laboratories have uncovered thousands of sites that undergo RNA editing, which underscores the critical importance of this regulatory mechanism in biology and medicine. To find these sites, technologies are required to decipher the sequences of nucleic acids, so that one can compare the sequences of DNA and RNA from the same sample and then identify the differences. In this review, we take a journey back in time in order to illustrate how identification of editing sites has matured over time.

The advent of DNA sequencing technologies

The 1970s was a remarkable decade for the study of nucleic acids. It was during this period of time when various methods were introduced to identify the bases in any given stretch of DNA molecule. Two approaches stood out. The Maxam–Gilbert method, also known as chemical sequencing, relies on base-specific treatments and cleavages to generate a series of different sized fragments (Maxam and Gilbert, 1977) (Fig. 8.1A). The Sanger method, also known as chain terminator sequencing, relies on the presence of a low concentration of dideoxynucleotides to synthesize DNA molecules of different lengths (Sanger *et al.*, 1977b) (Fig. 8.1B). Regardless of the exact method used, all these different DNA sequencing techniques allowed scientists to examine for the first time the nucleotide sequences of genes and small genomes. For example, the 5-kb genome of

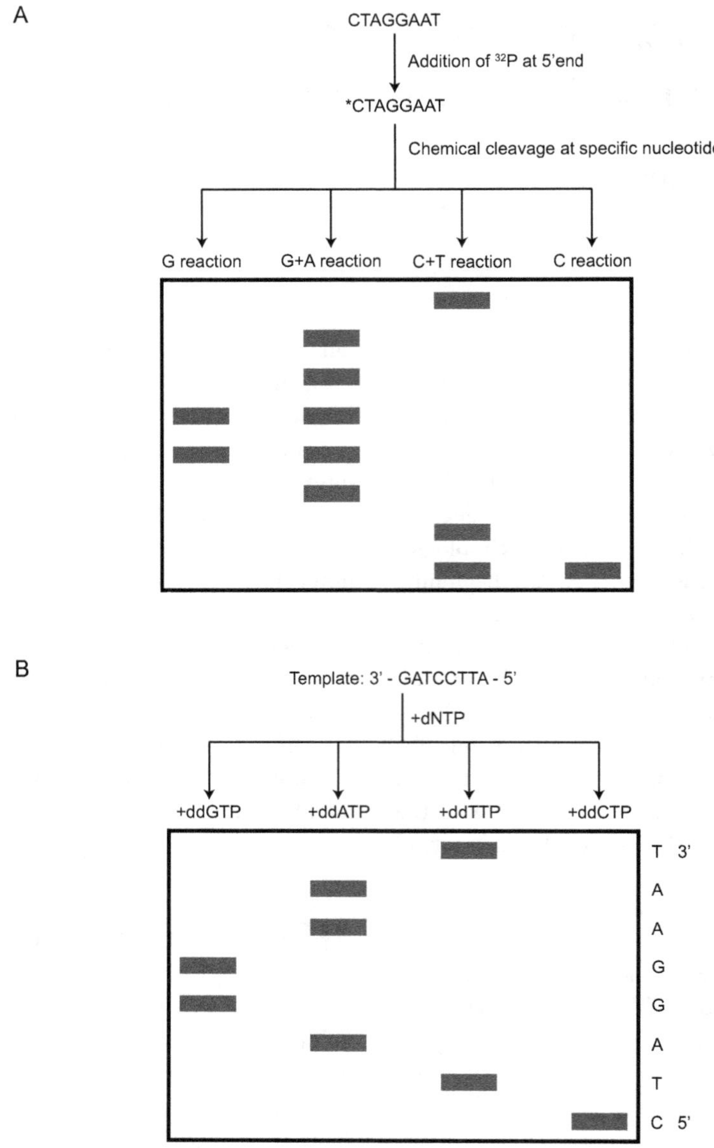

Figure 8.1 Early development DNA sequencing technologies. (A) Allan Maxam and Walter Gilbert introduced in the 1970s a method that relied on chemical treatment of DNA and subsequent cleavage at particular bases. Specifically, after radioactive labelling, the DNA is modified in four separate reactions. The G base is methylated by dimethyl sulphate, the purines (both A and G bases) are depurinated using formic acid, the pyrimidines (both C and T bases) are methylated using hydrazine, and only the C base is methylated using hydrazine and sodium chloride. The treated DNAs are then cleaved by hot piperidine and the fragments are size separated in a denaturing acrylamide gel. Finally, the DNA fragments are visualized by exposing the gel to X-ray film for autoradiography. (B) Frederick Sanger developed a method based on the use of dideoxynucleotides (ddNTPs) in addition to the standard nucleotides (dNTPs). Compared with dNTPs, ddNTPs contain a hydrogen group on the 3′ carbon instead of a hydroxyl group. Hence, when a ddNTP is incorporated into a sequence, the chain cannot be extended any further as a phosphodiester bond cannot form between the ddNTP and the next nucleotide. A DNA sample is aliquoted into four individual sequencing reactions, each of which contains a different dideoxynucleotide (ddGTP, ddATP, ddTTP, ddCTP) in addition to the normal nucleotides and DNA polymerase. The ddNTPs are in limited amounts, so that only a small fraction of DNA molecules cannot be extended further at each base and the reaction can still proceed for the other molecules. In this way, fragments of different lengths are generated. The ddNTPs are also radioactively or fluorescently labelled for detection.

the bacteriophage ΦX174 was the first complete DNA genome to be published (Sanger *et al.*, 1977a).

DNA sequencing became even more accessible when it was automated in the mid-1980s. Prior to automation, it was a labour-intensive experiment that required hazardous radioactivity and the manual running of polyacrylamide gels. However, radioactivity was eventually replaced by fluorescence, where a different coloured fluorophore was used for each of the bases, and the first automatic sequencer was announced in 1986 (Smith *et al.*, 1986). Since then, the sequencing machine has improved tremendously, such that today, the typical read length of each primer extension reaction is between 800 to 1000 basepairs. Armed with an automatic sequencer, scientists could begin to examine the sequences of genes and cDNAs with relative ease and this opened the door to the discovery of RNA editing sites.

The early days – one gene at a time

The first vertebrate RNA editing site was discovered in the transcript of the apolipoprotein B (apoB) gene (Powell *et al.*, 1987). At that time, two forms of the protein (apoB100 and apoB48) were known to exist, but no one knew exactly how they were related to each other. Examination of cDNA clones from the small intestine and liver as well as comparison with genomic DNA revealed that a cytosine in the pre-mRNA was converted to a uracil only in the small intestine to generate a premature STOP codon. Hence, as the CAA codon had been changed to UAA in the small intestine, the length of the protein in the human liver was 4563 amino acids long, while the length of the protein in the human small intestine contained only 2179 amino acids, 48% of the full length. The two isoforms perform different functions in lipid transport: apoB100 is a key component of low-density lipoprotein particles in the liver, while apoB48 is necessary for the formation of chylomicrons in the small intestine. Subsequently, APOBEC1, the enzyme that performed the C-to-U editing of apoB RNA, was cloned from rat small intestine (Teng *et al.*, 1993) and was further shown to require additional auxiliary proteins to carry out its editing function (Wedekind *et al.*, 2003).

A few years after the discovery of C-to-U editing in apoB, another type of substitutional editing was discovered – conversion of adenosine to inosine. Inosine is read by reverse transcriptases and the translation machinery as guanosine and it also basepairs with a cytosine. Interestingly, A-to-I editing was initially identified as an activity that unwound double-stranded RNA (Bass and Weintraub, 1987; Rebagliati and Melton, 1987). This unwinding activity was not far from the truth as it was later established that the enzymes that catalysed the A-to-I reaction, adenosine deaminases acting on RNA or ADARs, indeed recognize double-stranded RNA substrates and convert perfect A:U basepairing to imperfect I:U basepairing, thereby disrupting the double-stranded structure. While the first demonstration of A-to-I editing was based on chromatographic analysis of *in vitro* synthesized RNA duplexes (Bass and Weintraub, 1988), the first *in vivo* examples of A-to-I editing in mammals were found in the glutamate receptor subunits, GluA2 (previously called GluR2), GluK1 (previously called GluR5), and GluK2 (previously called GluR6), and were discovered largely through careful sequence analysis of cDNA and genomic DNA (Sommer *et al.*, 1991).

It soon became clear that A-to-I RNA editing is by far the most common type of substitutional editing, even though C-to-U editing was uncovered first. In mammals, five adenosines in the transcripts encoding the 2C subtype of serotonin receptor, 5-HT_{2C}R, have been found to be edited, three of which result in amino acid changes (Burns *et al.*, 1997). In *Xenopus*, sense and antisense transcripts from the basic fibroblast growth factor (bFGF) gene can hybridize to each other in the nucleus to form a perfect RNA duplex, resulting in the conversion of multiple adenosines to inosines within the duplex (Kimelman and Kirschner, 1989; Saccomanno and Bass, 1999). In *Drosophila*, specific A-to-I editing was first reported for a calcium channel α1 subunit gene (Smith *et al.*, 1996), while promiscuous A-to-I editing was first observed in transcripts from the 4f-rnp gene (Petschek *et al.*, 1996). Even viruses have been found to exploit A-to-I editing. For example, a single A-to-I editing event in the hepatitis delta virus causes a STOP codon to be converted to a tryptophan, thereby permitting the virus to generate two proteins from only one open reading frame (Polson *et al.*, 1996). Hence, by the end of the last century, the field of RNA editing was buzzing with excitement – how many RNA editing sites, in particular the A-to-I type, are there in nature?

Comparative genomics and bioinformatic screens of EST and SNP databases – moderate throughput approaches

By the turn of the twenty-first century, the pace of sequencing had accelerated so rapidly that billions of DNA nucleotides had been deposited in various databases, awaiting further examination. Not only had whole genomes of different model organisms been assembled, including the yeast *Saccharomyces cerevisiae* (Goffeau *et al.*, 1996), the worm *Caenorhabditis elegans* (C. elegans Sequencing Consortium, 1998), the fruit fly *Drosophila melanogaster* (Adams *et al.*, 2000), and the mouse (Mouse Genome Sequencing Consortium *et al.*, 2002), numerous cDNA clones and libraries of expressed sequence tags (ESTs) had been sequenced as well. These rich sources of information served as the basis for a higher-throughput identification of RNA editing sites beyond the dozen or so sites that had been serendipitously discovered in the 1990s.

A comparative genomics study of 18 *Drosophila* species identified 16 previously unknown RNA editing targets in the nervous system (Hoopengardner *et al.*, 2003). The authors searched 914 genes for highly conserved sequences and then performed reverse transcription–polymerase chain reaction (PCR) to check for the presence of guanosine in place of adenosine. The rationale for this approach was that any mutation, even in synonymous positions, may destroy the double-stranded RNA structure that is required for dADAR to bind. The 16 editing sites were further confirmed by sequencing cDNA from dADAR-null flies, which only contained pure adenosine signal, unlike cDNA from wildtype flies, which contained a mix of adenosine and guanosine signals.

Beyond *Drosophila*, evolutionary conservation can also be applied to find RNA editing sites in vertebrates and in particular mammals. Mismatches between RNA and DNA that are due to genomic variability between individuals or are simply a result of sequencing errors are much less likely to be conserved between different species. In addition, editing sites that are functionally important are expected to be highly conserved. A search for conserved genomic loci where an A-to-G mismatch appears in both human and mouse at the same position uncovered four novel editing targets on top of the known editing substrates encoding the glutamate receptor subunits (Levanon *et al.*, 2005). The new sites were validated not only in human and mouse but were further verified in chicken.

To leverage on the large collection of human ESTs and cDNAs, a computational pipeline was developed to identify genetic variants that are likely to be due to RNA editing from these sequences (Levanon *et al.*, 2004). The key to the methodology is to search for mismatches only in potential double-stranded regions, as ADAR substrates typically form dsRNA stems. Other important features of the pipeline include discarding poor quality sequences, removing mismatches that are supported by less than 5% of available sequences, and filtering known single nucleotide polymorphisms (SNPs). The outcome of this computational approach is the mapping of more than 10,000 A-to-I editing sites in ~ 1600 different genes, with the majority of sites occurring in non-coding regions of the genome and in particular within *Alu* repeats. The finding that numerous clustered editing sites occur within the large number of *Alu* elements in the human genome was independently confirmed by a separate study that examined high quality full-length cDNAs for groups of A-to-G mismatches (Kim *et al.*, 2004).

Owing to their clinical relevance, genomic sequence variations have been subjected to intense investigation over the past decade. Since the completion of the Human Genome Project (International Human Genome Sequencing Consortium, 2004), scientists have focused on understanding the areas of the genome that vary between individuals. In particular, many genome-wide association studies have linked a variety of SNPs to a wide range of diseases. Currently, the largest archive of polymorphisms is dbSNP (Sherry *et al.*, 2001) and the latest version, build 137, also includes genetic variants identified in phase 2 of the 1000 Genomes Project, which is a collaborative effort between multiple research institutions to sequence the genomes of a large number of people in order to generate a comprehensive catalogue of human genetic variation (1000 Genomes Project Consortium, 2010).

While the majority of SNPs are discovered by analysing genomic DNA from many individuals, a substantial fraction of polymorphisms are annotated based on analysis of expressed sequence data, also known as cDNA SNPs (Buetow *et al.*, 1999; Irizarry *et al.*, 2000). In principle, variants identified in a transcript may be a reflection of true variants in the corresponding DNA from which the RNA is transcribed, but they may also be due to gene duplications, sequencing errors, or RNA editing. Indeed, many cDNA SNPs in public databases are apparently not validated (Jiang *et al.*, 2003), which raises the question of how accurate the SNP databases actually are. This motivated two separate research groups to examine the databases for mis-annotated SNPs (Eisenberg *et al.*, 2005; Gommans *et al.*, 2008). The first group found that 102 expressed SNPs, 96 of which were located in *Alu* elements, showed strong characteristics of A-to-I editing and they proceeded to validate seven sites by Sanger sequencing (Eisenberg *et al.*, 2005), while the second group developed a scoring system based on an ADAR motif, conservation, and double-stranded character and they too managed to validate some of the highest ranking A-to-I recoding sites (Gommans *et al.*, 2008). Collectively, these studies demonstrated that variants identified in a transcript may not necessarily be due to variation in the DNA and that bona fide RNA editing sites may reside in SNP databases.

Entering a new era of ultra-high-throughput sequencing

The goal of the genomics community at the beginning of the century was to develop technologies for affordable multiplex sequencing, that is, to sequence thousands, if not millions, of DNA oligomers simultaneously and cheaply. This will enable whole-genome sequencing of many more organisms and will also bring personalized medicine closer to reality. In 2005, the first commercial next-generation sequencing platform was announced by 454 Life Sciences Corporation, which has since been acquired by Roche. The 454 system uses emulsion PCR and pyrosequencing, which is a sequencing-by-synthesis technique that relies on detection of pyrophosphate release upon nucleotide incorporation, and the earliest version was able to sequence 25 million bases at 99% accuracy (Margulies *et al.*, 2005). At around the same time, an academic laboratory reported the construction of a bead-based multiplex DNA sequencer using off-the-shelf instrumentation and reagents and sequenced the genome of an evolved bacteria strain at less than one error per million bases (Shendure *et al.*, 2005).

Within the short span of 7 years or so, a dazzling suite of ultra-high-throughput sequencing technologies is now available on the market and even more are expected to be announced in the near future. Most of them are essentially different implementations of cyclic array sequencing, whereby a two-dimensional array of distinct DNA clusters are enzymatically manipulated and imaged repeatedly (Shendure and Ji, 2008; Shendure *et al.*, 2011) (Fig. 8.2). Each technology or platform offers its own advantages and disadvantages but all of them are capable of generating a huge amount of data and can be leveraged for the study of RNA editing.

Applications of next-generation sequencing technologies to RNA editing

One of the earliest demonstrations of how new high-throughput technologies may be applied to the identification of RNA editing sites comes from sequencing pioneer George Church's laboratory (Li *et al.*, 2009). In this study, the authors combed through all human mRNAs and ESTs to look for good quality A-to-G mismatches from the reference genome and designed padlock probes to capture the shortlisted sites from human genomic DNA and cDNA samples. From over 8 million mRNAs and ESTs, they filtered out low quality sequences, poorly aligned

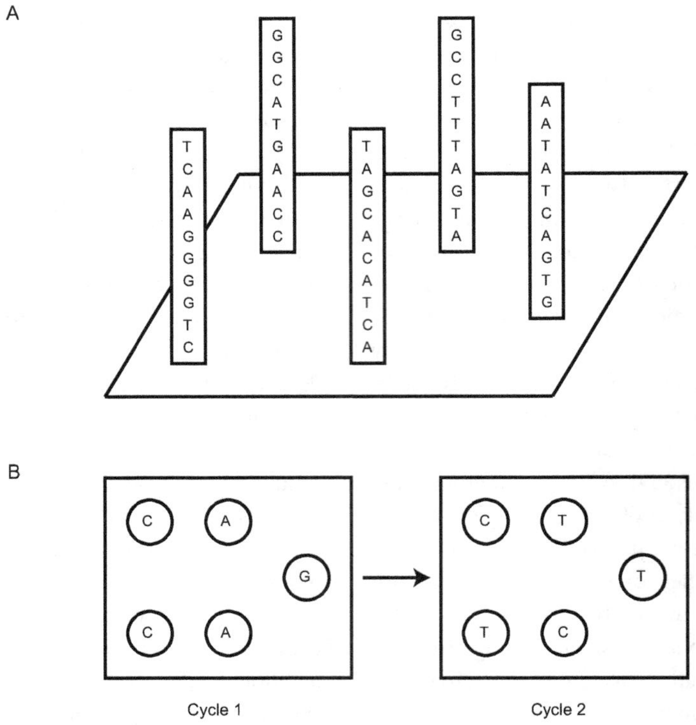

Figure 8.2 Ultra-high-throughput cyclic array sequencing. (A) The majority of current next-generation sequencing technologies follow a general paradigm known as cyclic array sequencing. Millions of individual DNA features are physically immobilized on a two-dimensional array. Each feature may be a single DNA molecule or a cluster of identical molecules. The features are sufficiently close to one another, so that as many of them can be packed on the array as possible, but they are not so close that they cannot be resolved as distinct units. In addition, they may be immobilized in an orderly manner or they may be scattered randomly. Since the DNA molecules are fixed to a surface, a single reagent volume can be used to simultaneously manipulate all the features in parallel. The sequencing procedure is cyclic because an enzymatic reaction coupled to some readout is repeatedly applied to identify the base at each specific position in every DNA feature on the array. In the popular Illumina sequencing platform, sequencing by synthesis is employed and the dNTPs used are fluorescently labelled, with each dNTP species labelled with a different colour. In each cycle after a single base extension has occurred, the array or flow cell is imaged in the four colours in order to interrogate the base identity at every DNA cluster. (B) In the example shown, there are five DNA features on the array. During the first cycle, the base closest to the surface is identified for all the features. During the second cycle, the next base for every feature is reported. The whole process is repeated until the end of all the DNA molecules is reached.

sequences, repetitive regions, and sequences that contained non A-to-G mismatches and eventually designed probes for ~36,000 sites. After target capture and DNA sequencing (Fig. 8.3), a few hundred RNA editing sites were identified, thereby expanding the number of known human editing sites in non-repetitive regions by about 10-fold.

Besides target capture, chemical methods coupled with deep sequencing may be used to identify RNA editing sites. To detect A-to-I edited positions, the inosine in the RNA may be reacted with acrylonitrile to form N^1-cyanoethylinosine, which blocks the progress of reverse transcriptases (Sakurai *et al.*, 2010). Hence, in untreated samples, an edited site in the RNA will show up as a mixture of adenosines and guanosines during sequencing, while in treated samples, the same position will only contain adenosines because the cyanoethylated inosine cannot be reverse transcribed (Fig. 8.4). By applying this technique, also termed inosine chemical erasing (ICE),

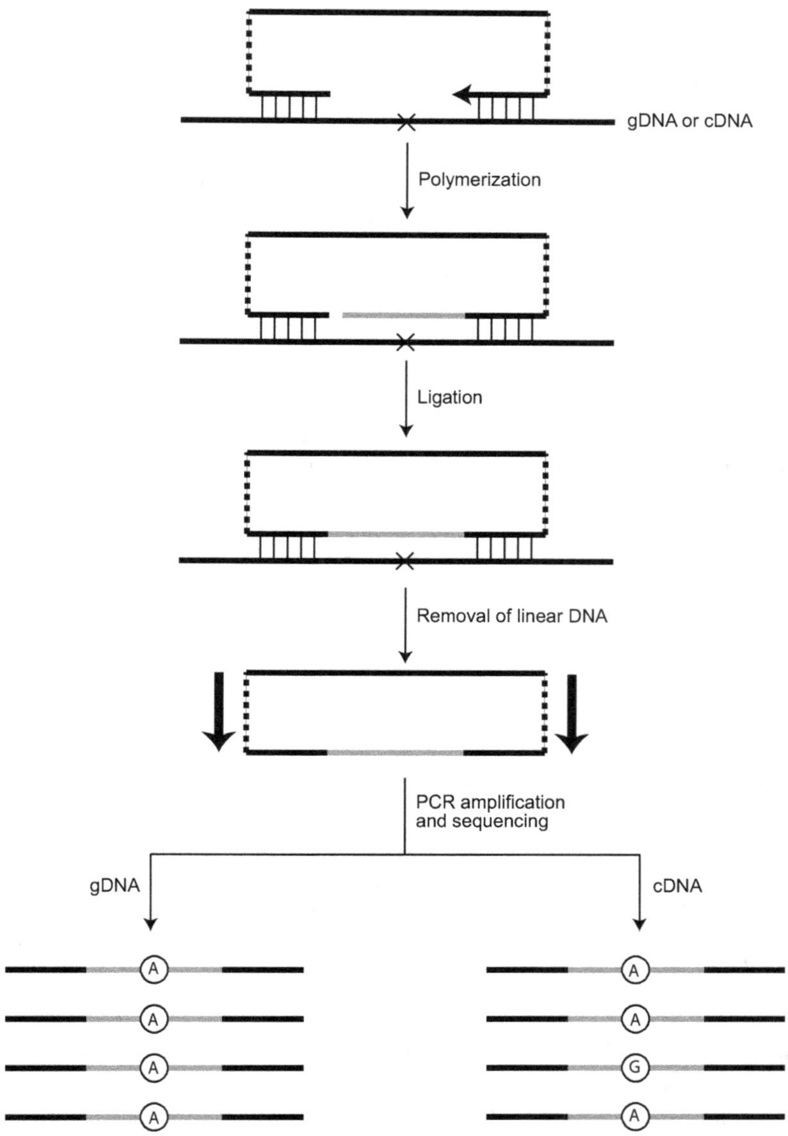

Figure 8.3 Padlock probe capture and sequencing of DNA molecules-of-interest. Padlock probes, which are also known as molecular inversion probes, are designed to target regions flanking both sides of a locus-of-interest. In the schematic, the locus-of-interest is a putative editing site, as indicated by a cross. The sequences of the probes that hybridize to the flanking regions are known as a ligation arm or an extension arm. After incubating the probes with genomic DNA (gDNA) or DNA reverse transcribed from RNA (cDNA) for over 24 hours to allow the hybridization to occur, DNA polymerase or Stoffel fragment is added to copy the captured sequences into the probes. DNA ligase is then added to circularize each probe, before exonucleases are used to remove linear DNA (both the gDNA or cDNA template and any uncircularized probe). Common primers, as indicated by the thick black arrows, are used to amplify the captured sites and the PCR products are then subjected to sequencing.

researchers uncovered ~5000 RNA editing sites in the transcriptome of a human brain.

In a comprehensive search for editing targets of the cytidine deaminase APOBEC1, researchers used RNA sequencing (RNA-Seq) to compare the transcriptomes of jejunal epithelial cells from wild-type mice and congenic Apobec1 null mice (Rosenberg *et al.*, 2011). They found 39 sites that

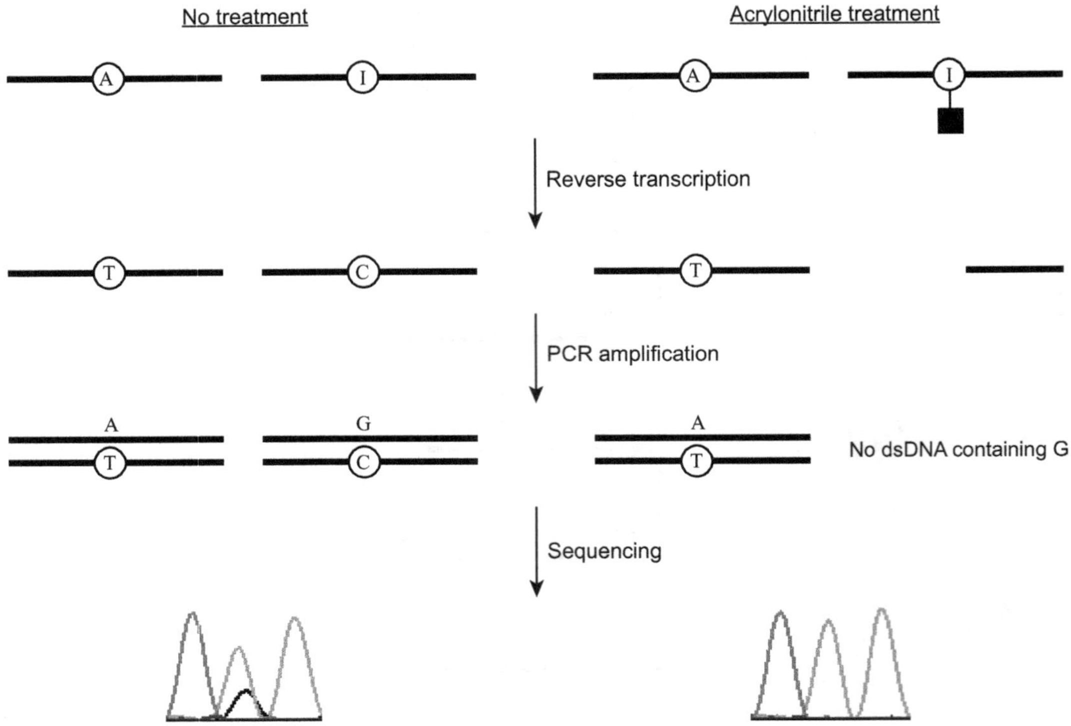

Figure 8.4 A chemical method to identify inosines. A RNA sample is divided into two aliquots, such that one aliquot receives no chemical treatment, while the other aliquot is treated with the chemical acrylonitrile in order to modify all the inosines. Acrylonitrile, which has the formula C_3H_3N, reacts with inosine in mildly alkaline conditions to form N^1-cyanoethylinosine through a type of nucleophilic conjugate addition. The presence of the cyanoethyl group, as indicated by a solid black box in the schematic, inhibits the progression of the reverse transcriptase. Hence, after reverse transcription, PCR amplification, and sequencing, both the A base and the G base can be detected at an A-to-I editing site for the untreated aliquot, but only the A base would be present for the chemically treated aliquot.

contained both C and T in the wild-type cDNA but only a C in Apobec1$^{-/-}$ cDNA, which indicated that these sites may be edited by Apobec1. 33 out of the 39 sites could be further validated by Sanger sequencing, including the apoB gene. Interestingly, with the exception of the edited coding site in apoB, all the remaining 32 sites are located in 3′UTRs. Nevertheless, regardless of the location, the small number of C-to-U sites uncovered in the study suggests that C-to-U editing is much less common than A-to-I editing.

The most powerful application of next-generation sequencing technologies lies in *de novo* unbiased discovery of RNA editing sites. As any typical next-generation sequencing run generates millions of reads, which is approximately equivalent to an entire EST database, it permits sensitive detection of editing events. By comparing RNA sequencing (RNA-Seq) data and whole genome sequencing data that were generated from the same sample, one can identify positions in the transcripts that are different from the corresponding genomic DNA loci from which the RNA molecules are derived. In a detailed analysis of the *Drosophila* transcriptome, 972 A-to-I editing sites within the transcripts of 597 genes were identified (Graveley *et al.*, 2011). Interestingly, most of the edited positions in *Drosophila* result in amino acid changes, unlike in mammals, where the vast majority of A-to-I editing occurs in non-coding regions.

Various research groups have also attempted to find new editing sites in different human cell lines. Perhaps surprisingly, one report claims that there are widespread RNA–DNA differences in B-cell lines, including thousands of non-A-to-G mismatches (Li *et al.*, 2011). Subsequently, other studies show that while the majority of RNA–DNA mismatches appear to be A-to-G disparities, there are also a significant number of non-canonical differences (Ju *et al.*, 2011; Bahn *et al.*, 2012; Peng *et al.*, 2012). Taken together, these potentially exciting results suggest that there are other types of substitutional RNA editing in mammals besides A-to-I and C-to-U editing. However, before relying on these results, one has to carefully examine the analysis pipeline used to identify the non-canonical editing sites.

Some important analysis considerations

Conceptually, bioinformatic identification of editing events from next-generation sequencing data appears to be straightforward – simply 'compare' the RNA-Seq reads with the genome sequencing reads and identify positions that do not agree between the two datasets. Typically, sequencing reads are first mapped to a reference genome, hence variants are first identified in the RNA and also in the DNA with respect to the reference genome and then the RNA editing candidates will be the set of RNA variants that are absent from the set of DNA variants.

In practice, there are several important issues to consider during the analysis process. If they are not properly handled, one can easily miss out on true editing sites (false negatives) or wrongly identify editing events (false positives). Here, we summarize some of the important issues to take note of.

Selective editing versus hyperediting

ADARs exhibit two different modes of adenosine deamination *in vivo* – highly specific RNA editing and hyperediting. Specific editing typically occurs in non-repetitive regions and only a single (or at most a small number of) adenosine(s) within an RNA is deaminated. On the other hand, hyperediting typically occurs in repetitive regions, such as *Alu* repeats, and numerous adenosines are edited in clusters. Most of the known A-to-I sites occur in repetitive regions via hyperediting.

Many reads will not align perfectly to the reference genome due to sequencing errors, SNPs in the genome, and RNA editing. Typically, one allows a small number of mismatches between each read and the reference genome but the exact number of mismatches allowed is somewhat arbitrary and can have different outcomes. When the number of permitted mismatches is small, reads may be placed more accurately but there can be a bias towards mapping of reads that contain reference bases instead of edited bases. Hence, one will only be able to identify transcripts that are selectively edited at a few, or even just one, site. On the other hand, when the number of permitted mismatches is larger, one can discover more clustered editing sites, such as the numerous A-to-I modifications observed in *Alu* repeats (Athanasiadis *et al.*, 2004; Kim *et al.*, 2004; Levanon *et al.*, 2004). However, there are likely to be more mapping errors as well.

One possible strategy to discover hyperedited RNAs is to perform '*in silico* editing' prior to mapping. For example, one can transform all adenosines in the genome and sequencing reads to guanosines, which effectively reduces the number of DNA bases from four to three, and then perform the mapping. This transformation has been effectively applied to discover 14,000 editing sites in just 760 transcripts (Carmi *et al.*, 2011).

Intrinsic mapping errors

Different mapping tools rely on different algorithms to place sequencing reads. For example, SHRiMP (Rumble *et al.*, 2009) uses spaced seeds for similarity searches and the Smith–Waterman algorithm to align reads, whereas Bowtie (Langmead *et al.*, 2009) and BWA (Li and Durbin, 2009) use schemes that are based on a reversible transformation of a block of input text, known as the Burrows–Wheeler transform. Hence, each tool has its own intrinsic vulnerabilities and may make different types of mapping errors. To overcome this issue, one can use two or more mappers to align the sequencing reads, call editing sites based on the various alignments, and then find the robust sites that are always detected regardless of which mapper is used. We find that mappers that do not support insertions and deletions (indels) tend to perform worse than mappers that do. Furthermore, additional filtering steps can be taken to remove systematic mapping artefacts. For example, putative editing sites that show an extreme degree of variation in the RNA (that is, an editing level of 100%) are likely to be false and may be discarded, since it is doubtful that editing is completely efficient at any site.

Read ends

Extra care should be taken during the analysis of both ends of each read. In general, sequencing quality tends to decline towards the 3′ end of a read, so one may need to trim off the poor quality bases as they are not reliable. On the 5′ end, a different problem exists. During the preparation of most sequencing libraries, the first strand synthesis (reverse transcription) step involves the use of random primers. When a random primer, which is typically six bases long, binds to a RNA molecule, the reverse transcriptase can then synthesize the cDNA from the primer. However, the primer may not bind perfectly all the time. For example, five out of six nucleotides may bind to the transcript and the polymerase can still extend the first strand. When this happens, one can mistakenly call the error introduced by the random primer as an editing site. To overcome this problem, one should discard all RNA variants that fall in the first six bases of a read (assuming that hexamers are used).

SNPs and somatic mutations in the genome

As mentioned earlier in the chapter, a variant in a RNA molecule can be due to RNA editing, a SNP in the genome, somatic mutations, or sequencing errors. To decrease the likelihood of sequencing errors, one can require that the variant is observed in more than one read or that a putative editing site is detected in more than one sample. Ideally, one should also have the corresponding genome or exome data from the same sample that the RNA-Seq library was generated from. For each putative editing site, the genomic position must be homozygous. While the criteria for homozygosity can be somewhat arbitrary, the consensus is that a coverage of at least 10 or 20 reads is required and that all the genome or exome sequencing reads must show the same nucleotide at the editing site. In principle, this should be sufficient to eliminate SNPs in the genome or somatic mutations. However, in reality, it is cost-prohibitive to sequence every genome deep enough so that there is adequate coverage at every nucleotide position. The distribution of sequencing reads across the genome is also non-uniform. Hence, one should filter off all the known genomic SNPs that are annotated in databases or discovered through various large scale genome sequencing or exome sequencing projects, such as the 1000 Genomes Project. Although the catalogue of known SNPs is still incomplete and is currently biased against certain ethnicities, there are ongoing efforts to sample the genomes of racially diverse or remotely located human populations so that in the near future, subtraction of catalogued genomic SNPs from variants detected in RNA molecules would result in a strong enrichment of genuine RNA editing sites.

Splice junctions

Alternative splicing is a widespread post-transcriptional mechanism that increases transcriptome diversity. It occurs in animals and plants and can even be found in yeast (Graveley, 2001) and bacteria (Robart *et al.*, 2004). In mammals, the vast majority of multiexon genes are believed to undergo alternative splicing (Pan *et al.*, 2008; Wang *et al.*, 2008) and different isoforms of the same gene have been found to perform different functions (Gabut *et al.*, 2011). Intriguingly, a few studies have shown that A-to-I editing can create or destroy splice sites, thereby creating novel isoforms (Rueter *et al.*, 1999; Beghini *et al.*, 2000).

Unfortunately, identifying RNA editing sites close to a splice junction is not an easy task. Although one can align reads to the transcriptome in addition to the reference genome, accurate mapping of reads spanning exon-exon boundaries is still challenging for various reasons. First, the lengths of introns are highly variable and can span from less than 100 bases to more than 100,000 bases. Second, some splice junctions may not be correctly annotated. Hence, a fraction of mappings to the transcriptome (RefSeq, ENSEMBL, or Gencode annotations) may be wrong. Third, even though 98% of annotated junctions utilize the canonical splicing signals, there is still a sizeable number of junctions that rely on non-canonical splicing sequences. When a read spans a non-canonical junction and a variant is detected at the junction, it can be difficult to know whether the variant represents a true editing event or whether the junction is simply incorrect. Fourth, there may exist two isoforms that differ from each other by only a few bases at the splice junction. Often, mapping to one isoform results in an editing site, while mapping to the other isoform abolishes the variant. Hence, given all these issues, RNA editing candidates identified close to a splice site can easily be artefacts and they should be examined carefully.

Gene duplications and pseudogenes

Correct read placement is essential for identifying bona fide RNA editing sites, but read mapping can be complicated by the presence of paralogues. In a true editing event, two RNA molecules that are identical except for a particular variant position must originate from the same genomic DNA locus. In a false editing call, the two RNA transcripts actually originate from duplicated genes that differ from each other at the exact variant position. To check for duplicated genes, one can use BLAST (Altschul *et al.*, 1990) or BLAT (Kent, 2002) to determine if sequences flanking putative editing sites can be found anywhere else in the genome. One can also use BLAST or BLAT to confirm whether each sequencing read covering an editing site is correctly mapped or whether the read only maps to one unique position in the genome. In addition, RNA editing candidates should be checked against databases of pseudogenes (Karro *et al.*, 2007), many of which have been found to be transcribed (Zheng *et al.*, 2007). By implementing all these different filters, one can remove many artefacts resulting from gene duplications and pseudogenes.

The devil is in the details

Given that numerous caveats lie in the identification of RNA editing sites from next-generation sequencing data, we should carefully re-examine the claim that there are widespread RNA–DNA sequence differences in humans of all twelve possible substitutional types (Li *et al.*, 2011). Four different laboratories have since published analyses that appear to debunk, or at the very least challenge, the claim. In the first report, researchers found that the majority of the putative editing sites reported by Li *et al.* (2011) can be explained by faithful transcription of paralogous sequences in the genome (Schrider *et al.*, 2011). Subsequently, three back-to-back technical comments highlighted multiple artefacts that contributed to the false calling of editing sites by Li *et al.* (2011) (Kleinman and Majewski, 2012; Lin *et al.*, 2012; Pickrell *et al.*, 2012). An unacceptably large fraction of putative editing sites is very likely to be due to incorrect mapping because they occur only at read ends or the supporting reads align only in one direction (Fig. 8.5). Specifically, approximately 87% of the sites fail a position bias test or a strand bias test ($P < 0.05$). In addition, the DNA sequencing coverage was very low at numerous sites, which introduces uncertainty in the genotype. Notably, many putative RNA–DNA differences are actually known SNPs annotated in existing databases. Some putative differences also arise from genomic loci known to be copy number variable in humans, such as the HLA-DQB2 gene. Taken together, the results from careful analysis of the data used by Li *et al.* (2011) by different research groups indicate that the vast majority of the alleged editing sites are false positives.

Our laboratory has developed a computational pipeline to accurately identify RNA editing sites from deep RNA-Seq and genome sequencing data (Ramaswami *et al.*, 2012). We used BWA (Li and Durbin, 2009) to map reads to both the reference genome and also to the annotated transcriptome. After filtering duplicate reads, we singled out sites that contained variants in at least two RNA-Seq reads and appeared to be homozygous in the genomic DNA based on the sequencing data. To ensure that no variation in the human genome may confound our results, we removed all SNPs annotated in dbSNP (except cDNA SNPs), the 1000 Genomes Project, and other exome sequencing projects. Furthermore, to avoid false positives arising from imperfect random hexamer priming, we discarded the first six bases of each read after mapping. With these steps, we found that over 90% of the sites identified in *Alu* regions were A-to-G changes, which are indicative of

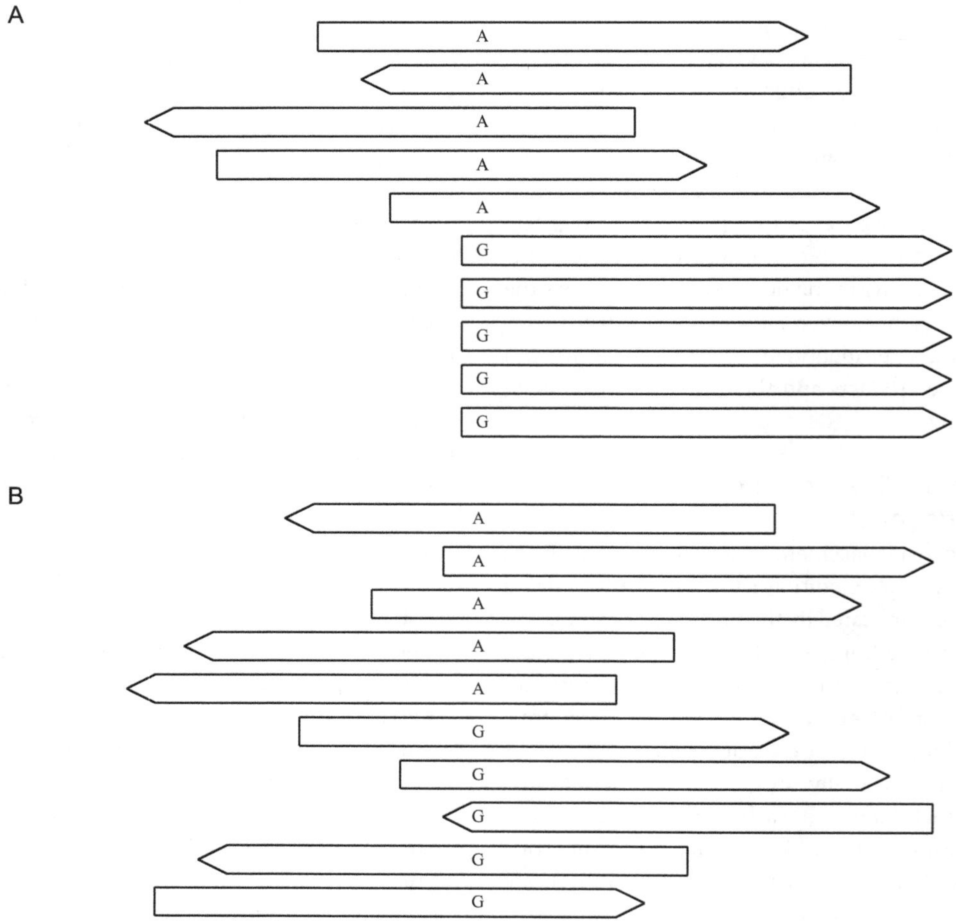

Figure 8.5 False and true positive alignments of sequencing reads. (A) The schematic depicts an example of mis-alignment of sequencing reads. In this example, the reference base in the genome at a particular site is A, while the non-reference or mismatched base is G. As shown, all the reads containing the G base are oriented in one direction only (strand bias). Furthermore, the variant site is present only at read ends and nowhere else (position bias). Here, all the G mismatches occur at the 5′ end and may be caused by imperfect priming during the first strand cDNA synthesis step of the library generation process. (B) The schematic illustrates a typical example of accurate read placement. Reads containing either the reference A base or the non-reference G base appear to be randomly oriented. In addition, the mismatch can occur at any position along a read.

A-to-I editing. Since repetitive regions tend to be hyperedited by ADAR, it is not surprising that the overwhelming majority of the sites discovered in these regions are A-to-I modifications. What is more challenging to find, however, are the true editing sites in non-repetitive regions. To address the challenge, we imposed more stringent criteria to identity RNA editing candidates in non-repetitive regions. We required that there are at least three RNA-Seq reads that contained the variant nucleotide and that the minimum editing level is at least 10%. We also removed intronic sites that are within four bases of splice junctions and also all sites that are in homopolymer runs greater than four basepairs. Finally, we used BLAT (Kent, 2002) to eliminate sites that lie in duplicated regions of the genome. With these additional steps, we found that over 70% of the RNA editing candidates in non-repetitive regions are A-to-I mismatches. Hence, our analysis confirmed that A-to-I editing is by far the most prevalent type of editing in humans and while other non-canonical types of editing may still exist, they are likely to be very infrequent and may occur only in specific biological contexts.

Other methodologies, besides ours, can also be used to accurately identify RNA editing sites. The Sanger Institute sequenced the genomes of 17 inbred mouse strains (Keane *et al.*, 2011) as well as the transcriptomes of the brains from 15 of these strains and developed a distinct pipeline for calling editing sites (Daneck *et al.*, 2012). The reads were first mapped using BWA (Li and Durbin, 2009) and variants were called using a statistical framework built into a set of utilities known as samtools/bcftools (Li, 2011). Those reads that contained RNA editing candidates were realigned in a splice-aware manner using a different tool known as exonerate (Slater and Birney, 2005) and reads that were not uniquely mapped were discarded. In order to remove artefacts, additional filters were imposed. For a position to qualify as an editing site, there must be a minimum cumulative depth of 10 RNA-Seq reads and no mismatch in genomic reads. The variant must also be observed in at least two biological replicates. After correcting for strand bias and variant distance bias, the authors identified 5579 editing sites, of which only 2.4% were non-A-to-I or T-to-C mismatches. Finally, to take into account the fact that ADAR frequently modifies multiple bases that are close to one another, any RNA mismatches within 10 bp from an editing site were iteratively added together with clusters of eight or more mismatches of the same type within a 20-bp window. All the sites included in this manner were A-to-I changes and the number of RNA editing sites was increased to 7389. In summary, the analysis performed by the Sanger Institute further verified that A-to-I modification is the most common type of substitutional RNA editing in mammals and that other non-canonical types of editing are most likely to be rare, if they even exist.

Déjà vu – back to the beginning

The world-renowned physicist Richard Feynman once said, 'It doesn't matter how beautiful your *theory* is, it doesn't matter how smart you are. If it doesn't agree with *experiment*, it's wrong.' While next-generation sequencing technologies have enabled scientists to discover an abundance of novel RNA editing candidates, whether in coding regions or in non-coding regions, one should proceed cautiously by experimentally confirming the editing sites. For example, even though computational filters can help to remove artefacts caused by gene duplications and pseudogenes, there can still be unknown duplications in the genome, either due to imperfect assembly or individual-to-individual variability. One approach to validate the sites identified from high-throughput datasets is to perform traditional Sanger sequencing or clonal sequencing. Other experimental methods include mass spectrometry to identify peptides that contain the predicted amino acid substitution or *in vitro* biochemical assays to determine if a synthetic RNA containing the site-of-interest can be edited in the presence of cell extract or purified enzyme. In a surreal way, we are now back to the beginning, where careful experimental examination of each RNA editing candidate site has to be undertaken.

A crystal ball gaze into the future

While a direct comparison of DNA and RNA sequencing data generated from the same sample is conceptually the correct way of identifying RNA editing sites, next-generation sequencing technologies are very often used to study the transcriptome and hence, there is a plethora of RNA-Seq datasets available today without the corresponding genome or exome sequences. Can we leverage on this wealth of information for the purpose of discovering even more editing sites? Relative to a reference genome, variations present in a RNA-Seq dataset may be caused by different reasons, namely SNPs, somatic mutations, transcription errors, sequencing errors, or RNA editing. The same transcription or sequencing errors are highly unlikely to recur in multiple independent samples, particularly if the RNA was sequenced on different platforms. Similarly, *de novo* somatic mutations are not likely to occur repeatedly in various biological contexts. Hence, RNA variants that are present in many different datasets are most probably due to SNPs or RNA editing. In addition, there are ongoing international efforts to sequence the genomes of humans from diverse races and ethnicities. With the careful curation of SNPs, we may be able to subtract away the vast majority of SNPs from the recurring RNA variants, thereby leaving us with those variants that are most likely due to RNA editing. The development of such a computational pipeline in the future will greatly enhance our ability to uncover editing sites from RNA-Seq datasets only.

Besides annotating which nucleotide position(s) in a RNA molecule can be edited, it is imperative to know what the editing levels are at every known site, especially since changes in editing levels can result in devastating diseases. For example, while the editing of the glutamate receptor subunit GluA2 is essentially 100% efficient in motor neurons dissected from healthy individuals as judged by restriction digest patterns on an agarose gel, the editing of GluA2 in motor neurons from ALS patients is strikingly incomplete and can range from a mean of 38–75% (Kawahara *et al.*, 2004). More broadly, while RNA editing sites are usually first identified in a particular context, for example in a specific tissue or cell line or in one individual, the subsequent questions to address are whether the sites are also edited in other contexts and whether they are edited to the same extent in different situations. An application of ultra-high-throughput sequencing technologies lies in accurate measurement of RNA editing levels. One cannot simply rely on examining chromatograms from Sanger sequencing because even though the height of the peaks can provide a reasonable estimate of the editing level (Eggington *et al.*, 2011), they are not sufficiently quantitative and they also do not reveal low levels of editing. An alternative to peak height estimation is clonal sequencing, whereby one transforms vectors containing amplicons with the editing site into bacteria and then sequences individual colonies. However, this approach is labour-intensive and cannot be easily scaled up. High throughput sequencing provides the ability to examine multiple editing sites in different contexts simultaneously. By counting the number of reads corresponding to a particular allele, one obtains a highly accurate measure of the editing level of a particular site. An example of an 'editotyping' study is to determine how the editing of known sites changes over development. By analysing 28 different sites in the mouse brain at different developmental stages using 454 sequencing, researchers found that in general, the editing level is lowest during early embryogenesis and highest during adulthood (Wahlstedt *et al.*, 2009). In the future, more comprehensive studies are needed to determine the editing levels of the thousands of sites that have been discovered to date.

As more and more RNA editing sites are identified (Fig. 8.6), particularly of the A-to-I type, the next big question is whether these editing events are functional, and if so, what their specific roles are. How do recoding sites alter the properties of the proteins involved? Does editing in the 3′UTRs affect binding by microRNAs? Does dysregulation of editing contribute to diseases? In addition, we can also start to determine the similarities and differences in editing between various species. Does RNA editing contribute to higher cognitive abilities in humans compared with other primates?

Of course, one of the biggest and most controversial puzzles in the field still remains to be resolved: Do other types of substitutional RNA editing besides A-to-I and C-to-U exist in mammals? While there have been recent reports of their existence (Li *et al.*, 2011; Bahn *et al.*, 2012; Peng *et al.*, 2012), we and others have found serious flaws in the analysis, which cast shadows of doubts on the findings. In the future, new sequencing technologies that offer longer read lengths or directly sequence RNA may help to resolve this issue. If the non-canonical editing types do exist, it will be crucial to understand their mechanisms and functional impact on biology and medicine.

References

1000 Genomes Project Consortium. (2010). A map of human genome variation from population-scale sequencing. Nature *467*, 1061–1073.

Adams, M.D., Celniker, S.E., Holt, R.A., Evans, C.A., Gocayne, J.D., Amanatides, P.G., Scherer, S.E., Li, P.W., Hoskins, R.A., Galle, R.F., *et al.* (2000). The genome sequence of *Drosophila melanogaster*. Science *287*, 2185–2195.

Altschul, S.F., Gish, W., Miller, W., Myers, E.W., and Lipman, D.J. (1990). Basic local alignment search tool. J. Mol. Biol. *215*, 403–410.

Athanasiadis, A., Rich, A., and Maas, S. (2004). Widespread A-to-I RNA editing of Alu-containing

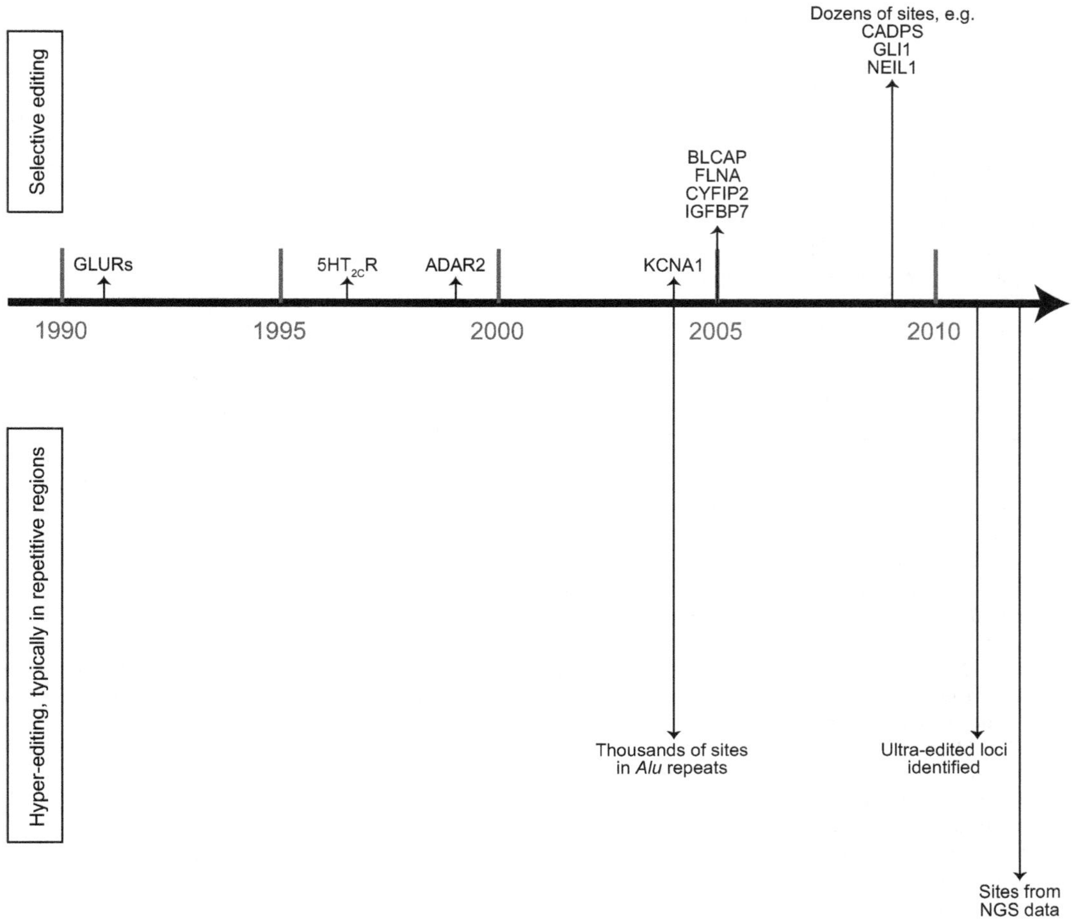

Figure 8.6 Timeline illustrating the rate of A-to-I editing site discovery in mammals. The first few mammalian RNA editing sites were found almost serendipitously, with researchers puzzled by why different cDNA clones of the same gene can contain different nucleotides at particular positions. The finding by Peter Seeburg's laboratory that the glutamate receptor subunits are edited set the stage for further discovery of A-to-I RNA modifications (Sommer *et al.*, 1991). At the beginning, editing sites were uncovered one at a time, with notable examples in the serotonin-2C receptor 5-HT_{2C}R (Burns *et al.*, 1997), the ADAR2 editing enzyme itself (Rueter *et al.*, 1999), and the potassium channel KCNA1 (Bhalla *et al.*, 2004). As higher throughput resources and technologies were developed, the rate of editing site discovery accelerated. The transcripts BLCAP, FLNA, CYFIP2, and IGFBP7 were found to be edited using a comparative genomics approach (Levanon *et al.*, 2005), while dozens of novel recoding sites were uncovered by targeted capture followed by high-throughput sequencing (Li *et al.*, 2009). Interestingly, whereas protein recoding sites proved to be challenging to find, clustered editing sites that typically reside in repetitive regions turned out to be much more abundant. Analysis of expressed sequences resulted in the mapping of over 10,000 A-to-I sites in the human transcriptome, the majority of which are in *Alu* repeats (Athanasiadis *et al.*, 2004; Kim *et al.*, 2004a; Levanon *et al.*, 2004), as well as the identification of hundreds of ESTs that cannot be aligned to the reference genome because they are ultra-edited (Carmi *et al.*, 2011). In addition, the recent availability of large datasets generated using next-generation sequencing (NGS) technologies have opened the floodgates for the detection of RNA editing sites (Ju *et al.*, 2011; Bahn *et al.*, 2012; Daneck *et al.*, 2012; Peng *et al.*, 2012; Ramaswami *et al.*, 2012).

mRNAs in the human transcriptome. PLoS Biol. *2,* e391.

Bahn, J.H., Lee, J.H., Li, G., Greer, C., Peng, G., and Xiao, X. (2012). Accurate identification of A-to-I RNA editing in human by transcriptome sequencing. Genome Res. *22,* 142–150.

Bass, B.L., and Weintraub, H. (1988). An unwinding activity that covalently modifies its double-stranded RNA substrate. Cell *55,* 1089–1098.

Bass, B.L., and Weintraub, H. (1987). A developmentally regulated activity that unwinds RNA duplexes. Cell *48,* 607–613.

Beghini, A., Ripamonti, C.B., Peterlongo, P., Roversi, G., Cairoli, R., Morra, E., and Larizza, L. (2000). RNA hyperediting and alternative splicing of hematopoietic cell phosphatase (PTPN6) gene in acute myeloid leukemia. Hum. Mol. Genet. *9,* 2297–2304.

Bhalla, T., Rosenthal, J.J., Holmgren, M., and Reenan, R. (2004). Control of human potassium channel inactivation by editing of a small mRNA hairpin. Nat. Struct. Mol. Biol. *11,* 950–956.

Buetow, K.H., Edmonson, M.N., and Cassidy, A.B. (1999). Reliable identification of large numbers of candidate SNPs from public EST data. Nat. Genet. *21,* 323–325.

Burns, C.M., Chu, H., Rueter, S.M., Hutchinson, L.K., Canton, H., Sanders-Bush, E., and Emeson, R.B. (1997). Regulation of serotonin-2C receptor G-protein coupling by RNA editing. Nature *387,* 303–308.

C. elegans Sequencing Consortium. (1998). Genome sequence of the nematode *C. elegans*: a platform for investigating biology. Science *282,* 2012–2018.

Carmi, S., Borukhov, I., and Levanon, E.Y. (2011). Identification of widespread ultra-edited human RNAs. PLoS Genet. *7,* e1002317.

Daneck, P., Nellaker, C., McIntyre, R.E., Buendia-Buendia, J.E., Bumpstead, S., Ponting, C.P., Flint, J., Durbin, R., Keane, T.M., and Adams, D.J. (2012). High levels of RNA editing site conservation amongst 15 laboratory mouse strains. Genome Biol. *13,* R26.

Eggington, J.M., Greene, T., and Bass, B.L. (2011). Predicting sites of ADAR editing in double-stranded RNA. Nat. Commun. *2,* 319.

Eisenberg, E., Adamsky, K., Cohen, L., Amariglio, N., Hirshberg, A., Rechavi, G., and Levanon, E.Y. (2005). Identification of RNA editing sites in the SNP database. Nucleic Acids Res. *33,* 4612–4617.

Gabut, M., Samavarchi-Tehrani, P., Wang, X., Slobodeniuc, V., O'Hanlon, D., Sung, H.K., Alvarez, M., Talukder, S., Pan, Q., Mazzoni, E.O., *et al.* (2011). An alternative splicing switch regulates embryonic stem cell pluripotency and reprogramming. Cell *147,* 132–146.

Goffeau, A., Barrell, B.G., Bussey, H., Davis, R.W., Dujon, B., Feldmann, H., Galibert, F., Hoheisel, J.D., Jacq, C., Johnston, M., *et al.* (1996). Life with 6000 genes. Science *274,* 546, 563–567.

Gommans, W.M., Tatalias, N.E., Sie, C.P., Dupuis, D., Vendetti, N., Smith, L., Kaushal, R., and Maas, S. (2008). Screening of human SNP database identifies recoding sites of A-to-I RNA editing. RNA *14,* 2074–2085.

Graveley, B.R. (2001). Alternative splicing: increasing diversity in the proteomic world. Trends Genet. *17,* 100–107.

Graveley, B.R., Brooks, A.N., Carlson, J.W., Duff, M.O., Landolin, J.M., Yang, L., Artieri, C.G., van Baren, M.J., Boley, N., Booth, B.W., *et al.* (2011). The developmental transcriptome of *Drosophila melanogaster*. Nature *471,* 473–479.

Hoopengardner, B., Bhalla, T., Staber, C., and Reenan, R. (2003). Nervous system targets of RNA editing identified by comparative genomics. Science *301,* 832–836.

International Human Genome Sequencing Consortium. (2004). Finishing the euchromatic sequence of the human genome. Nature *431,* 931–945.

Irizarry, K., Kustanovich, V., Li, C., Brown, N., Nelson, S., Wong, W., and Lee, C.J. (2000). Genome-wide analysis of single-nucleotide polymorphisms in human expressed sequences. Nat. Genet. *26,* 233–236.

Jiang, R., Duan, J., Windemuth, A., Stephens, J.C., Judson, R., and Xu, C. (2003). Genome-wide evaluation of the public SNP databases. Pharmacogenomics *4,* 779–789.

Ju, Y.S., Kim, J.I., Kim, S., Hong, D., Park, H., Shin, J.Y., Lee, S., Lee, W.C., Kim, S., Yu, S.B., *et al.* (2011). Extensive genomic and transcriptional diversity identified through massively parallel DNA and RNA sequencing of eighteen Korean individuals. Nat. Genet. *43,* 745–752.

Karro, J.E., Yan, Y., Zheng, D., Zhang, Z., Carriero, N., Cayting, P., Harrrison, P., and Gerstein, M. (2007). Pseudogene.org: a comprehensive database and comparison platform for pseudogene annotation. Nucleic Acids Res. *35,* D55–D60.

Kawahara, Y., Ito, K., Sun, H., Aizawa, H., Kanazawa, I., and Kwak, S. (2004). Glutamate receptors: RNA editing and death of motor neurons. Nature *427,* 801.

Keane, T.M., Goodstadt, L., Danecek, P., White, M.A., Wong, K., Yalcin, B., Heger, A., Agam, A., Slater, G., Goodson, M., *et al.* (2011). Mouse genomic variation and its effect on phenotypes and gene regulation. Nature *477,* 289–294.

Kent, W.J. (2002). BLAT – the BLAST-like alignment tool. Genome Res. *12,* 656–664.

Kim, D.D., Kim, T.T., Walsh, T., Kobayashi, Y., Matise, T.C., Buyske, S., and Gabriel, A. (2004). Widespread RNA editing of embedded alu elements in the human transcriptome. Genome Res. *14,* 1719–1725.

Kimelman, D., and Kirschner, M.W. (1989). An antisense mRNA directs the covalent modification of the transcript encoding fibroblast growth factor in *Xenopus oocytes*. Cell *59,* 687–696.

Kleinman, C.L., and Majewski, J. (2012). Comment on 'Widespread RNA and DNA sequence differences in the human transcriptome'. Science *335,* 1302; author reply 1302.

Langmead, B., Trapnell, C., Pop, M., and Salzberg, S.L. (2009). Ultrafast and memory-efficient alignment of short DNA sequences to the human genome. Genome Biol. *10,* R25.

Levanon, E.Y., Eisenberg, E., Yelin, R., Nemzer, S., Hallegger, M., Shemesh, R., Fligelman, Z.Y., Shoshan,

A., Pollock, S.R., Sztybel, D., *et al.* (2004). Systematic identification of abundant A-to-I editing sites in the human transcriptome. Nat. Biotechnol. *22,* 1001–1005.

Levanon, E.Y., Hallegger, M., Kinar, Y., Shemesh, R., Djinovic-Carugo, K., Rechavi, G., Jantsch, M.F., and Eisenberg, E. (2005). Evolutionarily conserved human targets of adenosine to inosine RNA editing. Nucleic Acids Res. *33,* 1162–1168.

Li, H. (2011). A statistical framework for SNP calling, mutation discovery, association mapping and population genetical parameter estimation from sequencing data. Bioinformatics *27,* 2987–2993.

Li, H., and Durbin, R. (2009). Fast and accurate short read alignment with Burrows-Wheeler transform. Bioinformatics *25,* 1754–1760.

Li, J.B., Levanon, E.Y., Yoon, J.K., Aach, J., Xie, B., Leproust, E., Zhang, K., Gao, Y., and Church, G.M. (2009). Genome-wide identification of human RNA editing sites by parallel DNA capturing and sequencing. Science *324,* 1210–1213.

Li, M., Wang, I.X., Li, Y., Bruzel, A., Richards, A.L., Toung, J.M., and Cheung, V.G. (2011). Widespread RNA and DNA sequence differences in the human transcriptome. Science *333,* 53–58.

Lin, W., Piskol, R., Tan, M.H., and Li, J.B. (2012). Comment on 'Widespread RNA and DNA sequence differences in the human transcriptome'. Science *335,* 1302; author reply 1302.

Margulies, M., Egholm, M., Altman, W.E., Attiya, S., Bader, J.S., Bemben, L.A., Berka, J., Braverman, M.S., Chen, Y.J., Chen, Z., *et al.* (2005). Genome sequencing in microfabricated high-density picolitre reactors. Nature *437,* 376–380.

Maxam, A.M., and Gilbert, W. (1977). A new method for sequencing DNA. Proc. Natl. Acad. Sci. U.S.A. *74,* 560–564.

Mouse Genome Sequencing Consortium, Waterston, R.H., Lindblad-Toh, K., Birney, E., Rogers, J., Abril, J.F., Agarwal, P., Agarwala, R., Ainscough, R., Alexandersson, M., *et al.* (2002). Initial sequencing and comparative analysis of the mouse genome. Nature *420,* 520–562.

Pan, Q., Shai, O., Lee, L.J., Frey, B.J., and Blencowe, B.J. (2008). Deep surveying of alternative splicing complexity in the human transcriptome by high-throughput sequencing. Nat. Genet. *40,* 1413–1415.

Peng, Z., Cheng, Y., Tan, B.C., Kang, L., Tian, Z., Zhu, Y., Zhang, W., Liang, Y., Hu, X., Tan, X., *et al.* (2012). Comprehensive analysis of RNA-Seq data reveals extensive RNA editing in a human transcriptome. Nat. Biotechnol. *30,* 253–260.

Petschek, J.P., Mermer, M.J., Scheckelhoff, M.R., Simone, A.A., and Vaughn, J.C. (1996). RNA editing in *Drosophila* 4f-rnp gene nuclear transcripts by multiple A-to-G conversions. J. Mol. Biol. *259,* 885–890.

Pickrell, J.K., Gilad, Y., and Pritchard, J.K. (2012). Comment on 'Widespread RNA and DNA sequence differences in the human transcriptome'. Science *335,* 1302; author reply 1302.

Polson, A.G., Bass, B.L., and Casey, J.L. (1996). RNA editing of hepatitis delta virus antigenome by dsRNA-adenosine deaminase. Nature *380,* 454–456.

Powell, L.M., Wallis, S.C., Pease, R.J., Edwards, Y.H., Knott, T.J., and Scott, J. (1987). A novel form of tissue-specific RNA processing produces apolipoprotein-B48 in intestine. Cell *50,* 831–840.

Ramaswami, G., Lin, W., Piskol, R., Tan, M.H., Davis, C., and Li, J.B. (2012). Accurate identification of human Alu and non-Alu RNA editing sites. Nat. Methods *9,* 579–581.

Rebagliati, M.R., and Melton, D.A. (1987). Antisense RNA injections in fertilized frog eggs reveal an RNA duplex unwinding activity. Cell *48,* 599–605.

Robart, A.R., Montgomery, N.K., Smith, K.L., and Zimmerly, S. (2004). Principles of 3′ splice site selection and alternative splicing for an unusual group II intron from *Bacillus anthracis.* RNA *10,* 854–862.

Rosenberg, B.R., Hamilton, C.E., Mwangi, M.M., Dewell, S., and Papavasiliou, F.N. (2011). Transcriptome-wide sequencing reveals numerous APOBEC1 mRNA editing targets in transcript 3′ UTRs. Nat. Struct. Mol. Biol. *18,* 230–236.

Rueter, S.M., Dawson, T.R., and Emeson, R.B. (1999). Regulation of alternative splicing by RNA editing. Nature *399,* 75–80.

Rumble, S.M., Lacroute, P., Dalca, A.V., Fiume, M., Sidow, A., and Brudno, M. (2009). SHRiMP: accurate mapping of short color-space reads. PLoS Comput. Biol. *5,* e1000386.

Saccomanno, L., and Bass, B.L. (1999). A minor fraction of basic fibroblast growth factor mRNA is deaminated in Xenopus stage VI and matured oocytes. RNA *5,* 39–48.

Sakurai, M., Yano, T., Kawabata, H., Ueda, H., and Suzuki, T. (2010). Inosine cyanoethylation identifies A-to-I RNA editing sites in the human transcriptome. Nat. Chem. Biol. *6,* 733–740.

Sanger, F., Air, G.M., Barrell, B.G., Brown, N.L., Coulson, A.R., Fiddes, C.A., Hutchison, C.A., Slocombe, P.M., and Smith, M. (1977a). Nucleotide sequence of bacteriophage phi X174 DNA. Nature *265,* 687–695.

Sanger, F., Nicklen, S., and Coulson, A.R. (1977b). DNA sequencing with chain-terminating inhibitors. Proc. Natl. Acad. Sci. U. S. A. *74,* 5463–5467.

Schrider, D.R., Gout, J.F., and Hahn, M.W. (2011). Very few RNA and DNA sequence differences in the human transcriptome. PLoS One *6,* e25842.

Shendure, J., and Ji, H. (2008). Next-generation DNA sequencing. Nat. Biotechnol. *26,* 1135–1145.

Shendure, J., Porreca, G.J., Reppas, N.B., Lin, X., McCutcheon, J.P., Rosenbaum, A.M., Wang, M.D., Zhang, K., Mitra, R.D., and Church, G.M. (2005). Accurate multiplex polony sequencing of an evolved bacterial genome. Science *309,* 1728–1732.

Shendure, J.A., Porreca, G.J., Church, G.M., Gardner, A.F., Hendrickson, C.L., Kieleczawa, J., and Slatko, B.E. (2011). Overview of DNA sequencing strategies. Curr. Protoc. Mol. Biol. *Chapter 7,* Unit7.1.

Sherry, S.T., Ward, M.H., Kholodov, M., Baker, J., Phan, L., Smigielski, E.M., and Sirotkin, K. (2001). dbSNP: the

NCBI database of genetic variation. Nucleic Acids Res. *29,* 308–311.

Slater, G.S., and Birney, E. (2005). Automated generation of heuristics for biological sequence comparison. BMC Bioinformatics *6,* 31.

Smith, L.A., Wang, X., Peixoto, A.A., Neumann, E.K., Hall, L.M., and Hall, J.C. (1996). A *Drosophila* calcium channel alpha1 subunit gene maps to a genetic locus associated with behavioral and visual defects. J. Neurosci. *16,* 7868–7879.

Smith, L.M., Sanders, J.Z., Kaiser, R.J., Hughes, P., Dodd, C., Connell, C.R., Heiner, C., Kent, S.B., and Hood, L.E. (1986). Fluorescence detection in automated DNA sequence analysis. Nature *321,* 674–679.

Sommer, B., Kohler, M., Sprengel, R., and Seeburg, P.H. (1991). RNA editing in brain controls a determinant of ion flow in glutamate-gated channels. Cell *67,* 11–19.

Teng, B., Burant, C.F., and Davidson, N.O. (1993). Molecular cloning of an apolipoprotein B messenger RNA editing protein. Science *260,* 1816–1819.

Wahlstedt, H., Daniel, C., Enstero, M., and Ohman, M. (2009). Large-scale mRNA sequencing determines global regulation of RNA editing during brain development. Genome Res. *19,* 978–986.

Wang, E.T., Sandberg, R., Luo, S., Khrebtukova, I., Zhang, L., Mayr, C., Kingsmore, S.F., Schroth, G.P., and Burge, C.B. (2008). Alternative isoform regulation in human tissue transcriptomes. Nature *456,* 470–476.

Wedekind, J.E., Dance, G.S., Sowden, M.P., and Smith, H.C. (2003). Messenger RNA editing in mammals: new members of the APOBEC family seeking roles in the family business. Trends Genet. *19,* 207–216.

Zheng, D., Frankish, A., Baertsch, R., Kapranov, P., Reymond, A., Choo, S.W., Lu, Y., Denoeud, F., Antonarakis, S.E., Snyder, M., *et al.* (2007). Pseudogenes in the ENCODE regions: consensus annotation, analysis of transcription, and evolution. Genome Res. *17,* 839–851.

Regulation of Gene Expression Through Inosine-containing Untranslated Regions

9

Heather A. Hundley

Abstract

Adenosine deaminases that act on RNA (ADARs) 'edit' RNA by converting adenosines to inosines within double-stranded regions. Because adenosine and inosine have different basepairing properties, editing alters RNA structure. Furthermore, as inosine is recognized as guanosine by the translation and splicing machinery, editing can alter the protein coding potential and splicing patterns of mRNA. In fact, the ability to diversify the proteome is an important function of ADARs. However, it is now clear that editing within coding regions is rare compared to editing of non-coding regions, such as introns and untranslated regions (UTRs). Although editing of UTRs is widespread, the function of these modifications is unclear. Here, we review the diverse fates that have been reported for inosine-containing RNAs and specifically discuss whether editing is required for these outcomes.

Introduction

Proper control of gene expression is critical for normal development of all organisms. Post-transcriptional regulation is fundamental to proper gene expression and likely matches transcriptional control in importance and sophistication (Vogel *et al.*, 2010; Schwanhausser *et al.*, 2011). Within the nucleus, newly synthesized RNA undergoes many processing events, including splicing, polyadenylation and RNA editing, that contribute to both the efficiency and fidelity of gene expression (Moore and Proudfoot, 2009). RNA editing specifically alters RNA sequences and structures through modification, insertion or deletion of nucleotides (Gott and Emeson, 2000; Keegan *et al.*, 2001). In higher eukaryotes, the most prevalent type of RNA editing is a nucleotide modification of adenosine (A) to inosine (I) (Farajollahi and Maas, 2010; Nishikura, 2010). These deamination events are catalysed by the ADAR family (Goodman *et al.*, 2012), which are present in animals ranging from sea anemones to humans (Jin *et al.*, 2009). ADARs are required for normal development and behaviour, with defects in ADAR function linked to diseases such as epilepsy, schizophrenia, amyotrophic lateral sclerosis, and many types of cancer (Dominissini *et al.*, 2011; Galeano *et al.*, 2011). Despite their biological importance, the effects of ADARs and RNA editing on gene expression are largely unknown.

Majority of ADAR editing events are in non-coding regions

Current estimates predict that there are over 40,000 editing sites in the human transcriptome (Kiran and Baranov, 2010). Based on both the density and location of editing events within a particular RNA, ADAR targets can be broadly divided into two categories. The first contains mRNAs that are edited within coding regions at specific adenosines (Pullirsch and Jantsch, 2010; Hood and Emeson, 2012). These ADAR substrates were first identified when researchers noticed discrepancies between sequences encoded by the genome versus cDNA (Sommer *et al.*, 1991; Lomeli *et al.*, 1994; Burns *et al.*, 1997). Since inosine is reverse transcribed as guanosine (G), an A-to-G change between genomic DNA and cDNA is indicative of A-to-I editing (Fig. 9.1). Editing within coding regions can generate protein variants with single amino acid substitutions that affect the protein structure, function

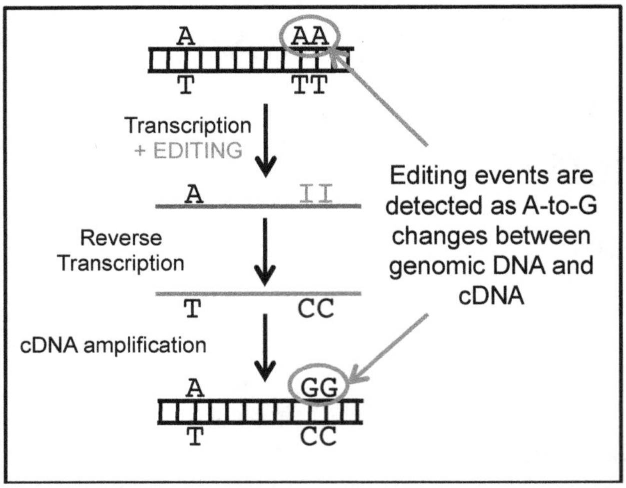

Figure 9.1 Identification of ADAR RNA editing sites. Transcription of genomic DNA (top molecule) gives rise to mRNA (blue line). Editing of the mRNA molecule changes certain adenosines (A) to inosines (I). When the cDNA (green line) is produced from the mRNA, reverse transcriptase recognizes inosines as guanosines (G). Amplification of the cDNA results in a DNA molecule that has A-to-G changes compared to the genomic DNA. A colour version of this figure is located in the plate section at the back of the book.

and pharmacology (Streit and Decher, 2011). Modification of coding potential is an important function of ADARs, especially in the nervous system where many of these substrates, mostly neurotransmitter receptors and ion channels, are primarily expressed (Jepson and Reenan, 2008). However, despite the functional importance of these editing events and transcriptome-wide searches for selective editing events, editing in coding regions is only known to occur for ~30 human genes (Li *et al.*, 2009).

The vast majority of editing sites comprise the second type of ADAR substrates, which are promiscuously edited and occur within non-coding regions of the genome (Hundley and Bass, 2010). The first indications that non-coding regions were important ADAR targets came from the use of a biochemical approach to isolate inosine-containing RNAs from both human brain and *Caenorhabditis elegans* (Morse and Bass, 1999; Morse *et al.*, 2002). Nearly all inosine-containing mRNAs isolated contained editing sites within the untranslated regions (UTRs). The biological importance of these editing events was not fully appreciated until four groups independently performed systematic searches of cDNA and expressed sequence tag (EST) libraries and uncovered over 10,000 editing sites within non-coding regions of the human transcriptome (Athanasiadis *et al.*, 2004; Blow *et al.*, 2004; Kim *et al.*, 2004; Levanon *et al.*, 2004). In particular, the majority of editing events occur within 3′ UTRs (54%), with 12% in 5′ UTRs and 33% in introns (Levanon *et al.*, 2004). It is important to note that the exact number of editing events within introns is difficult to predict as some sites annotated as introns may actually reside in alternative exons and most intronic sequences are not expected to be present in EST databases. Regardless, using the fact that non-coding editing sites localize to long double-stranded duplexes formed from inverted repeat sequences, additional bioinformatics analyses and DNA capture with parallel next-generation sequencing have expanded the number of ADAR editing sites 4-fold (Li *et al.*, 2009; Carmi *et al.*, 2011). Furthermore, with the recent advent of high-throughput RNA sequencing (RNA-Seq) and the ability to sequence multiple individual human genomes for a relatively low price, several groups have performed unbiased searches for differences between genomic DNA and expressed RNA to identify potential RNA editing sites (Li *et al.*, 2011; Bahn *et al.*, 2012; Peng *et al.*, 2012). Two of these large-scale RNA-Seq studies have made the provocative suggestion that there are many types of nucleotide modifications present in human mRNA and A-to-G changes represent only between 20% and 60% of the DNA–RNA differences in the human transcriptome (Li *et al.*, 2011; Bahn *et al.*, 2012); however, the bioinformatics approaches used in these studies are controversial (Schrider *et al.*, 2011; Peng *et al.*, 2012). Importantly, consistent with the initial computational screens for ADAR editing sites, the RNA-Seq study with the most stringent criteria for detecting DNA–RNA differences has indicated that over 90% of changes are indicative of A-to-I editing (Peng *et al.*, 2012). As sequencing technology advances in the coming years, additional ADAR substrate RNAs will likely be uncovered, however for now, it is clear that ADARs modify tens of thousands of adenosines in thousands of genes and result in widespread modification of non-coding regions of the human transcriptome.

Non-coding editing sites are located within inverted repetitive elements

Most non-coding editing sites are present within transcribed repetitive elements (Wahlstedt and Ohman, 2011). Repeat sequences are abundant in eukaryotic genomes, constituting nearly a fifth of the *C. elegans* genome and upwards of half the human genome (Britten and Kohne, 1968; Sulston and Brenner, 1974; Venter *et al.*, 2001). These repeat sequences are primarily composed of transposable elements (Devine *et al.*, 1997; Cordaux and Batzer, 2009). Although these elements have commonly been referred to as 'junk' or 'selfish' DNA (Doolittle and Sapienza, 1980; Orgel and Crick, 1980), the fact that repeats are interspersed with coding sequences in the genome, has led to speculation that these elements may play a regulatory function in gene expression (Britten and Davidson, 1969; Brosius, 1999). In fact, recent high-throughput sequencing studies have revealed that repeats are widely transcribed, with an overrepresentation of these elements in the 3′ UTRs of protein-coding mRNAs (Faulkner *et al.*, 2009). As UTRs are molecular 'hotspots' for elements that regulate gene expression, including mRNA stability, localization and translation (Conne *et al.*, 2000; Chatterjee and Pal, 2009), repetitive elements have the potential to affect many of these processes (Ponicsan *et al.*, 2010; Berger and Strub, 2011).

When two repetitive sequences that are inversely orientated are present within one mRNA, the inverted repeats can basepair to form long intramolecular double-stranded RNA (dsRNA) structures that are ideal ADAR substrates (Fig. 9.2). Although repetitive sequences are predicted to be major targets of ADARs across species, the frequency of A-to-I editing in humans is at least one order of magnitude greater than the editing observed in fly, chicken, mouse or rat transcriptomes (Eisenberg *et al.*, 2005). This high frequency of editing in humans is attributed, in part, to the abundance (over one million copies) of one repeat, the Alu element. Alus are primate-specific sequences that belong to the short

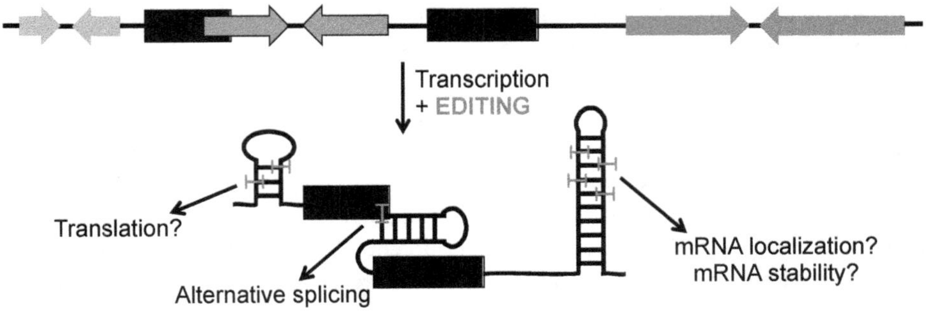

Figure 9.2 UTRs and introns are the primary targets of ADARs, but the function of these editing events is largely unknown. Inverted repetitive elements (arrows) present in genomic DNA (top molecule) give rise to long intramolecular dsRNA structures in pre-mRNA (bottom molecule). These repeats localize primarily to UTRs and introns, where they are predicted to affect gene expression. However, the repeats can also overlap with exons (black rectangles) and alter splicing of the pre-mRNA. A colour version of this figure is located in the plate section at the back of the book.

interspersed repetitive element (SINE) family (Batzer and Deininger, 2002). While the overall number of SINEs is equal in mouse and human, the dominance and high sequence similarity of Alu elements accounts for the higher editing levels in the human transcriptome (Eisenberg *et al.*, 2005). In fact, it has been predicted that ~ 85% of human pre-mRNAs contain at least two Alu elements in inverted orientation, which form dsRNA and undergo A-to-I editing (Athanasiadis *et al.*, 2004). Although Alus are prevalent in all primates, a recent study has indicated that editing of Alu elements in human brain is significantly higher than in non-human primates (Paz-Yaacov *et al.*, 2010). Although the exact mechanism for regulating editing levels is unknown, these data suggest that editing of non-coding regions by ADARs may be important for the development of higher brain function.

Inverted repeats are hyperedited by ADARs

Non-coding regions that are targeted by ADARs contain clusters of inosines, where up to 50% of nearby adenosines in a single transcript undergo editing (Wahlstedt and Ohman, 2011). This type of editing is referred to as hyperediting or promiscuous/non-selective editing. Hyperediting is thought to occur on long completely dsRNA, as introduction of mismatches and bulges into long dsRNA structures has been shown to reduce the extent of ADAR editing *in vitro* (Lehmann and Bass, 1999). The near-perfect complementarity of inverted repetitive elements provides an ideal substrate for ADAR binding at multiple sites across the dsRNA and, therefore, leads to more deamination within the inverted repeats. The higher level of editing in humans is in part due to the length of the ADAR substrate. Whereas the repeat sequences in lower organisms are typically 150 nt, inverted Alu elements form ~ 300 bp dsRNA structures (Eisenberg *et al.*, 2005). However, while most adenosines within long dsRNA have the potential to be deaminated, not all adenosines are targeted by ADARs. In fact, editing levels amongst neighbouring adenosines within an Alu repeat can vary from a few per cent to 60–80% (Greenberger *et al.*, 2010). It is striking that although inverted Alus contain ~ 67 adenosines, only eight particular adenosines account for two-thirds of the editing events in the human transcriptome (Barak *et al.*, 2009). Nevertheless, even these 'hotspots' are not deaminated in every transcript. Although the extent of editing at each adenosine is variable, the overall pattern of editing efficiency within the Alu is consistent between different human individuals (Greenberger *et al.*, 2010). In addition, global hypoediting of non-coding regions has been observed in many different types of tumours (Paz *et al.*, 2007; Galeano *et al.*, 2011). These data suggest that editing within non-coding regions is tightly regulated, however, the mechanism for selecting which adenosines undergo deamination is not well-understood (Wahlstedt and Ohman, 2011). In addition, while it is clear that non-coding editing generates tremendous diversity within the human transcriptome, the functional consequence of these editing events is largely a mystery (Farajollahi and Maas, 2010; Hundley and Bass, 2010). Editing within non-coding regions has been suggested to affect nearly every step of mRNA metabolism, including splicing, nuclear export, degradation and translation (Fig. 9.2). In this chapter, we will focus on how A-to-I editing of non-coding regions regulates these processes to ultimately affect gene expression.

Editing in introns affects splicing

One of the most definitive functions for editing within non-coding regions is the modulation of splicing. The splicing machinery interprets inosine as guanosine, therefore ADARs have the potential to regulate splicing by a number of mechanisms (Nishikura, 2010) (Fig. 9.3). First, A-to-I editing has the potential to create canonical splice site acceptor sequences (AG) from AA dinucleotides and splice donor sequences (GU) from AU dinucleotides. In addition, ADARs could eliminate splice acceptor sites by changing the acceptor sequence from AG to IG. Last, ADARs have the potential to modify the branch site adenosine, thus preventing intron removal. In fact, examples of ADAR regulation of splicing through all of the above mechanisms have been observed (Hogg *et al.*, 2011).

One of the first and perhaps most interesting examples of RNA editing altering splicing is the ability of mammalian ADAR2 to edit its own

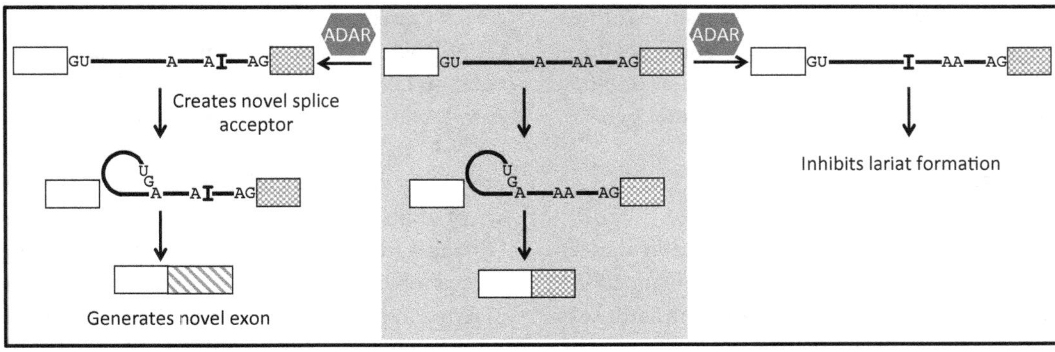

Figure 9.3 RNA editing affects splicing. The middle pathway (shaded grey) shows the canonical splicing pathway for two exons (white and chequerboard rectangles) of a pre-mRNA in the absence of ADAR (dark grey hexagon). In the first step, the splice donor sequence (GU) forms a lariat structure with the branch site adenosine (A in middle of molecule). In the second step, the intron (black line) is removed and the two exons are joined together to form the mature mRNA. Left pathway: Editing of the second adenosine (bold I) of an AA dinucleotide in the intron creates a novel splice acceptor site and generates a novel exon. Right pathway: Editing of the branch site adenosine (bold I) inhibits lariat formation and removal of the intron.

pre-mRNA (Rueter *et al.*, 1999). Editing of an AA dinucleotide creates a novel splice site acceptor that results in inclusion of 47-nt exon within the mRNA that when translated, generates a non-functional protein. Thus, the ability of ADAR2 to modulate splice site recognition creates an elegant negative feedback loop that controls ADAR2 levels in mammals.

As editing levels vary across mammalian tissues and throughout development (Paul and Bass, 1998; Wahlstedt *et al.*, 2009), alteration of splice sites by ADARs could result in tissue and/or development specific regulation of gene expression. In fact, tissue-dependent modulation of splicing has been clearly demonstrated for human nuclear prelaminin A recognition factor (NARF) (Lev-Maor *et al.*, 2007). A series of RNA editing events within the pre-mRNA of NARF both create a splice site acceptor and alter splicing enhancers to generate an mRNA that contains a novel Alu exon. The exon inclusion rates vary amongst different tissues, but correlate with the level of RNA editing known to occur in those tissues. For example, brain tissue, which has very high levels of A-to-I editing, showed the most inclusion of the Alu exon within NARF mRNA, while skeletal muscle, which has low levels of editing, had one of the lowest levels of exon inclusion.

The ability of ADARs to alter splicing and generate non-functional proteins, suggests that RNA editing, if not properly regulated, could affect disease manifestation and progression. In support of this idea, A-to-I editing and alternative splicing of the protein tyrosine phosphatase gene (*PTPN6*) was found to differ between healthy individuals and those suffering from acute myeloid leukaemia (AML) (Beghini *et al.*, 2000). In AML patients, hyperediting of a branch site adenosine resulted in intron retention within the mature mRNA, which was predicted to disrupt PTPN6 protein function. This suggests that editing and alternative splicing, together, may play an important role in leukaemogenesis, and is consistent with recent findings that ADAR1 is overexpressed in acute lymphoblastic leukaemias (ALL) (Ma *et al.*, 2011). Yet, these findings are in contrast to the general model that lower A-to-I editing is present in many types of tumours compared with the normal tissues (Paz *et al.*, 2007). However, it is important to note that even in the aforementioned study, some editing events, such as those in the intron of BRCA1 were found to be significantly up-regulated in tumour tissues compared to control tissues. Interestingly, this is reminiscent of studies of DNA methylation where global decreases in genomic methylation occur in various cancers, but increased DNA methylation have been demonstrated to occur at the promoters of various tumour-suppressor proteins (Ehrlich, 2002). Thus, it is critical to characterize editing on a gene-by-gene basis in order to uncover important effects, such as those on splicing, which result in the generation of novel mRNAs and likely effect gene expression at different points in development and/or tumorigenesis.

Localization of mRNAs with edited 3′ UTRs

The first hyperedited RNAs detected in the cell were of viral origin (Jantsch and Ohman, 2008). Interestingly, some of these RNAs, such as those of the polyoma virus, are sequestered in the nucleus, thus preventing viral gene expression (Kumar and Carmichael, 1997). While this implicates RNA editing as an important antiviral defence mechanism (Samuel, 2011), whether hyperedited cellular RNAs undergo a similar fate is controversial. Nuclear retention of two mRNAs, Nicolin 1 (NICN-1) in human cells and mouse CTN-RNA, which contain multiple inosines within 3′ UTR inverted repeat sequences has been reported (Prasanth *et al.*, 2005; Chen and Carmichael, 2008). However, other endogenous human and *C. elegans* mRNAs that possess edited 3′ UTRs have been detected within the cytoplasm on translating ribosomes (Hundley *et al.*, 2008). Furthermore, recent studies on several additional endogenous human mRNAs have not revealed any significant differences in localization between mRNAs containing edited 3′ UTRs and those lacking inverted repeats (Capshew *et al.*, 2012). Consistent with this, mouse embryonic fibroblasts (MEFs) lacking ADARs did not reveal significant differences in expression of reporter constructs containing inverted repeats compared to wild-type MEFs (M. Tajaddod, F. Huber, S. Schopoff, and M.F. Jantsch, unpublished results). Thus, it is clear that hyperediting of cellular mRNAs is not a general mechanism used to silence endogenous gene expression.

Although A-to-I editing within 3′ UTRs does not preclude an mRNA from reaching the cytoplasm, it is possible that some inosine-containing RNAs are retained in the nucleus, specifically within paraspeckles (Chen and Carmichael, 2012). Paraspeckles are subnuclear bodies that are present in most mammalian cells cultured *in vitro* (Fox and Lamond, 2010). CTN-RNA is an alternative splice isoform of the mCAT2 gene that is enriched in paraspeckles of mouse NIH3T3 cells (Prasanth *et al.*, 2005). The 3′ UTR of CTN-RNA is required for nuclear retention and contains inverted repeat elements, which undergo A-to-I editing at several positions. In addition, CTN-RNA co-localizes with $p54^{nrb}$, a protein previously demonstrated to bind and retain extensively edited synthetic dsRNA within nuclei (Zhang and Carmichael, 2001). These data and studies of human reporter mRNAs with edited inverted repeats led to a prevalent model that extensively edited transcripts are recognized by $p54^{nrb}$ and trapped in the nucleus to silence gene expression (Chen *et al.*, 2008). However, studies in human embryonic stem cells have revealed that neither extensive editing nor binding to $p54^{nrb}$ is sufficient for nuclear retention, and in fact, it is the formation of paraspeckles that is critical for regulating nuclear export (Chen and Carmichael, 2009). Interestingly, paraspeckles were proposed to form upon differentiation of the human embryonic stem cells into trophoblasts, suggesting that a biological function for paraspeckles may be in regulating gene expression during differentiation. In addition, recent data from the first analysis of paraspeckles in mice have suggested that paraspeckles are non-essential bodies that form in response to particular environmental cues (Nakagawa *et al.*, 2011). For example, the structures are largely absent from normal mouse tissue, but rapidly form upon culturing of MEFs *in vitro*, suggesting that paraspeckle formation may be a 'natural' response to certain changes in the extracellular environment. Together, these data strongly suggest that the biological function of paraspeckles may be to act as a nuclear reservoir for certain cellular mRNAs that will be utilized in response to various cellular conditions (Bass *et al.*, 2005; Fox and Lamond, 2010). However, it is unclear what elements in cellular mRNAs are required for localization to paraspeckles and the role, if any, that inosines play in nuclear retention.

Hyperediting targets mRNAs for degradation

Extensive editing by ADARs can result in specific degradation of the inosine-containing dsRNA. A nuclease activity that cleaves dsRNA at sites of alternating IU and UI basepairs has been detected in cytoplasmic extracts from various tissues

(Scadden and Smith, 2001). Replacement of IU and UI basepairs with the isomorphic GU and UG basepairs blocks cleavage, suggesting that a specific structure created by hyperediting is required for degradation. Characterization of the nuclease activity identified Tudor staphylococcal nuclease (Tudor-SN) as one component (Scadden, 2005). Tudor-SN is present in most eukaryotes and functions in various cellular processes including transcription and splicing (Friberg *et al.*, 2009). Interestingly, Tudor-SN was originally identified as a component of the RNAi-induced silencing complex (RISC) (Caudy *et al.*, 2003). As RNA editing and RNAi are competing pathways, it is intriguing to speculate that Tudor-SN serves as a general surveillance factor that either utilizes unmodified dsRNA for RNAi-mediated gene silencing or destroys inosine-containing dsRNA (Scadden, 2005).

As there are currently no known endogenous RNAs that are recognized by Tudor-SN, the biological significance of the inosine-specific nuclease activity is unclear. In fact, studies of endogenous ADAR substrate mRNAs in *C. elegans* lacking ADAR editing did not reveal any significant changes in mRNA levels compared to wild-type worms, suggesting these mRNAs are not targeted by Tudor-SN under normal growth conditions (Hundley *et al.*, 2008). However, there is one clear example of an endogenous inosine-containing RNA that undergoes cleavage in the cell, mouse CTN-RNA, but the cleavage only happens in response to cellular stress (Prasanth *et al.*, 2005). As discussed above, CTN-RNA is nuclear retained under normal growth conditions. However, when cells undergo various stresses, such as interferon-γ or lipopolysaccharide (LPS) stimulation, the inosine-containing 3′ UTR is cleaved, allowing for the remaining RNA to be exported to the cytoplasm and translated to produce mCAT2 protein. The mCAT2 protein is a cell surface receptor that is responsible for uptake of L-arginine, a precursor to nitric oxide (NO). The NO production pathway increases in response to stress and is key component of the innate immune response (Lowenstein and Padalko, 2004). Thus, the nuclear storage and cleavage of CTN-RNA to generate abundant mRNA for translation provides a rapid mechanism for post-transcriptional production of mCAT2 protein when cellular stress is encountered (Bass *et al.*, 2005; Das and Carmichael, 2005).

The studies of CTN-RNA have revealed a novel mode of controlling gene expression via nuclear retention and suggested a biological function for paraspeckles (Fox and Lamond, 2010). However, as neither the nuclease nor the cleavage site are known, the molecular mechanism for the processing of CTN-RNA, and the role editing plays in this process, is unknown. If inosine is required for cleavage, could Tudor-SN be the nuclease that regulates CTN-RNA? Although Tudor-SN was originally purified from cytoplasmic ribosomes (Caudy *et al.*, 2003), it has recently been demonstrated that Tudor-SN shuttles between the nucleus and cytoplasm (Weissbach and Scadden, 2012). However, upon treatment with arsenite or interferon induction, similar stresses as those that stimulate CTN–RNA cleavage, Tudor-SN localizes to cytoplasmic stress granules (SGs). Thus, it is unclear whether Tudor-SN and CTN-RNA co-localize at the time that the inosine-containing 3′ UTR is cleaved. Moreover, a recent bioinformatics approach to identify mRNAs regulated in a manner similar to CTN-RNA has suggested that inosine-containing 3′ UTRs are not cleaved *per se*, but instead undergo non-canonical splicing to remove the inverted repeats from the mRNA (Osenberg *et al.*, 2009). Importantly, it was noted that the nucleotides present at the 'splice' sites do not correspond to the alternating IU and UI basepairs recognized by Tudor-SN. While these data are intriguing and suggest that over 500 human and 100 mouse genes may be regulated in this manner, detailed experimental evidence for such transcripts is lacking. Furthermore, although the extent of editing in various tissues correlated with the number of expressed sequence tags (ESTs) lacking the inverted repeats, the dependence on inosine for transcript processing has not been tested. Regardless of the exact molecular mechanism, it is clear that multiple transcripts in the cell can be produced using non-canonical 3′ end processing (Wilusz and Spector, 2010). Future work to understand the dynamics between nuclear retention, RNA editing and post-transcriptional cleavage events will be critical for understanding how inosine-containing 3′ UTRs regulate gene expression.

Inverted repeats and editing in 3′ UTRs alter translation

Elements within 3′ UTRs are often recognized by *trans*-acting factors that regulate translation of the mRNAs (Wickens *et al.*, 1997; Spriggs *et al.*, 2010). Translational control is critical during development, in response to cellular stress and for proper neuronal function (Costa-Mattioli *et al.*, 2009; MacNicol and MacNicol, 2010; Spriggs *et al.*, 2010). Regulation of gene expression at the level of translation has drawn increasing attention over the last decade due in large part to the discovery that microRNAs (miRNAs) are present in most organisms, regulate many pathways, and function aberrantly in many diseases (Carthew and Sontheimer, 2009; Ghildiyal and Zamore, 2009; Mondol and Pasquinelli, 2012). Although the primary mechanism of action is unclear, in general, miRNAs regulate protein synthesis by repressing translation and/or deadenylation and degradation of mRNA (Fabian *et al.*, 2010). Although miRNAs are ~21 nt in length, repression of gene expression only requires basepairing between bases 2–8 of the miRNA (known as the 'seed' sequence) and the target mRNA (Bartel, 2009). Many recent studies have focused on the impact of ADARs and RNA editing on miRNA biogenesis and function (Chapter 6). However, it is important to consider that editing sites within 3′ UTRs can also affect miRNA-mediated gene regulation. In fact, bioinformatic analysis of A-to-I changes within human UTRs, revealed that one-fourth of the nearly 13,000 deamination events characterized create perfect complements to seed sequences of 30 different miRNA families, with miR-513, miR769-3p and miR450b-3p preferentially targeting deaminated sequences (Borchert *et al.*, 2009). Furthermore, the inosines within these miRNA-binding sites are the Alu adenosines that are most highly targeted by ADARs. Interestingly, co-expression of miR-513 and miR-769 with reporters bearing 'edited' (G substituted for I) or unedited 3′ UTRs from the DNA fragmentation

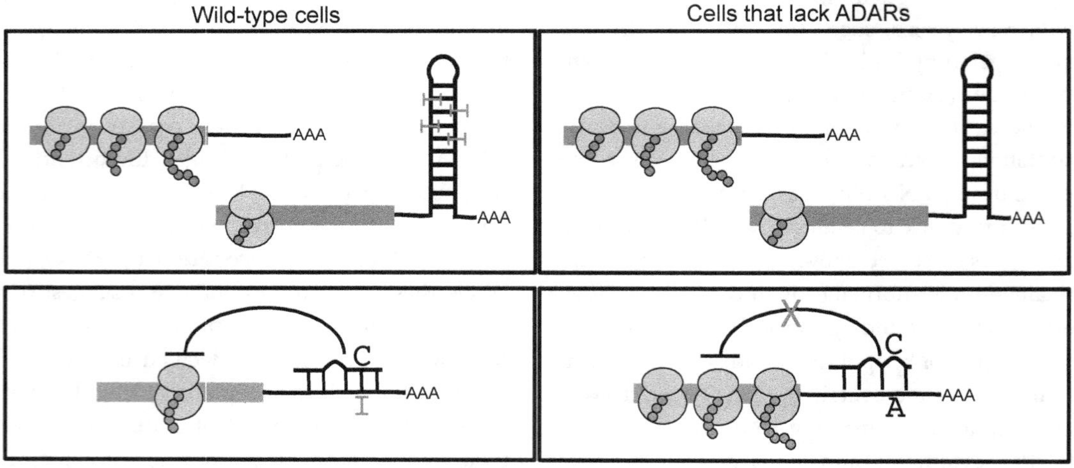

Figure 9.4 Effects of double-stranded 3′ UTRs and editing on translation. Twin circles (beige) represent ribosomes translating the coding regions (rectangles) of mRNA. (Upper panels) In wild-type cells (left panel), alternative splicing can generate mRNAs that have identical coding regions (purple rectangles) but either contain, or lack a double-stranded 3′ UTR. Comparison of these mRNAs has indicated that the double-stranded 3′ UTR reduces translation efficiency. Although the double-stranded 3′ UTRs are extensively edited (editing sites are represented by bold I), studies in worms and human cell lines lacking ADARs (right panel) have indicated that the effect on translation is independent of editing. (Lower panels) In wild-type cells (left panel), editing within 3′ UTRs can create novel miRNA-binding sites that result in decreased protein expression. In the absence of editing (right panel), the miRNA has decreased complementarity to the 3′ UTR and cannot repress translation. A colour version of this figure is located in the plate section at the back of the book.

factor alpha (*DFFA*) gene suggest that editing is required for miRNA-mediated repression of DFFA. Thus, an important biological function for non-coding editing may be to generate novel miRNA binding sites (Fig. 9.4). Furthermore, as unedited Alu sequences are probable targets for over 50 different miRNAs (Daskalova *et al.*, 2006; Smalheiser and Torvik, 2006), it seems plausible that editing will disrupt basepairing between these miRNAs and their target 3′ UTRs leading to aberrant gene regulation.

In addition to recognizing sequence elements, many *trans*-acting factors that modulate translation recognize short (20–40 bp) hairpin structures within 3′ UTRs (Svoboda and Di Cara, 2006). Interestingly, recent studies have suggested that the long (300–600 bp) dsRNA duplexes formed from inverted repeat elements can alter translation efficiency (Fig. 9.4). In *C. elegans*, comparisons of both endogenous and reporter mRNAs with double-stranded 3′ UTRs demonstrated that these mRNAs associate with a smaller number of translating ribosomes than mRNAs lacking these structures (Hundley *et al.*, 2008). However, comparisons between wild-type worms and those lacking ADARs did not show significant differences in ribosome-association. Consistent with this, recent studies of reporter mRNAs with inverted Alu elements from the *PSMB2* gene have shown a similar editing-independent effect on these dsRNA structures translation in human cell lines (Capshew *et al.*, 2012). Thus, while inosine-containing 3′ UTRs affect translation, it is unclear what role, if any, that editing plays in regulating translation.

Conclusions and outlook

It is now well established that the complexity of higher organisms depends on the ability to produce a large number of different transcripts from one gene locus (Baltimore, 2001). Transcript diversity is in large part generated by the variety of ways in which pre-mRNA is processed (Warf and Berglund, 2010). For example, in theory, alternative splicing can produce over 38,000 different isoforms of the *Drosophila Dscam* gene (Schmucker *et al.*, 2000). As the human genome contains over 40,000 A-to-I RNA editing sites, the transcript diversity generated by ADARs could even exceed that of alternative splicing (Barak *et al.*, 2009). Taking in to consideration that one Alu element can have ~ 30 editing sites and varying combinations of these modifications are present in UTRs, in theory, over 2 billion different transcripts could be generated from one locus. However, it is important to note that certain adenosines within non-coding regions have been observed to undergo deamination more often than others, therefore, the number of transcripts generated from editing in one Alu element is likely to be smaller than the estimate above. To uncover the true impact of ADARs, future studies need to utilize techniques such as RNA-Seq, that allow for an accurate assessment of the number of deamination events present within individual transcripts and, thus, transcript diversity.

As the majority of A-to-I changes occur within non-coding regions of the genome, the transcript diversity generated by RNA editing likely affects regulatory elements in the mRNA. In this regard, various effects on post-transcriptional gene expression have been described for inosine-containing RNAs. However, in most cases, whether or not the observed effects are a consequence of editing or are merely due to the dsRNA structure itself is unclear. In fact, the handful of studies that have addressed this point have indicated that editing is not required for the effects of inverted repeat UTRs to regulate post-transcriptional gene expression (Hundley and Bass, 2010). For example, in *C. elegans*, comparisons of endogenous and reporter mRNAs with inverted repeat-containing UTRs did not reveal any significant differences in mRNA levels or translation efficiency between wild-type worms and those lacking editing (Hundley *et al.*, 2008). However, it is important to note that these studies were carried out for only one or two developmental points and under normal growth conditions. Perhaps editing of non-coding regions plays an important role in regulating gene expression in response to specific stimuli? In this regard, many of the pathways associated with inosine-containing RNAs, such as paraspeckle formation, CTN-RNA cleavage, and Tudor localization, all seem to be regulated in response to cellular stresses (Prasanth *et al.*, 2005; Nakagawa *et al.*, 2011; Weissbach and Scadden,

2012). Future work to characterize the effects of non-coding editing on gene expression clearly needs to be performed under a variety of cellular conditions. Furthermore, although studies that measure global effects of editing on gene expression provide useful overviews of the changes present in the transcriptome, the diverse fates that have thus far been reported for inosine-containing UTRs suggest that careful analysis of individual transcripts will be critical to determine the effects of editing on gene expression.

References

Athanasiadis, A., Rich, A., and Maas, S. (2004). Widespread A-to-I RNA editing of Alu-containing mRNAs in the human transcriptome. PLoS Biol. *2*, e391.

Bahn, J.H., Lee, J.H., Li, G., Greer, C., Peng, G., and Xiao, X. (2012). Accurate identification of A-to-I RNA editing in human by transcriptome sequencing. Genome Res. *22*, 142–150.

Baltimore, D. (2001). Our genome unveiled. Nature *409*, 814–816.

Barak, M., Levanon, E.Y., Eisenberg, E., Paz, N., Rechavi, G., Church, G.M., and Mehr, R. (2009). Evidence for large diversity in the human transcriptome created by Alu RNA editing. Nucleic Acids Res. *37*, 6905–6915.

Bartel, D.P. (2009). MicroRNAs: target recognition and regulatory functions. Cell *136*, 215–233.

Bass, B.L., Hellwig, S., and Hundley, H.A. (2005). A nuclear RNA is cut out for translation. Cell *123*, 181–183.

Batzer, M.A., and Deininger, P.L. (2002). Alu repeats and human genomic diversity. Nat. Rev. Genet. *3*, 370–379.

Beghini, A., Ripamonti, C.B., Peterlongo, P., Roversi, G., Cairoli, R., Morra, E., and Larizza, L. (2000). RNA hyperediting and alternative splicing of hematopoietic cell phosphatase (PTPN6) gene in acute myeloid leukemia. Human Mol. Genet. *9*, 2297–2304.

Berger, A., and Strub, K. (2011). Multiple roles of alu-related noncoding RNAs. Prog. Mol. Subcell. Biol. *51*, 119–146.

Blow, M., Futreal, P.A., Wooster, R., and Stratton, M.R. (2004). A survey of RNA editing in human brain. Genome Res. *14*, 2379–2387.

Borchert, G.M., Gilmore, B.L., Spengler, R.M., Xing, Y., Lanier, W., Bhattacharya, D., and Davidson, B.L. (2009). Adenosine deamination in human transcripts generates novel microRNA binding sites. Human Mol. Genet. *18*, 4801–4807.

Britten, R.J., and Davidson, E.H. (1969). Gene regulation for higher cells: a theory. Science *165*, 349–357.

Britten, R.J., and Kohne, D.E. (1968). Repeated sequences in DNA. Hundreds of thousands of copies of DNA sequences have been incorporated into the genomes of higher organisms. Science *161*, 529–540.

Brosius, J. (1999). RNAs from all categories generate retrosequences that may be exapted as novel genes or regulatory elements. Gene *238*, 115–134.

Burns, C.M., Chu, H., Rueter, S.M., Hutchinson, L.K., Canton, H., Sanders-Bush, E., and Emeson, R.B. (1997). Regulation of serotonin-2C receptor G-protein coupling by RNA editing. Nature *387*, 303–308.

Capshew, C.R., Dusenbury, K.L., and Hundley, H.A. (2012). Inverted Alu ds RNA structures do not affect localization but can alter translation efficiency of human mRNAs independent of RNA editing. Nucleic Acids Res. *40*, 8637–8645.

Carmi, S., Borukhov, I., and Levanon, E.Y. (2011). Identification of widespread ultra-edited human RNAs. PLoS Genet. *7*, e1002317.

Carthew, R.W., and Sontheimer, E.J. (2009). Origins and mechanisms of miRNAs and siRNAs. Cell *136*, 642–655.

Caudy, A.A., Ketting, R.F., Hammond, S.M., Denli, A.M., Bathoorn, A.M., Tops, B.B., Silva, J.M., Myers, M.M., Hannon, G.J., and Plasterk, R.H. (2003). A micrococcal nuclease homologue in RNAi effector complexes. Nature *425*, 411–414.

Chatterjee, S., and Pal, J.K. (2009). Role of 5′ – and 3′-untranslated regions of mRNAs in human diseases. Biol. Cell *101*, 251–262.

Chen, L.L., and Carmichael, G.G. (2008). Gene regulation by SINES and inosines: biological consequences of A-to-I editing of Alu element inverted repeats. Cell Cycle *7*, 3294–3301.

Chen, L.L., and Carmichael, G.G. (2009). Altered nuclear retention of mRNAs containing inverted repeats in human embryonic stem cells: functional role of a nuclear noncoding RNA. Mol. Cell *35*, 467–478.

Chen, L.L., and Carmichael, G.G. (2012). Nuclear editing of mRNA 3′-UTRs. Curr. Top. Microbiol. Immunol. *353*, 111–121.

Chen, L.L., DeCerbo, J.N., and Carmichael, G.G. (2008). Alu element-mediated gene silencing. EMBO J. *27*, 1694–1705.

Conne, B., Stutz, A., and Vassalli, J.D. (2000). The 3′ untranslated region of messenger RNA: A molecular 'hotspot' for pathology? Nat. Med. *6*, 637–641.

Cordaux, R., and Batzer, M.A. (2009). The impact of retrotransposons on human genome evolution. Nat. Rev. Genet. *10*, 691–703.

Costa-Mattioli, M., Sossin, W.S., Klann, E., and Sonenberg, N. (2009). Translational control of long-lasting synaptic plasticity and memory. Neuron *61*, 10–26.

Das, A.K., and Carmichael, G.G. (2005). Unleash the messenger. Nat. Cell Biol. *7*, 1054–1055.

Daskalova, E., Baev, V., Rusinov, V., and Minkov, I. (2006). 3′UTR-located ALU elements: donors of potential miRNA target sites and mediators of network miRNA-based regulatory interactions. Evol. Bioinform. Online *2*, 103–120.

Devine, S.E., Chissoe, S.L., Eby, Y., Wilson, R.K., and Boeke, J.D. (1997). A transposon-based strategy for sequencing repetitive DNA in eukaryotic genomes. Genome Res. *7*, 551–563.

Dominissini, D., Moshitch-Moshkovitz, S., Amariglio, N., and Rechavi, G. (2011). Adenosine-to-inosine RNA editing meets cancer. Carcinogenesis *32*, 1569–1577.

Doolittle, W.F., and Sapienza, C. (1980). Selfish genes, the phenotype paradigm and genome evolution. Nature *284*, 601–603.

Ehrlich, M. (2002). DNA methylation in cancer: too much, but also too little. Oncogene *21*, 5400–5413.

Eisenberg, E., Nemzer, S., Kinar, Y., Sorek, R., Rechavi, G., and Levanon, E.Y. (2005). Is abundant A-to-I RNA editing primate-specific? Trends Genet. *21*, 77–81.

Fabian, M.R., Sonenberg, N., and Filipowicz, W. (2010). Regulation of mRNA translation and stability by microRNAs. Annu. Rev. Biochem. *79*, 351–379.

Farajollahi, S., and Maas, S. (2010). Molecular diversity through RNA editing: a balancing act. Trends Genet. *26*, 221–230.

Faulkner, G.J., Kimura, Y., Daub, C.O., Wani, S., Plessy, C., Irvine, K.M., Schroder, K., Cloonan, N., Steptoe, A.L., Lassmann, T., *et al.* (2009). The regulated retrotransposon transcriptome of mammalian cells. Nat. Genet. *41*, 563–571.

Fox, A.H., and Lamond, A.I. (2010). Paraspeckles. Cold Spring Harbor Perspect. Biol. *2*, a000687.

Friberg, A., Corsini, L., Mourao, A., and Sattler, M. (2009). Structure and ligand binding of the extended Tudor domain of *D. melanogaster* Tudor-SN. J. Mol. Biol. *387*, 921–934.

Galeano, F., Tomaselli, S., Locatelli, F., and Gallo, A. (2011). A-to-I RNA editing: The 'ADAR' side of human cancer. Semin. Cell Dev. Biol. *23*, 244–250.

Ghildiyal, M., and Zamore, P.D. (2009). Small silencing RNAs: an expanding universe. Nat. Rev. Genet. *10*, 94–108.

Goodman, R.A., Macbeth, M.R., and Beal, P.A. (2012). ADAR proteins: structure and catalytic mechanism. Curr. Top. Microbiol. Immunol. *353*, 1–33.

Gott, J.M., and Emeson, R.B. (2000). Functions and mechanisms of RNA editing. Annu. Rev. Genet. *34*, 499–531.

Greenberger, S., Levanon, E.Y., Paz-Yaacov, N., Barzilai, A., Safran, M., Osenberg, S., Amariglio, N., Rechavi, G., and Eisenberg, E. (2010). Consistent levels of A-to-I RNA editing across individuals in coding sequences and non-conserved Alu repeats. BMC Genomics *11*, 608.

Hogg, M., Paro, S., Keegan, L.P., and O'Connell, M.A. (2011). RNA editing by mammalian ADARs. Adv. Genet. 73, 87–120.

Hood, J.L., and Emeson, R.B. (2012). Editing of neurotransmitter receptor and ion channel RNAs in the nervous system. Curr. Top. Microbiol. Immunol. 353, 61–90.

Hundley, H.A., and Bass, B.L. (2010). ADAR editing in double-stranded UTRs and other noncoding RNA sequences. Trends Biochem. Sci. 35, 377–383.

Hundley, H.A., Krauchuk, A.A., and Bass, B.L. (2008). C. elegans and H. sapiens mRNAs with edited 3′ UTRs are present on polysomes. RNA *14*, 2050–2060.

Jantsch, M.F., and Ohman, M. (2008). RNA editing by adenosine deaminases that act on RNA (ADARs). In RNA Editing, Goringer, H.U., ed. (Springer, Berlin), pp. 51–84.

Jepson, J.E., and Reenan, R.A. (2008). RNA editing in regulating gene expression in the brain. Biochim. Biophys. Acta *1779*, 459–470.

Jin, Y., Zhang, W., and Li, Q. (2009). Origins and evolution of ADAR-mediated RNA editing. IUBMB Life *61*, 572–578.

Keegan, L.P., Gallo, A., and O'Connell, M.A. (2001). The many roles of an RNA editor. Nat. Rev. Genet. *2*, 869–878.

Kim, D.D., Kim, T.T., Walsh, T., Kobayashi, Y., Matise, T.C., Buyske, S., and Gabriel, A. (2004). Widespread RNA editing of embedded alu elements in the human transcriptome. Genome Res. *14*, 1719–1725.

Kiran, A., and Baranov, P.V. (2010). DARNED: a DAtabase of RNa EDiting in humans. Bioinformatics *26*, 1772–1776.

Kumar, M., and Carmichael, G.G. (1997). Nuclear antisense RNA induces extensive adenosine modifications and nuclear retention of target transcripts. Proc. Natl. Acad. Sci. U.S.A. *94*, 3542–3547.

Lehmann, K.A., and Bass, B.L. (1999). The importance of internal loops within RNA substrates of ADAR1. J. Mol. Biol. *291*, 1–13.

Lev-Maor, G., Sorek, R., Levanon, E.Y., Paz, N., Eisenberg, E., and Ast, G. (2007). RNA editing-mediated exon evolution. Genome Biol. *8*, R29.

Levanon, E.Y., Eisenberg, E., Yelin, R., Nemzer, S., Hallegger, M., Shemesh, R., Fligelman, Z.Y., Shoshan, A., Pollock, S.R., Sztybel, D., *et al.* (2004). Systematic identification of abundant A-to-I editing sites in the human transcriptome. Nat. Biotechnol. *22*, 1001–1005.

Li, J.B., Levanon, E.Y., Yoon, J.K., Aach, J., Xie, B., Leproust, E., Zhang, K., Gao, Y., and Church, G.M. (2009). Genome-wide identification of human RNA editing sites by parallel DNA capturing and sequencing. Science *324*, 1210–1213.

Li, M., Wang, I.X., Li, Y., Bruzel, A., Richards, A.L., Toung, J.M., and Cheung, V.G. (2011). Widespread RNA and DNA sequence differences in the human transcriptome. Science *333*, 53–58.

Lomeli, H., Mosbacher, J., Melcher, T., Hoger, T., Geiger, J.R., Kuner, T., Monyer, H., Higuchi, M., Bach, A., and Seeburg, P.H. (1994). Control of kinetic properties of AMPA receptor channels by nuclear RNA editing. Science *266*, 1709–1713.

Lowenstein, C.J., and Padalko, E. (2004). iNOS (NOS2) at a glance. J. Cell Sci. *117*, 2865–2867.

Ma, C.H., Chong, J.H., Guo, Y., Zeng, H.M., Liu, S.Y., Xu, L.L., Wei, J., Lin, Y.M., Zhu, X.F., and Zheng, G.G. (2011). Abnormal expression of ADAR1 isoforms in Chinese pediatric acute leukemias. Biochem. Biophys. Res. Commun. *406*, 245–251.

MacNicol, M.C., and MacNicol, A.M. (2010). Developmental timing of mRNA translation – integration of distinct regulatory elements. Mol. Reprod. Dev. 77, 662–669.

Mondol, V., and Pasquinelli, A.E. (2012). Let's make it happen: the role of let-7 microRNA in development. Curr. Top. Dev. Biol. *99*, 1–30.

Moore, M.J., and Proudfoot, N.J. (2009). Pre-mRNA processing reaches back to transcription and ahead to translation. Cell *136*, 688–700.

Morse, D.P., and Bass, B.L. (1999). Long RNA hairpins that contain inosine are present in *Caenorhabditis elegans* poly(A)+ RNA. Proc. Natl. Acad. Sci. U.S.A. *96*, 6048–6053.

Morse, D.P., Aruscavage, P.J., and Bass, B.L. (2002). RNA hairpins in noncoding regions of human brain and *Caenorhabditis elegans* mRNA are edited by adenosine deaminases that act on RNA. Proc. Natl. Acad. Sci. U.S.A. *99*, 7906–7911.

Nakagawa, S., Naganuma, T., Shioi, G., and Hirose, T. (2011). Paraspeckles are subpopulation-specific nuclear bodies that are not essential in mice. J. Cell Biol. *193*, 31–39.

Nishikura, K. (2010). Functions and regulation of RNA editing by ADAR deaminases. Annu. Rev. Biochem. *79*, 321–349.

Orgel, L.E., and Crick, F.H. (1980). Selfish DNA: the ultimate parasite. Nature *284*, 604–607.

Osenberg, S., Dominissini, D., Rechavi, G., and Eisenberg, E. (2009). Widespread cleavage of A-to-I hyperediting substrates. RNA *15*, 1632–1639.

Paul, M.S., and Bass, B.L. (1998). Inosine exists in mRNA at tissue-specific levels and is most abundant in brain mRNA. EMBO J. *17*, 1120–1127.

Paz, N., Levanon, E.Y., Amariglio, N., Heimberger, A.B., Ram, Z., Constantini, S., Barbash, Z.S., Adamsky, K., Safran, M., Hirschberg, A., *et al.* (2007). Altered adenosine-to-inosine RNA editing in human cancer. Genome Res. *17*, 1586–1595.

Paz-Yaacov, N., Levanon, E.Y., Nevo, E., Kinar, Y., Harmelin, A., Jacob-Hirsch, J., Amariglio, N., Eisenberg, E., and Rechavi, G. (2010). Adenosine-to-inosine RNA editing shapes transcriptome diversity in primates. Proc. Natl. Acad. Sci. U.S.A. *107*, 12174–12179.

Peng, Z., Cheng, Y., Tan, B.C., Kang, L., Tian, Z., Zhu, Y., Zhang, W., Liang, Y., Hu, X., Tan, X., *et al.* (2012). Comprehensive analysis of RNA-Seq data reveals extensive RNA editing in a human transcriptome. Nat Biotechnol. *30*, 253–260.

Ponicsan, S.L., Kugel, J.F., and Goodrich, J.A. (2010). Genomic gems: SINE RNAs regulate mRNA production. Curr. Opin. Genet. Dev. *20*, 149–155.

Prasanth, K.V., Prasanth, S.G., Xuan, Z., Hearn, S., Freier, S.M., Bennett, C.F., Zhang, M.Q., and Spector, D.L. (2005). Regulating gene expression through RNA nuclear retention. Cell *123*, 249–263.

Pullirsch, D., and Jantsch, M.F. (2010). Proteome diversification by adenosine to inosine RNA editing. RNA Biol. *7*, 205–212.

Rueter, S.M., Dawson, T.R., and Emeson, R.B. (1999). Regulation of alternative splicing by RNA editing. Nature *399*, 75–80.

Samuel, C.E. (2011). Adenosine deaminases acting on RNA (ADARs) are both antiviral and proviral. Virology *411*, 180–193.

Scadden, A.D. (2005). The RISC subunit Tudor-SN binds to hyperedited double-stranded RNA and promotes its cleavage. Nat. Struct. Mol. Biol. *12*, 489–496.

Scadden, A.D., and Smith, C.W. (2001). Specific cleavage of hyperedited dsRNAs. EMBO J. *20*, 4243–4252.

Schmucker, D., Clemens, J.C., Shu, H., Worby, C.A., Xiao, J., Muda, M., Dixon, J.E., and Zipursky, S.L. (2000). *Drosophila* Dscam is an axon guidance receptor exhibiting extraordinary molecular diversity. Cell *101*, 671–684.

Schrider, D.R., Gout, J.F., and Hahn, M.W. (2011). Very few RNA and DNA sequence differences in the human transcriptome. PloS One *6*, e25842.

Schwanhausser, B., Busse, D., Li, N., Dittmar, G., Schuchhardt, J., Wolf, J., Chen, W., and Selbach, M. (2011). Global quantification of mammalian gene expression control. Nature *473*, 337–342.

Smalheiser, N.R., and Torvik, V.I. (2006). Alu elements within human mRNAs are probable microRNA targets. Trends Genet. *22*, 532–536.

Sommer, B., Kohler, M., Sprengel, R., and Seeburg, P.H. (1991). RNA editing in brain controls a determinant of ion flow in glutamate-gated channels. Cell *67*, 11–19.

Spriggs, K.A., Bushell, M., and Willis, A.E. (2010). Translational regulation of gene expression during conditions of cell stress. Mol. Cell *40*, 228–237.

Streit, A.K., and Decher, N. (2011). A-to-I RNA editing modulates the pharmacology of neuronal ion channels and receptors. Biochemistry Biokhimiia *76*, 890–899.

Sulston, J.E., and Brenner, S. (1974). The DNA of *Caenorhabditis elegans*. Genetics *77*, 95–104.

Svoboda, P., and Di Cara, A. (2006). Hairpin RNA: a secondary structure of primary importance. Cel. Mol. Life Sci. *63*, 901–908.

Venter, J.C., Adams, M.D., Myers, E.W., Li, P.W., Mural, R.J., Sutton, G.G., Smith, H.O., Yandell, M., Evans, C.A., Holt, R.A., *et al.* (2001). The sequence of the human genome. Science *291*, 1304–1351.

Vogel, C., Abreu Rde, S., Ko, D., Le, S.Y., Shapiro, B.A., Burns, S.C., Sandhu, D., Boutz, D.R., Marcotte, E.M., and Penalva, L.O. (2010). Sequence signatures and mRNA concentration can explain two-thirds of protein abundance variation in a human cell line. Mol. Syst. Biol. *6*, 400.

Wahlstedt, H., Daniel, C., Enstero, M., and Ohman, M. (2009). Large-scale mRNA sequencing determines global regulation of RNA editing during brain development. Genome Res. *19*, 978–986.

Wahlstedt, H., and Ohman, M. (2011). Site-selective versus promiscuous A-to-I editing. Wiley Interdiscip. Rev. RNA *2*, 761–771.

Warf, M.B., and Berglund, J.A. (2010). Role of RNA structure in regulating pre-mRNA splicing. Trends Biochem. Sci. *35*, 169–178.

Weissbach, R., and Scadden, A.D. (2012). Tudor-SN and ADAR1 are components of cytoplasmic stress granules. RNA *18*, 462–471.

Wickens, M., Anderson, P., and Jackson, R.J. (1997). Life and death in the cytoplasm: messages from the 3′ end. Curr. Opin. Genet. Dev. *7*, 220–232.

Wilusz, J.E., and Spector, D.L. (2010). An unexpected ending: noncanonical 3′ end processing mechanisms. RNA *16*, 259–266.

Zhang, Z., and Carmichael, G.G. (2001). The fate of dsRNA in the nucleus: a p54(nrb)-containing complex mediates the nuclear retention of promiscuously A-to-I edited RNAs. Cell *106*, 465–475.

ADAR and the Balance Game Between Virus Infection and Innate Immune Cell Response

10

Sara Tomaselli[1], Federica Galeano[1], Franco Locatelli and Angela Gallo

Abstract

All viruses that have dsRNA structures at any stages of their life cycle may potentially undergo RNA editing mediated by ADAR enzymes. Indeed, a number of reports describe A-to-I sequence changes in viral genomes and/or transcripts that are consistent with ADAR activity. These modifications can appear as either hyperediting during persistent viral infections or specific RNA editing events in viral dsRNAs.

It is now well established that ADAR enzymes can affect virus interaction with their host in both an editing-dependent and -independent manner, with ADARs playing for both sides: the host and the virus. Despite the discovery of editing events on viral RNA dates back to thirty years ago, the biological consequences of A-to-I changes during viral infection is still far to be completely elucidated. In particular, the proviral role played by ADAR1, partly due to PKR inhibition, together with its antiviral effect following hyperediting events put in evidence the complex role played by RNA editing in the regulation of viral infections and innate immune response.

Introduction

The number of virus families as well as their different replication and expression strategies is countless. However, viruses share the necessity to use the host cellular machinery for their RNA and protein synthesis. The presence of viral RNA molecules inside the host cell is the starting point common to all viruses: either viruses carry an RNA genome or viruses replicate through RNA intermediates. These viral genomes/transcripts can form double-stranded RNA (dsRNA) structures, which are quickly recognized by many cellular systems, including the family of adenosine deaminase that act on RNA (ADAR) enzymes. Therefore, viral dsRNAs can undergo editing, allowing ADARs to play an important part in the virus life cycle. Indeed, nucleotide changes, consistent with A-to-I editing events mediated by ADAR enzymes, have been discovered in several viral RNAs.

The first report of adenosine (A) residues mutated to guanosine (G) residues in a viral genome dates back to 1984, when O'Hara and co-authors described this 'biased hypermutation' event in the RNA of vesicular stomatitis virus (VSV) (O'Hara *et al.*, 1984), some years prior to the discovery of ADAR enzymes. Thenceforth, ADARs have been found implicated in the RNA editing of several viruses.

ADARs can carry out highly site-selective editing events (i.e. in KSHV, EBV and HDV), as well as random hyperediting, defined as editing at up to 50% of all adenosine residues of viral transcripts (i.e. in MV and MPyV).

Who is the major player in editing of viral RNAs is one of the first key questions that has been raised, as both ADAR1 (the full-length p150 and the N-terminally truncated p110 isoform) and ADAR2 have an active deaminase domain. One determining factor could be the cellular localization: only the long form of ADAR1 (p150) is present in both cytoplasm and nucleus,

[1]These authors contributed equally to this work.

whilst ADAR1 p110 and ADAR2 are mainly nuclear proteins. As most viruses replicate in the cytoplasm, the cytoplasmic ADAR1 p150 is likely the main player in editing these viruses. Additionally, the observation that interferon (IFN) – the first cytokine discovered – induces up-regulation of ADAR1 p150 (Patterson and Samuel, 1995) raised the hypothesis that this isoform functions in host defence mechanisms against viral infection and inflammation. This hypothesis found support in the discovery that the physical interaction between ADAR1 and nuclear factor 90 (NF90) proteins positively affects the NF90-mediated gene expression, stimulating the activation of the cellular antiviral cascade and interferon-β activation (Nie *et al.*, 2005). A second line of evidence for a role of ADARs in antiviral response came from several works showing hyperediting in viral RNA genomes and transcripts during persistent infections (Samuel, 2011). Subsequent to these findings on ADAR1, it is not surprising that, at the beginning, A-to-I RNA editing was linked to an antiviral function. However, although ADAR editing activity has been demonstrated on a variety of viral RNAs (Samuel, 2011), an ADAR1 antiviral role has been definitively established only for HCV (Taylor *et al.*, 2005).

Recent lines of research have pointed out that ADAR enzymes do not play only an antiviral role. Specifically, they can act both as proviral and antiviral factors.

Somewhat unexpectedly, ADAR1 has emerged as a promoter for replication of many viruses (HIV-1, MV, VSV, HDV) during acute infections (Clerzius *et al.*, 2011). This proviral role of ADAR1 appears to occur via two mechanisms: RNA editing activity and inhibition of the double-stranded RNA-dependent protein kinase (PKR) (George *et al.*, 2009; Clerzius *et al.*, 2011; Pfaller *et al.*, 2011). The two mechanisms can act independently but simultaneously within the host cell to increase viral replication.

PKR was first recognized as a 68-kDa protein induced by interferon (Meurs *et al.*, 1990). It is a serine/threonine kinase, whose activation is a critical component of antiviral cell defence and cell growth pathways (Garcia *et al.*, 2006). Two N-terminal dsRNA-binding domains (dsRBDs) give PKR the role of a dsRNA sensor to signal pathogenic invasions. The presence of dsRNAs in the cell promotes dimerization and activation of PKR that, through the C-terminal catalytic kinase domain, phosphorylates a variety of substrates, such as the alpha subunit of the translation eukaryotic initiation factor 2 (eIF-2α) (Lemaire *et al.*, 2008). Phosphorylation of eIF-2α on serine 51 negatively alters the efficiency of translation initiation, reducing cellular as well as viral protein synthesis (Sadler and Williams, 2007).

Viruses have developed different mechanisms to counteract PKR activity that include the following: the expression of dsRNAs or viral proteins directly suppressing PKR activity, promoting PKR degradation and/or eIF-2α dephosphorylation. Viruses have also learned to exploit host cellular factors to negatively regulate PKR activation, like the PKR activator (PACT), the TAR binding protein (TRBP) and ADAR1 (Clerzius *et al.*, 2011).

ADAR1 plays a role as PKR suppressor in several acute infections, resulting in an enhanced viral replication of the vesicular stomatitis virus (VSV), measles virus (MV) and human immunodeficiency virus type 1 (HIV-1) (Samuel, 2011). ADAR1 binds to the first dsRBD of PKR, reducing its kinase activity and suppressing eIF-2α phosphorylation, by an editing-independent mechanism (Nie *et al.*, 2007).

The existence of INF-inducible proteins (ADAR1/PKR) with opposite effects seems puzzling, and highlights the existence of a complex and fine-tuned balance between viral spread and host cell defence. Besides, for some viruses ADARs could act in both a proviral and antiviral manner, depending on the cellular/time context (i.e. for MV, HDV). This is not surprising, under the point of view that each cell has to achieve an equilibrium to avoid death due to either viral replication or to apoptosis induced by hyperactive immune response (Gelinas *et al.*, 2011).

Herein, we gather the current knowledge of A-to-I RNA editing and viruses, classified according to their family and genome type (Table 10.1 and Fig. 10.1).

Table 10.1 Viruses undergoing RNA editing events by ADARs

Family/genus	Virus	Genome	Host	Editing	ADAR enzyme	ADAR effect
Polyomaviridae	MpyV	Circular dsDNA	Mouse	Hyperediting	Ad1 p110 Ad1	Antiviral Proviral
Herpesviridae	EBV	Linear dsDNA	Humans	pri-miRBART6	ND	Viral latency
	KSHV	Linear dsDNA	Humans	Kaposin A and miR-K10	Ad1 p110	Latent/lytic switch
Flaviviridae	HCV	ss(+)RNA	Humans	Hyperediting	Ad1 p150	Antiviral
Retroviridae	RAV-1	ss(+)RNA	Birds	Hyperediting	ND	ND
	ALV	ss(+)RNA	Birds	Hyperediting	ND	ND
	HIV-1	ss(+)RNA	Humans	5′UTR, Tat, Rev, Env (RRE)	Ad1 and Ad2 Ad1	Proviral Antiviral
Arenaviridae	LCMV	ss(–)RNA	Mammals	Hyperediting	Ad1 p150	Antiviral(?)
Bunyaviridae	RVFV	ss(–)RNA	Humans	Hyperediting	ND	ND
Deltavirus	HDV	ss(–)RNA	Humans	Amber/W	Ad1 p110	Proviral
Orthomyxoviridae	Influenza A virus	ss(–)RNA	Mammals, birds	Hyperediting	Ad1 p150	Antiviral
Paramyxoviridae	MV	ss(–)RNA	Humans	Hyperediting	Ad1 p150 Ad1	Antiviral Proviral
	HRSV	ss(–)RNA	Humans	Hyperediting	ND	Proviral
	HPIV3	ss(–)RNA	Humans	Hyperediting	ND	ND
Rabdoviridae	VSV	ss(–)RNA	Plants, insects, mammals	Hyperediting	Ad1 p150	Proviral
	Drosophila sigma virus	ss(–)RNA	Drosophila	Hyperediting	dADAR	ND

ND, not determined.

Polyomaviridae: murine polyomavirus (MPyV)

The members of the Polyomaviridae family are naked viruses, with an icosahedric capsid and a small circular double-stranded DNA (dsDNA) genome (~5 kbp). Polyomaviruses can infect both animals and humans, and are often associated with tumour development.

Murine polyoma virus (MPyV) was the first polyomavirus to be discovered (Gross, 1953). After the viral attachment through the interaction between the viral protein 1 (VP1) and the sialic acid residues of cell gangliosides, virus is endocytosed and transported into the nucleus, where it uses the host's machinery for gene expression and DNA replication. The polyoma virus life cycle is divided in two distinct phases: the early phase, which occurs immediately after the infection and before viral replication, and the late phase, which begins after the onset of DNA synthesis. The early and late-genes are transcribed from opposite strands of the viral genome. Spliced early RNA transcripts encode for non-structural proteins, known as large, middle and small T-antigens, which have a crucial role in regulating the viral life cycle, by promoting DNA synthesis. As polyoma virus needs host cell machinery to replicate itself, the host cell has to be in S-phase of cell cycle. This is achieved through the activation of cell signalling pathways (such as p53, pRB, MAPK pathways) by the viral T-antigens. This abnormal cell proliferation is at the basis of oncogenic transformation (Fluck and Schaffhausen, 2009). After replication, late-gene expression occurs, producing structural proteins (VP1–3) necessary for virion assembly.

During the viral life cycle, there is an accurate regulation of the early-late phase 'switch'. At early times after infection, genes encoded from the

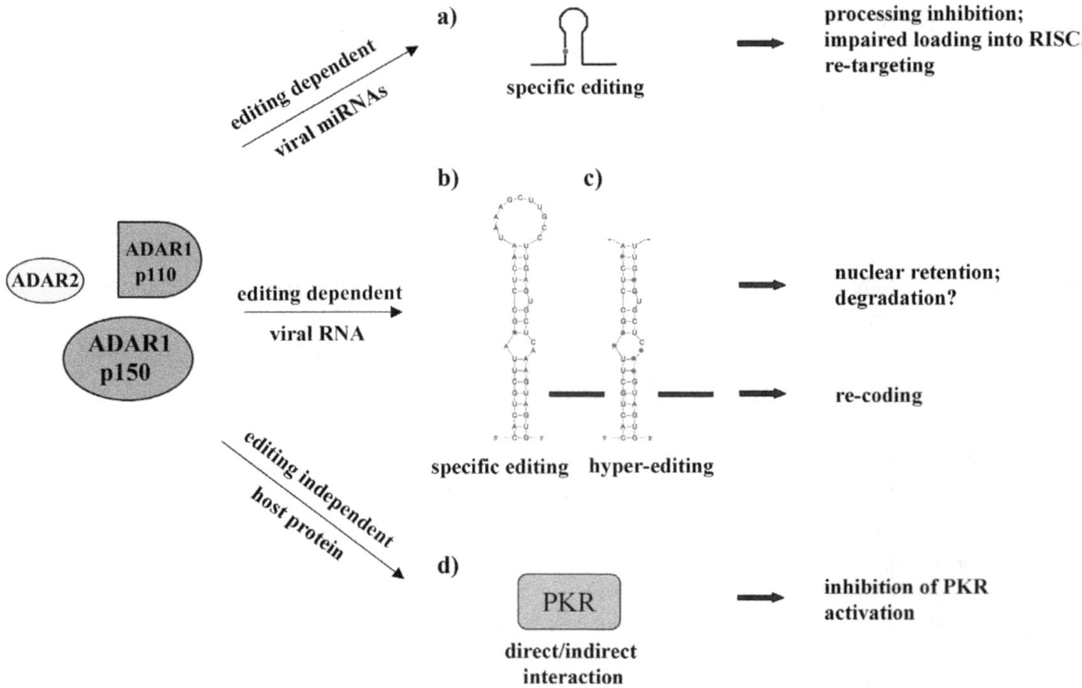

Figure 10.1 Possible ADAR-mediate effects after viral infection: (a) editing of viral miRNAs; editing of viral transcripts (b) at specific sites (such as 5′-UTR of HIV) or (c) at multiple non-specific sites (hyperediting); (d) direct/indirect interaction with host proteins (such as PKR). A colour version of this figure is located in the plate section at the back of the book.

early transcription units accumulate predominantly (late:early ratio of 1:10); however, after viral DNA synthesis, expression of early genes dramatically decreases with a remarkable increase of late-gene RNA levels (late:early ratio of 50:1) (Gu *et al.*, 2009). This temporal transition is independent from changes in early/late promoter efficiency (Hyde-DeRuyscher and Carmichael, 1988; Liu and Carmichael, 1993) and resulted to be regulated post-transcriptionally by, at least in part, RNA editing processes (Kumar and Carmichael, 1997; Gu *et al.*, 2009).

At late times, primary late-strand transcripts are inefficiently polyadenylated at their 3′ ends and RNA polymerase II encircles the genome multiple times, generating long multimeric RNAs, with regions of complementarity to early transcripts. These multigenomic late RNAs, acting as natural antisense regulators within the nucleus, have the potential to efficiently inhibit early gene expression in mouse polyoma virus (Liu and Carmichael, 1994). The molecular bases of early strand RNA repression were described a few years later: the consequent formation of early/late dsRNA hybrids into the nucleus enables RNA editing phenomena to occur (Kumar and Carmichael, 1997). Analysis of early strand RNAs isolated in the late phase of infection revealed widespread nucleotide modifications, consistent with adenosine deamination mediated by ADARs. About 40–60% of adenosines in early-strand RNAs were modified to inosines/guanosines. This A-to-I hyperediting seems to be responsible for the retention of the modified RNAs in the nucleus (Kumar and Carmichael, 1997).

Later, new studies demonstrated that the production of multigenomic primary RNA late-strands, generated by inefficient polyadenylation and transcription termination, is also due to RNA editing. First, key regulatory elements, involved in the transition from efficient (early phase) to inefficient (late phase) late-strand transcription termination, were found in the region of the poly-adenylation signal (Gu *et al.*, 2006). Additionally, the authors noticed that the poly(A) regions of early and late transcripts overlap, generating dsRNA structures that represent an excellent substrate for ADAR enzymes (Gu *et al.*, 2009). Indeed, in these overlapping regions a high content of inosines was found that is able to alter polyadenylation signals leading to transcription read-through in MPyV. These multigenomic late RNAs are highly stable, leading to up-regulation of late transcript levels with the consequent switch from early to late phase (Gu *et al.*, 2009).

The functional significance of RNA editing in regulating the polyoma virus life cycle has been proven by *ADAR1* knock down in MPyV-permissive NIH3T3 cells. A 4-fold reduction of ADAR1, with a consequent absence of editing, was causative of a defective early-to-late switch pattern of the virus (Gu *et al.*, 2009).

Noteworthy is the hypothesis proposed by the authors that similar mechanisms may also occur in eukaryotic/mammalian cells under physiological conditions, representing a new form of gene regulation mediated by RNA editing. Indeed, evidence has emerged that nuclear retention of promiscuously edited RNAs is a general phenomenon involving a multiprotein complex composed of the inosine-specific RNA binding protein p54nrb, the splicing factor PSF and the inner nuclear matrix structural protein matrin 3 (Zhang and Carmichael, 2001). This multi-protein complex seems to anchor hyperedited RNAs to the nuclear matrix, thus keeping them away from the translational machinery.

Recently, George and Samuel explored the role of ADARs in MPyV infection using mouse embryonic fibroblast (MEF) cells isolated from knockout of either *Adar1* or *Adar2* mice (George and Samuel, 2011). The hypothesis was that if ADAR-mediated hyperediting was a critical component for the gene expression regulation in MPyV, then the genetic depletion of either *Adar1* or *Adar2* should alter virus growth. Actually, this was not observed: viral growth (and viral protein expression) was comparable between $Adar1^{-/-}$ and $Adar2^{-/-}$ MEF cells and, unexpectedly, more efficient than in wild-type cells. According to these findings, the authors supposed that the formation of dsRNA structures, rather than the editing events, is the critical component for gene regulation. Additionally, only the absence of Adar1, and in particular of its constitutive isoform p110, was found to enhance MPyV-induced cytotoxicity of MEF cells. Surprisingly, this protective role of Adar1 p110 seems to be independent of its editing activity. The authors concluded that both Adar1 and Adar2 are important for suppress MPyV growth in MEFs, whereas only Adar1 p110 modulates cell response to MPyV infection (George and Samuel, 2011).

Even though these findings are intriguing, the fact that they are partially in contrast with the previous studies seems puzzling.

Taken together, these data point out that the role of A-to-I editing, and in particular of ADAR1, is crucial in mouse polyoma virus infection; nevertheless, the exact functions and mechanisms of ADARs need to be further elucidated.

Herpesviridae

Herpesviridae family members are composed of a linear dsDNA genome, an icosahedric capsid, a tegument containing viral proteins and an envelope. Herpesviruses infect cells through the interaction with specific receptors on the cell surface; after the internalization of the virions, viral DNAs migrate to the cell nucleus where they are transcribed to RNA by recruiting the host cellular transcription machinery. All herpesviruses have two distinct life cycles: latent and lytic replication phase. During acute (lytic) infection, the virus sequentially expresses its entire repertoire of genes. The symptomatic infection is characterized by an enhanced replication and production of virions that spread infection from cell to cell, often leading to their death. After a self-limited period of lytic infection, in some specific host cells the virus can enter into a latent state expressing only selected latent genes, evading host immune surveillance mechanisms and establishing an asymptomatic, persistent infection.

According to the site of latency, herpesviruses are divided into three subfamilies: α, β and γ. The latter includes HHV-4 (Epstein–Barr virus) and HHV-8 (Kaposi's sarcoma-associated herpesvirus), with B cells as site of latency. Both of these viruses encode RNA transcripts that undergo highly site-selective A-to-I editing events.

Epstein–Barr virus (EBV or HHV-4)

Epstein–Barr virus (EBV), also known as human herpesvirus 4 (HHV-4), is characterized by two different infection strategies with distinct programmes of gene expression: an acute lytic phase (with the production of viral proteins and virions) and a persistent latent phase. The exclusive host of EBV is human and its primary cell targets are B cells and epithelial cells. EBV causes infectious mononucleosis during its lytic infection; in addition, EBV latent viral infection is associated with a variety of lymphoma (Burkitt lymphoma, Hodgkin disease) and some epithelial cancers (nasopharyngeal carcinoma).

HHV-4 has a linear dsDNA genome of 172 kbp coding for about 100 genes, of which only 9–11 are transcribed during virus latency. Three different types of latency have been identified (I–III) and the transition from lytic infection to distinctive latency states is finely regulated by both viral genes and viral microRNAs. Indeed, EBV codes also for more than 20 microRNAs implicated in the attenuation of host antiviral immune responses and in the regulation of transition from latent to lytic replication (Cullen, 2009; Skalsky and Cullen, 2010). The EBV miRNAs seem to target both viral and host cell genes (Cullen, 2009).

Among the EBV miRNAs, four primary microRNAs (pri-miRNAs) were found to undergo site-specific A-to-I editing events: pri-miR-BHRF1-1, pri-miRBART6, pri-miR-BART8 and pri-miR-BART16 (Iizasa *et al.*, 2010). Pri-miR-BART6 showed high levels of editing at the +20 site of the 3p strand in three latently EBV-infected cell lines (GM607 B lymphoblastoid cells, Daudi Burkitt lymphoma cells and C666-1 nasopharyngeal carcinoma cells) (Lizasa *et al.*, 2010). Western blotting analysis revealed an abundant expression of ADAR1 p150 and p110 and a lack of ADAR2 in these cells, suggesting that ADAR1 is the enzyme responsible for these editing events.

To investigate the effect of editing at the +20 site of miR-BART6, the authors analysed its maturation process. Interestingly, a combination of A-to-I editing at the +20 site and a naturally occurring mutation (three U residues deletion) in the pri-miR-BART6, as found in two of the three cell lines analysed (Daudi and C666-1 cells), was able to completely inhibit Drosha cleavage, with a total lack of pre- and mature miRNA production. However, editing alone (without the 3bp deletion) did not affect miR-BART6 biogenesis (as found in the GM607 cell line), but is able to reduce the correct loading of miR-BART6-5p onto the microRNA-induced silencing complex (miRISC). Remarkably, this is the first report of A-to-I editing of a pri-miRNA affecting miRISC loading (Iizasa *et al.*, 2010).

To further understand the biological significance of pri-miR-BART6 editing, the authors investigated the role of this miRNA in the viral life cycle. They discovered that miR-BART6-5p targets human Dicer, with a consequent modulation of expression of many viral and host-cell miRNAs and, in turns, genes. In particular, miR-BART6, by silencing of Dicer, suppresses the viral EBNA2 and LMP1 proteins required for transition from type I–II latency to type III. It also reduces expression of Zta and Rta viral proteins, which are essential for lytic replication. In this way, miR-BART6 maintains EBV-infected cells in a particular state of latency (I–II), by suppressing both the lytic replication and the transition to type III latency, a more immunoresponse-prone state of the viral infection. miR-BART6 also down-regulates host miRNAs necessary for the immune response.

Editing of pri-miR-BART6, through different mechanisms, reduces the activity of mature miR-BART6. Therefore, A-to-I editing, together with naturally occurring mutations, may be an adaptive mechanism to antagonize miR-BART6 function, playing a crucial role in the regulation of EBV life cycle and cell immune response.

Kaposi's sarcoma-associated herpesvirus (KSHV or HHV-8)

Kaposi's sarcoma-associated herpesvirus (KSHV) is one of the seven currently known human oncoviruses (i.e. KSHV, HCV, HBV, HTLV-1, HPV, HIV, EBV) (Talbot and Crawford, 2004). According to the International Committee on Taxonomy of Viruses (ICTV), the KSHV formal name is human herpesvirus 8 (HHV-8). HHV-8 has a linear dsDNA genome of about 160–170 kbp. After infection, the genome is circularized and enters into the cell as episomal DNA using the cellular replication machinery to replicate itself. Usually the virus remains in a latent state expressing limited genes, such as the viral latency-associated nuclear antigen (LANA), without integrating into host chromosomes and with no virion production. However, various signals, such as inflammation, may provoke the virus to enter into lytic replication. Epigenetic regulation seems to be essential for the activation/inactivation of latent genes (Ohsaki and Ueda, 2012). The latently infected tumour cells overexpress viral proteins and some of them are directly implicated in oncogenesis, because of their ability to bind to the tumour-suppressor protein p53 and to the retinoblastoma protein pRb (Cathomas, 2003). Indeed, HHV-8 is the etiologic agent responsible for the Kaposi's sarcoma (KS), a cancer of lymphatic endothelium forming vascular channels that fill with blood cells, giving the tumour its characteristic bruise-like appearance. In addition to KS, HHV-8 is also closely associated with other tumours, such as the rare primary effusion lymphoma (PEL), originating from preterminal B cells, and the multicentric Castleman's disease (MCD), a plasmacytic lymphadenopathy.

The K12 (kaposin) RNA transcript is expressed abundantly during the latent phase of HHV-8 infection and is also induced during lytic viral replication. It gives rise to three isoforms of the kaposin protein (A, B and C) and to the miR K-10, thanks to different translation starting codons (Sadler *et al.*, 1999). Interestingly, the kaposin A ORF (open reading frame), that includes both kaposin A mRNA and the miR-K10, contributes to the tumorigenic potential of the HHV-8 virus (Muralidhar *et al.*, 1998, 2000; Damania, 2004).

A specific A-to-I nucleotide variant, consistent with an ADAR-mediated RNA editing event, has been reported at position 117,990 of the viral transcript isolated from PEL tumours and cell lines infected with HHV-8 (Pfeffer *et al.*, 2005; Gandy *et al.*, 2007). This alteration modifies both the coding sequence of the kaposin A protein (changing the serine to a glycine at amino acid 38) and the position 2 at the 5′ end of the mature miR-K10 (affecting the seed sequence and potentially its mRNA target genes) (Pfeffer *et al.*, 2005; Gandy *et al.*, 2007). An expression analysis of RNA derived from PEL cells showed that *ADAR1* mRNA levels are 10- to 20-fold higher than *ADAR2*. ADAR1 p110 resulted to be able to edit the K12 transcript at position 117,990 in a highly specific manner by *in vitro* assay (Gandy *et al.*, 2007). However, ADAR2 and ADAR1 p150 involvement were not tested.

Remarkably, the authors observed that this A-to-I editing has a functional significance, with a key role in the replication strategy of HHV-8 and in its tumorigenic potential. In particular, PEL cells harbouring latent viral infection produce a K12 transcript mostly unedited and with transforming activity, whilst RNA editing of the same transcript is particular frequent during lytic viral replication and seems to eliminate its oncogenic potential (Gandy *et al.*, 2007).

It would be interesting to investigate whether editing events also occur in K12 transcripts of KS tumours.

Since the biological role of ADAR-mediated editing events in cellular miRNAs has been clearly disclosed, it would be not surprising that ADARs may have a significant impact on viral life cycle by editing of viral miRNAs. Nevertheless, to date the A-to-I editing events of EBV pri-miR-BART6 and of KSHV miR-K10 are the only examples of ADAR-mediated editing of viral miRNAs. Further research could better investigate this issue.

Flaviviridae: hepatitis C virus (HCV)

The name Flaviviridae (from the Latin *flavus*, yellow) is due to the first isolated member of this family, the yellow fever virus. The members of Flaviviridae family are single-stranded positive-sense RNA (ss(+)RNA) viruses, provided with envelope and an icosahedric capsid (Choo *et al.*, 1991).

Hepatitis C virus (HCV) is the unique member of the *Hepacivirus* genus (Maniloff, 1995) and was first isolated in 1989 as a non-A, non-B hepatitis agent (Choo *et al.*, 1989). The virus is transmitted parentally (blood–blood contact), sexually and vertically and to date more than 170 million people are estimated to be infected (Alter and Seeff, 2000; Raza *et al.*, 2007). HCV is recognized as the major cause of acute and chronic liver disease worldwide (Clement *et al.*, 2009); indeed, the acute infection becomes lifelong persistent in more than 80% of cases and may results in chronic liver disease, cirrhosis and hepatocellular carcinoma (Levrero, 2006).

The HCV genome consists of an uncapped, positive, ssRNA molecule of 9.6 kbp, which is characterized by the presence of conserved secondary structures and is predicted to contain a high rate of internal basepairing (Tuplin *et al.*, 2002; Simmonds, 2004). The translation of HCV genome generates a single precursor polypeptide, which is cleaved by host cell and viral proteases to produce 10 functional proteins (Bartenschlager *et al.*, 2004). One third of the HCV genome encodes for the capside/core protein C, the glycoproteins E1 and E2 and the p7 protein. The remainder of the HCV genome encodes for the non-structural proteins: NS2, NS3, NS4A, NS4B, NS5A e NS5B (Bartenschlager *et al.*, 2004). HCV infection initiates by the viral binding to several receptors, including unidentified liver-specific molecules. Upon binding, virions are internalized and released in the host cell cytoplasm, where viral RNA translation occurs. The subsequent *de novo* replication produces genome-length negative strand RNA molecules, which are found in partially or fully double-stranded replicative intermediates (Ali *et al.*, 2002). These negative RNAs serve as template for multiple rounds of positive-sense RNA synthesis (Lohmann *et al.*, 1999). The mature virions are then assembled and the infectious progeny buds from the host cells.

The first work reporting A-to-I editing events in the HCV genome came from Taylor and colleagues in 2005. Beyond the early interferon induction and PKR activation (measured through eIF-2α phosphorylation), the authors detected A-to-I mutations in RNA from interferon-treated Huh7 hepatoma cells stably transfected with HCV RNA replicon (HuhBB7 cells) (Taylor *et al.*, 2005). As ADAR1 is the only ADAR enzyme with an interferon inducible isoform (p150), they overexpressed adenovirus-associated (VA) RNA$_I$, known to inhibit ADAR1 (Lei *et al.*, 1998), in the same cells. Upon interferon treatment plus VA, Taylor *et al.* observed a strong reduction of A-to-I mutations, suggesting that these modifications were consistent with ADAR1-mediated editing events. Moreover, by a silencing approach, the authors elegantly demonstrated that A-to-I modifications were catalysed exclusively by ADAR1 and that the knockdown of this enzyme stimulated replicon expression in HuhBB7 cells. Taylor and coworkers did not find a specific editing site in the HCV genome and they argued that ADAR1 may catalyse a random hyperediting phenomenon and suggested that hypermutated viral RNAs may be degraded by an inosine-specific RNase (Scadden and Smith, 1997, 2001) or inefficiently replicated due to alterations of important (regulatory?) sequences.

In a recent work, Welzel and colleagues observed that a synonymous G > A SNP within exon 9 of *ADAR1* was associated with a sustained virologic response (SVR) in European American patients (Welzel *et al.*, 2009). Such evidence suggests an important role for ADAR1 during HCV infection, even if not yet fully elucidated.

Only in the last years the generation of a HCV cell culture infection model based on JFH-1HCV and human hepatoma HuH7 cells allowed researchers to better investigate the pathways involved in innate antiviral responses of host infected cells.

By using the novel cell infection model, two independent studies stated that HCV infection triggers PKR activation and eIF-2α phosphorylation, leading to a general reduction of host protein synthesis, whereas HCV translation results unaffected (Garaigorta and Chisari, 2009; Arnaud *et al.*, 2010). Taken together these data demonstrate how HCV is able to divert PKR from its antiviral role, thus strongly decreasing interferon-mediated innate immune response and interferon expression (Arnaud *et al.*, 2010).

Interestingly, it has been shown that ADAR1 can suppress PKR activation in cells infected with negative-stranded RNA viruses, such as vesicular stomatitis virus (VSV) and measles virus (MV) (Nie *et al.*, 2007; George *et al.*, 2009). In this context, ADAR1 results to be proviral, as PKR inhibition leads to suppression of host and viral translation, with a great reduction in virus replication. Based on these findings and those reported above (Garaigorta and Chisari, 2009; Arnaud *et al.*, 2010), it could be claimed that ADAR1-mediated PKR inhibition may result in an antiviral effect against HCV infection, because it should allow interferon-induced host immune response.

In strict correlation with these considerations,

Alisi and coworkers found that the ADAR1 p150 isoform is up-regulated in Huh7.5 cells when using the HCV infection and replication JFH1 model, whilst the constitutive ADAR1 p110 is not significantly modified (Alisi *et al.*, 2011). At the same time, another group found that apolipoprotein B messenger RNA editing enzyme catalytic polypeptide-like 3G (APOBEC3G) is induced as an antiviral factor against HCV infection (Peng *et al.*, 2011). Taken together these data demonstrate that at least two editing enzymes are activated as host defensive factors against HCV infection, introducing an additional step of complexity.

In conclusion, an antiviral role of ADAR1 in HCV infection, dependent on its editing activity, can be likely stated. ADAR1 p150 random editing activity may induce RNA degradation or mutations in coding sequences thus directly preventing HCV genome replication and viral protein synthesis. Moreover, the viral dsRNA editing-dependent degradation may also inhibit PKR activation.

Retroviridae

The members of the Retroviridae family consist of enveloped virions containing two copies of their RNA genome and the reverse transcriptase (RT) and integrase enzymes essential for replication. The retroviral genome consists of two ss(+)RNA molecules (about 7–10 kbp long), with a 5′ cap and 3′ poly(A) end sequences, that can be reverse transcribed by the viral RT in the cytoplasm and then randomly integrated into the host cell genome by the integrase enzyme. At this point the virus is called provirus and undergoes the usual transcription and translation processes to express viral genes (*gag, pol, env*). *Gag* (group-specific antigen) codes for core and structural proteins, *pol* (polymerase) codes for RT, protease and integrase, *env* (envelope) codes for the retroviral coat proteins. The viral RT is an RNA-dependent DNA polymerase that lacks the proofreading function; consequently, retroviruses have a very high mutation rate.

Alpharetrovirus: rous-associated virus type 1 (RAV-1) and avian leucosis virus (ALV)

Sequence changes, probably corresponding to A-to-I mutations by ADARs, have been also described for two alpharetroviruses (Retroviridae family): rous-associated virus type 1 (RAV-1) (Felder *et al.*, 1994) and avian leucosis virus (ALV) (Hajjar and Linial, 1995).

The first hint that A-to-I RNA editing could have fundamental implications in retroviruses derives from Felder *et al.*, who detected A-to-G mutations in the genome of a *c-mil/c-raf* transducing RAV-1 (designated IC4). IC4 was found to carry a highly mutated U3 long terminal repeat (LTR), in which 48% of A is converted to G (Felder *et al.*, 1994). These mutations were widespread in the U3 sequence, leading to changes in several viral regulatory elements. This hyperediting resulted in an inactive polyadenylation signal (PAS) of the 3′ LTR, allowing readthrough transcription of downstream cellular sequences. These readthrough transcripts may undergo alternative splicing, whereas viral transcription and replication are not impaired (Felder *et al.*, 1994).

One year later, an avian retroviral recombinant provirus (referred to as TBJ1R2), generated during *in vitro* passage, was described to contain a 150-nt inverted repeat with about 50% of A mutated to G. The authors proposed that these changes were the result of ADAR activity and that would likely render the virus replication defective (Hajjar and Linial, 1995).

Nevertheless, at that time dsRNA-modifying enzymes had been just discovered (Bass and Weintraub, 1988; Wagner *et al.*, 1989) and additional studies were not carried out to demonstrate the real involvement of ADARs in RAV-1 and ALV hyperediting phenomena.

Human immunodeficiency virus type I (HIV-1)

Human immunodeficiency virus type 1 (HIV-1) is a lentivirus (a member of the Retroviridae family), discovered in 1983 (Barre-Sinoussi *et al.*, 1983; Gallo *et al.*, 1983). It is the causative agent of acquired immunodeficiency syndrome (AIDS) (Weiss, 1993), a severe and fatal disease characterized by progressive failure of the immune system. The primary targets of HIV-1 are the immune system components: T-cells, monocytes, macrophages and dendritic cells. When the CD4+ T-lymphocyte count drops below 200/μl, the stage of clinical AIDS has been reached and the characteristic life-threatening opportunistic infections and neoplasms of AIDS start to appear.

The viral genome is composed of two single positive strands of RNA (9749 bp long) that code for nine genes: in addition to the three major structural genes (*gag, pol* and *env*), there are other 'accessory' genes unique to HIV (the transactivators *tat, rev, vpr* and the regulators *vif, nef, vpu*). The mature HIV-1 virions have a conical capsid composed of the viral protein p24 (from *gag* gene, typical of lentiviruses) and a lipid envelope containing the antigen gp120 (from *env* gene), which is responsible for tropism to CD4+ receptors. Within the HIV RNA there are conserved secondary structures, such as the trans-activating responsive (TAR) element, located within the LTR at the 5′ and 3′ ends, and the Rev response element (RRE), located within the *env* gene. These structures seem to be important for the HIV life cycle by altering the function of protease and reverse transcriptase (RT).

Brenda Bass's group provided the first evidence that A-to-I RNA editing plays a role in HIV-1 infection (Sharmeen *et al.*, 1991). They showed that the intramolecular TAR stem–loop structure is a substrate for A-to-I editing when injected into the nuclei of *Xenopus laevis* oocytes. In particular, 5–10% adenosines of TAR resulted to be converted to inosines, but only in the presence of Tat, a potent viral activator of HIV-1 gene expression.

Despite that the biological role of these editing events has yet to be established, this finding raised the question whether RNA editing within HIV-1 transcripts occurs also in HIV-1 infected T-cells and to which extent this may contribute to the different steps of the viral life cycle. The observation that ADAR1 mRNA and protein levels increased following HIV infection of primary T-cells supported the hypothesis of ADAR1 involvement in HIV-1 replication (Phuphuakrat *et al.*, 2008; Clerzius *et al.*, 2009). In particular, these subsequent studies provided evidence that HIV-1 takes advantage of ADAR1 for viral replication by editing-dependent and – independent mechanisms (Phuphuakrat *et al.*, 2008; Clerzius *et al.*, 2009; Doria *et al.*, 2009).

ADAR1 has been found to have a proviral role, enhancing HIV-1 expression. Overexpression of wild-type ADAR1, but not of a catalytic mutant lacking deaminase activity, up-regulated p24 protein expression in producer cells (COS-7, 293T) (Phuphuakrat *et al.*, 2008). On the contrary, 293T cells, silenced for ADAR1 by siRNA methodology and transfected with HIV-1, showed a decreased production of the Gag precursor p55 and the capsid protein p24. The authors tested also the infectivity of the progeny virions, finding no differences between viruses produced by cells transfected with active/inactive ADAR1 or siADAR1 and the controls. The two double-stranded regions TAR and RRE of the HIV–1 genome were sequenced and site-specific A-to-G changes were found around the RRE region within the *env* gene (nucleotide positions 8164 to 8166) (Phuphuakrat *et al.*, 2008). Strangely, similar changes were not found in TAR, as previously reported using the model of *Xenopus laevis* oocyte nuclei (Sharmeen *et al.*, 1991). The biological significance of *env* editing was explored by creating a HIV-1 genome with three site-specific edited sites, mimicking the naturally edited HIV-1 RNA. This mutant construct was expressed more efficiently than the wild type, underscoring the proviral effect of RNA editing in HIV-1 (Phuphuakrat *et al.*, 2008).

Further studies dissected the mechanisms of the ADAR1 proviral effect and the contribution of its editing activity, leading to partially discrepant conclusions. The authors showed that, in contrast to the findings of Phuphuakrat *et al.* (2008), both the active (wild-type) and inactive (E/A) form of ADAR1 are able to strongly increase the overall accumulation of viral proteins (such as gp120, p24 and Nef) in HIV-1 producer cells (Clerzius *et al.*, 2009; Doria *et al.*, 2009). However, the ADAR1-mediated editing activity is crucial to increase the assembly/release of progeny virions and their infectious potential (Doria *et al.*, 2009). New editing events were found both in coding (Rev and Tat) and non-coding (5′UTR) regions of viral RNA. Specifically, four out five editing events within the 5′UTR were located within the poly(A) signal. Rev and Tat sequences were found to be edited at six and one sites, respectively, with five out of six editing events in the Rev sequence leading to amino acid changes in crucial regions for its regulative function (Doria *et al.*, 2009). However, how these editing events mediate the effects observed on HIV-1 replication is still an open question that deserves additional investigations.

In addition to the editing-dependent effect of ADAR1 on HIV-1 replication, another editing-independent mechanism has been well characterized. As previously mentioned, ADAR1 physically interacts with PKR, leading to PKR inactivation (Nie *et al.*, 2007). This ADAR1-mediated PKR inhibition is one of the mechanisms used by HIV-1 to impair the host antiviral response (Clerzius *et al.*, 2009; Doria *et al.*, 2009). Reduced levels of both phosphorylated PKR and its target eIF-2α were observed in 293T cells co-transfected with HIV-1 and with either the active or the inactive form of ADAR1, providing a plausible explanation for the editing-independent capacity of ADAR1 to stimulate HIV-1 protein expression (Clerzius *et al.*, 2009; Doria *et al.*, 2009). In particular, ADAR1 p150 and p110 expression is enhanced during HIV infection of T-cells (Jurkat). The binding of ADAR1 to PKR at the peak of HIV-1 infection is responsible for the PKR inhibition and correlates with increased viral protein production (p55 and p24) (Clerzius *et al.*, 2009). The ADAR1-mediated PKR suppression requires at least one Z-DNA binding domain (Z-DBD) and the three dsRNA binding domains (dsRBDs) of ADAR1, whilst the deaminase activity is not necessary (Clerzius *et al.*, 2009).

However, additional mechanisms by which ADAR1 can positively modulate HIV-1 replication are highly likely to exist and need further studies.

In contrast to the evidence just described, a recent work by Biswas *et al.* (2012) has proposed that ADAR1 is actually an antiviral factor against HIV-1. After overexpressing and silencing ADAR1 in different cells, the authors stated that ADAR1 inhibits HIV-1 protein accumulation without affecting HIV-1 RNA synthesis. This antiviral effect has been attributed to site-specific editing events in *env* and *rev* coding sequences. The ADAR1-induced A-to-G mutations, mainly located in the RRE binding domain, resulted in a defective binding/activity of Rev. As a consequence, the Rev-mediated transport activity of primary transcripts *gag*, *pol*, *env* from the nucleus to cytoplasm was inhibited and protein synthesis affected. There is no clear understanding at this stage of why these results contrast with previous data. However, unfortunately, the PKR activity has not been tested in this latter setting of experiments.

Another member of the ADAR family, ADAR2, has been recently proposed to act as a proviral factor (Doria *et al.*, 2011). Notably, HIV-1 target cells (Jurkat cells) stably silenced for ADAR2 showed a significant impairment of viral replication. Consistently, ADAR2 overexpression in producer cells (293T) enhanced p24 protein synthesis by an editing-independent mechanism, likely through an inhibitory effect on PKR. Indeed, reduced levels of phosphorylated eIF2α were found in cells expressing the active and inactive form of ADAR2. This effect was similar, even if less pronounced, to the ones already observed with ADAR1. The multistep process of viral particle release was positively modulated by ADAR2 in an editing-dependent manner as well (Doria *et al.*, 2011). However, in contrast to ADAR1, ADAR2 does not increase the infectious potential of HIV-1. ADAR2 can target adenosines similar to those of ADAR1 (i.e. within the 5′UTR of viral RNA), whilst Rev and Tat sequences were found edited exclusively by ADAR1. The different effects on HIV-1 between the two editing enzymes suggest that ADAR1 may use additional mechanisms to promote the infectious potential of the virus. Overall, ADAR2 shares some but not all the proviral functions of ADAR1.

How ADARs exert their proviral effect during HIV-1 replication and which substrates are involved (endogenous and viral transcripts) has yet to be fully discovered. To answer this point, a deep-sequencing analysis would help to identify all the specific editing events critical for ADAR-mediated HIV-1 modulation.

Notably, ADAR1 action seems to be in contrast to the activity of the DNA editing enzyme APOBEC3G, a potent host restriction factor that impairs retrovirus growth by cytidine deamination of retroviral DNA (Chiu and Greene, 2008). It would be interesting to explore whether direct or indirect interactions occur between these two editing enzymes with opposite effects on viral replication.

Arenaviridae: lymphocytic choriomeningitis virus (LCMV)

The members of the Arenaviridae family are enveloped viruses, provided with a helical capsid and a negative single-stranded RNA (ss(–)RNA) genome.

Lymphocytic choriomeningitis virus (LCMV) belongs to the Old World Arenaviruses and was first isolated in 1933. Persistent infection of primary host (*Mus Musculus*) is generally asymptomatic, whilst the transmission to humans can cause neurological diseases, leading to death in 1% of cases. The viral genome consists of two RNA segments, the S and the L molecules, containing an intergenic region, which can fold into single or double stem–loop structures (Franze-Fernandez *et al.*, 1987), and is responsible for transcription termination of the genome units. The S RNA encodes for the viral glycoprotein precursor (GPC), which is post-translationally cleaved by the cellular protease SP1 into the envelope proteins (GP-1 and GP-2), and the nucleocapsid protein (NP). The L molecule encodes for the viral RNA dependent RNA polymerase (RdRP) and a small zinc finger protein (Z), a structural component of the mature virion.

The LCMV infection is initiated by the viral binding to a specific cell receptor, the Alfa-dystroglycan, through the envelope glycoprotein GP-1 (Cao *et al.*, 1998). Upon receptor binding, virions are internalized by a membrane fusion mechanism, driven by the GP-2 protein, and the ribonucleoproteins (RNPs) are released in the host cell cytoplasm, where viral RNA transcription/translation and replication occur (Eschli *et al.*, 2006; Morin *et al.*, 2010). LCMV uses an ambisense coding strategy to synthesize two different polypeptides in opposite orientations from the same genomic RNA segment (Buchmeier, 2002). In particular, a primary transcription process generates NP and L mRNAs, whilst GP and Z mRNA are subsequently produced by a copy of the full-length antigenome RNA, which also serves as a replicative intermediate. Upon replication of the genome and after all the structural proteins and the envelope glycoproteins are newly synthesized, the mature virons are assembled and infectious progeny buds from the host cells, both processes depending on functional Z and GPC proteins.

The presence of an intermediate molecule consisting of dsRNA during the replication process makes the LCMV genome an excellent hypothetical substrate for ADAR enzymes. Indeed, the A-to-G hypermutation pattern was detected in the LCMV RNA genome in infected cell lines and in acutely infected mice (Zahn *et al.*, 2007). In particular, Zahn and colleagues observed that a clear A-to-G mutation bias emerged in the late phases of LCMV WE strain infection, both in murine L929 cell line (7 days post infection) and mice (5–8 days post infection). They focused the analysis on the glycoprotein GPC precursor within the S segment and found that about 38% of the adenosines in the short stretch were converted to guanosines. The frequency of mutation in this region was higher than expected if calculated on the basis of spontaneous events occurring during viral replication. Indeed, the authors observed an up-regulation of PKR and the ADAR1 p150 isoform (but not the constitutive ADAR1 p110) as a result of interferon induction due to LCMV infection. Moreover, as some viruses are found to express RNAs or proteins that inhibit ADAR1 activity (such as the adenovirus VAI RNA and the Vaccinia virus E3L protein), the authors performed transfection experiments in SH-SY5Y neuroblastoma cells, demonstrating that LCMV is not able to affect ADAR1 p150 editing activity measured on an exogenous substrate carrying the GluR-B hot spot +60 editing site. Taken together, these data show that the A-to-G hypermutation pattern in LCMV is consistent with ADAR-mediated editing events (Zahn *et al.*, 2007).

Editing at multiple sites may generate amino acid substitutions or premature stop codons. Accordingly, the authors found that about 50% of mutations in the GPC RNA segment lead to the production of mutated glycoproteins, often affecting their functionality. In particular, some mutations prevent or reduce GPC precursor cleavage by cellular SP1; other mutations affect the GPC glycosylation pattern thus impairing GPC migration and envelope expression. Owing to the great role played by GPC and GP-1/GP-2 glycoproteins in the initial (cellular receptor binding and entry) and final (virion assembly and budding) steps of infection, the authors conclude that the massive ADAR1 p150 mediated A-to-G hypermutation in LCMV impairs viral infectivity, suggesting an antiviral role for ADAR1 p150 in interferon-induced innate defence against viruses.

In this context, the recent study by Ward and collaborators about the effect of ADAR1 in several viral infections (Ward *et al.*, 2011) is interesting. In particular, the authors did not find any ability of ADAR1 p150 in inhibiting LCMV 13 strain replication.

The discrepancies between the reported findings suggest that new studies are needed to better dissect the role of A-to-I hyperediting on the LCMV genome and the potential implications on the virus life cycle.

Bunyaviridae: Rift Valley fever virus (RVFV)

Bunyaviridae family members are enveloped viruses, provided with a helical capsid and a segmented ss(–)RNA genome. Rift valley fever virus (RVFV) belongs to the *Phlebovirus* genus and was first characterized in 1931 by Daubney and colleagues in a farm of the Rift Valley, Kenya. It is transmitted to animals and humans mainly through *Aedes, Culex* and *Anopheles* mosquitos (Papa *et al.*, 2010; Seufi and Galal, 2010). Symptoms in humans range from photophobia and headaches to encephalitis, retinitis and high lethal haemorrhagic fever in 1% of the cases (Adam *et al.*, 2010).

The RVFV genome consists of three RNA segments (S, M and L) which show high nucleotide complementary at their 5′ and 3′ ends. Watson–Crick basepairing between these two regions leads to the formation of a dsRNA secondary structure and explains why the RNA genome segments are present as circular molecules in the cytoplasm of infected cells (Pettersson and von Bonsdorff, 1975). The L and M segments share the same negative polarity, whilst the S RNA segment uses the ambisense coding strategy (Giorgi *et al.*, 1991). The L molecule encodes the viral component of the RNA dependent RNA polymerase (RdRP), which is responsible for mRNA transcription and genome replication. The M segment encodes the Gn and Gc glycoproteins; moreover, the M molecule encodes two non-structural polypeptides of 78 kDa and 14 kDa, named NSm1 and NSm2 respectively, whose functions have not yet been fully elucidated. The genomic strand of S RNA encodes the nucleocapsid proteins (N) whilst the antigenomic one codes for another type of non-structural protein, the 31-kDa NSs (Walter and Barr, 2011).

All steps of the RVFV life cycle take place in the cytoplasm of the infected cell and display some peculiar features. The RVFV is able to enter the host cell via an endocytic pathway mediated by viral glycoproteins that likely bind dendritic DC-SIGN receptors (Lozach *et al.*, 2011). Upon virus entry, the ribonucleoproteins (RNPs) are released in the host cell cytoplasm, where the viral RNAs are transcribed and replicated by the RdRP. When the genome is replicated and all the structural proteins and the envelope glycoproteins are translated, the mature virions are assembled. Bunyaviruses assembly involves a characteristic mechanism consisting of the organization of tube-like factories around the Golgi complex, by which viruses couple genome replication to RNP assembly. The final step in the RVFV life cycle is the budding of the infectious virions from the host cells, a process probably mediated by cytoskeletal proteins (Walter and Barr, 2011).

As it occurs for all viruses, the Bunyaviruses entry activates host cell immune response, which starts with the induction of interferon-mediated protein expression. The RVFV strategies to escape the immune response are well characterized; it is known that the NS proteins play a contrasting role against the antiviral response by at least three main mechanisms. First, it has been shown that the NS proteins interact with host cell transcription factors SAP30 and YY1, generating peculiar filamentous structures in the nuclei of infected cells (Le May *et al.*, 2008). Secondly, the NS proteins can inhibit the formation of another transcription factor, TFIIH, by competitive binding to the p44 subunit with an overall suppression of host RNA synthesis (Le May *et al.*, 2008). Finally, RVFV NS proteins are able to suppress interferon β induction via promoter inhibition (Le May *et al.*, 2008) and proteosoma-mediated PKR degradation (Habjan *et al.*, 2009). All together these mechanisms contribute to weak host cell antiviral response, thus making NS proteins one of the key factors for RVFV pathogenesis *in vivo*.

Recently, Suspène and collaborators developed the 3DI-PCR technique, which allows the detection of GC-rich ADAR-edited RNAs. They used this approach to selectively amplify edited PCR products from RVFV-infected cells and they found extensive A-to-G hypermutations in the L segment of the RVFV RNA genome (Suspene *et al.*, 2008). The authors claim that this is the first reported case of hyperediting in the RVFV genome, although they did not identify the ADAR enzyme being responsible or its biological consequences. However, the enhancement of A-to-I events upon interferon response suggests that the ADAR1 p150 isoform may be involved. In this context, it is important to remember that RVFV has the ability to strongly suppress the interferon cascade via NS proteins, thus making the picture extremely interesting.

Deltavirus genus: hepatitis Δ virus (HDV)

First discovered in the mid-1970s as a novel HBV antigen in HBV-infected patients, hepatitis Δ virus (HDV) was recognized as a different pathogen only 3 years later (Rizzetto *et al.*, 1980). Because of its particular features, HDV does not belong to any virus family; however, it is classified as the only member of the Deltavirus genus (Pascarella and Negro, 2010). HDV is a small defective virus requiring HBV for its life cycle and, as a consequence, it can infect only persons that are HBsAg positive, in which it can lead to a more severe liver disease and eventually to death (Romeo *et al.*, 2009).

HDV is a circular ss(–)RNA virus, provided with an envelope, made up by the three HBV envelope proteins (HBAg) (Sureau, 2006) and host lipids, and a nucleocapsid formed by the HDV genome molecule and HDV long and small antigen proteins (HDAg) (Wang *et al.*, 1986). The HDV genome was completely sequenced in 1986 by Wang and colleagues. They found that it corresponds to one open reading frame, encoding only the two HDAg proteins that exist as small (SHDAg) and large (LHDAg) isoforms (Wang *et al.*, 1986). The first one is produced early during infection and is essential for genome replication (Kuo *et al.*, 1989); the latter appears later and acts as an inhibitor of viral replication and is necessary to drive virion assembly (Chang *et al.*, 1991; Chen *et al.*, 1992).

Unique among viruses, the HDV circular genome has 74% of internal basepairing and can fold into an unbranched, double-stranded, rod-like structure (Wang *et al.*, 1986).

HDV can infect only hepatocytes and thus it can only replicate in the liver. Recently, it has been observed that LHDAg specifically interacts with clathrin heavy chain, thus suggesting that the clathrin-mediated endocytic pathway may be the route of HDV entry (Huang *et al.*, 2007). Upon entry, the virus is uncoated and transported to the nucleus thanks to a specific signal in HDAg (Chou *et al.*, 1998) and the host transcriptional machinery is used for viral replication (Lai, 2005). The transcription process produces mRNAs, that are exported to the cytoplasm and translated by host cell factors, and antigenome transcripts, used as template for replication of new genome molecules. Newly synthesized SHDAgs return to the nucleus where they sustain viral replication (Yamaguchi *et al.*, 2001). LHDAgs are generated later during infection and they also return to the nucleus where they inhibit replication and associate with SHDAgs and the HDV RNA genome to form viral ribonucleoproteins (RNPs). RNPs are exported to the cytoplasm, where virion assembly occurs. This process is mediated by LHDAg and involves the association of the RNP with HBAg leading to the generation of enveloped, infectious particles (Wang *et al.*, 1991). The final step in the HDV life cycle is the budding of virions from hepatocytes through a still unclear mechanism that may involve clathrin-mediated exocytosis (Pascarella and Negro, 2010).

As mentioned above, both the two viral HDAg proteins are essential for a correct and complete life cycle. They are expressed at different time-points during HDV replication and have different and partially counteracting functions. SHDAg is produced early during infection and is necessary to support RNA replication (Kuo *et al.*, 1989), whilst LHDAg is synthesized later and drives virion packaging and assembly. Moreover, LHDAg is also able to prevent RNA replication, acting as a dominant negative inhibitor (Chang *et al.*, 1991; Chen *et al.*, 1992). These considerations suggest that HDAgs are critical for the switch from the early to the late phase of infection, i.e. from viral RNA synthesis to virus particle formation.

An event mediated by host enzymatic activity allows the generation of these two functionally distinct proteins from the same HDV coding sequence. In 1986, Wang and coworkers identified heterogenicity at nucleotide 1012 of the HDV genome, affecting HDAg protein length (Wang *et al.*, 1986). Indeed, the clones containing the UAG (amber) stop codon encoded the SHDAg protein, whilst the ones containing the UGG tryptophan codon encoded the LHDAg protein, which is 19 amino acids longer at the carboxyl terminus (Wang *et al.*, 1992). It was subsequently demonstrated that this heterogenicity is due to a hydrolytic deamination event catalysed by the host ADAR1 p110 editing enzyme (Jayan and Casey, 2002b). Since this editing changes the codon from amber to tryptophan (W), the corresponding editing site is named the amber/W site (Polson *et al.*, 1996).

The ADAR1 p110-mediated amber/W modification represents one of the best examples of a site-specific editing event among viral RNA substrates. Moreover, since ADAR1 p110 activity modulates the expression of the key HDV proteins and consequently the switch from early to late phase of infection, this enzyme plays an essential proviral role in the HDV life cycle (Wong and Lazinski, 2002; Casey, 2006).

In addition, several studies have demonstrated that decreased editing levels prevent virion assembly (Jayan and Casey, 2002b, 2005), whilst increased editing levels lead to a strong RNA synthesis inhibition via LHDAg up-regulation (Jayan and Casey, 2002a). Furthermore, as the editing event occurs on antigenome molecules that are used as template for viral genome replication, some HDV particles contain edited copies thus resulting in non-infectious virions because only the LHDAg can be produced (Wong and Lazinski, 2002). According to these data, we can argue that mis-regulated RNA editing may have an antiviral role against HDV infection.

Regulation of RNA editing by HDV is reached by preventing random editing events beyond the specific amber/W one. In fact, it seems that HDV can prevent biased changes likely by characteristic structural and sequence features of its genome (Chen *et al.*, 2010).

The editing at the amber/W site seems a slow process which starts 4 days post infection (4% of editing) and reaches the maximum level (30%) only 14 days post infection. Several viral features may be responsible for this control. First, low level of basepairing at the 3′ of the amber/W site makes this site suboptimal for ADAR1 (Jayan and Casey, 2005). Second, the substrate availability is limited by the binding of HDAg proteins to the viral RNA (HDV genotype 1) (Defenbaugh *et al.*, 2009) or by the dynamic conversion of viral RNA from unbranched (uneditable) to branched (editable) conformation (HDV genotype 3) (Linnstaedt *et al.*, 2009). Finally, a negative feedback regulates RNA editing: as editing events occur only during replication, LHDAg prevents the editing by inhibiting genome synthesis (Sato *et al.*, 2004).

Taken together these observations highlight how HDV has developed several strategies to redirect a host cell activity (ADAR1) towards its own advantage.

Orthomyxoviridae: influenza A virus

Orthomyxoviridae members are enveloped viruses, provided with a helical nucleocapsid and segmented, linear, ss(−)RNA genome. Orthomyxoviruses infect a wide host range, which comprise invertebrates and vertebrates, including birds, mammals and humans (Medina and Garcia-Sastre, 2011).

The influenza A genus is a very heterogeneous virus population, comprising several strains subtyped on the basis of the antigenic properties of haemagglutinin (HA) and neuraminidase (NA) glycoproteins. These viruses are responsible for seasonal and pandemic influenza in humans, highly contagious respiratory diseases characterized by headache, fever, pneumonia and, eventually, death (Neumann *et al.*, 2009).

The influenza A genome consists of eight uncapped, ss(−)RNA molecules, tightly associated with viral polymerase and nucleoproteins thus forming vRNPs. Each vRNP acts as an independent unit during transcription and replication leading to the production of 11 viral proteins. The already mentioned HA and NA are the two main envelope glycoproteins and mediate viral entry and virion release respectively. M2 is the third glycoprotein surface factor and plays an important role in virus entry as well as in virion assembly and budding. Just beneath the viral envelope, the matrix protein M1 connects the viral lipid membrane to the ribonucleoprotein core and provides structural support to the virions. The ribonucleoprotein core is made up of RNA segments covered with the RNA polymerase complex formed by PB1, PB2 and PA proteins and with the NP nucleocapsid protein. In addition, there are two non-structural proteins: NS1, which is a critical factor counteracting the host immune response, and NS2 (or NEP), which is responsible for the nuclear export of viral RNAs during replication. Several influenza A strains produce an additional protein, the PB1-F2, which derives from a second reading frame within the PB1 sequence and plays a key role in the induction of host cell apoptosis (Rossman and Lamb, 2011).

Bioinformatics studies provided evidence for several regions of potentially conserved secondary structures within the positive strand of each viral segment (Gog *et al.*, 2007). It has been proposed that secondary and tertiary structures may be important in regulating the splicing of viral RNA molecules and in general the viral packaging (Moss *et al.*, 2011). Upon receptor attachment, the virus enters the host cell by an endocytic pathway; the subsequent M2-mediated fusion between host endosomal membrane and viral envelope releases the nucleocapsid in the cytoplasm. Differently from most RNA viruses, Influenza A virus transcription and replication occur in the host cell nucleus. These steps of the viral life cycle thus depend on the cellular nucleus-cytoplasm transport mechanism, likely mediated by a Crm1 dependent pathway (Boulo *et al.*, 2007). In the nucleus, transcription of RNPs is driven by viral RNA polymerase; the 5′ capped and 3′ polyadenylated viral mRNAs are translated by the host cell machinery in the cytoplasm. The complementary positive RNA molecules generated during viral transcription and replication are used for protein synthesis and as templates to increase vRNAs production (Medina and Garcia-Sastre, 2011). Influenza A infection induces an extremely powerful interferon-mediated antiviral response. An excellent study from Tenoever and colleagues showed that the IKKε kinase, a key factor in coordinating the INF enhanceosome, is indispensable for the correct activation of some interferon-stimulated genes (ISGs) (Tenoever *et al.*, 2007). IKKε$^{-/-}$ mice infected with the influenza A WSN strain were more sensitive to infection. An array analysis performed on total RNA extracted from primary lung samples revealed that a subset of ISGs was poorly induced upon infection. The *Adar1* gene was one of them and the functional consequence of its reduced expression could be related to a strong decrease in hyperediting of the M segment of the influenza A genome. In particular, in wild-type mice the editing level was about 30% within the M segment, whilst in the IKKε$^{-/-}$ animals the editing level decreased to about 5%. Taken together, these data show that IKKε is responsible for the induction of ISGs, such as *Adar1*, which in turn exert an antiviral activity against influenza A infection. Even if the biological significance of Adar1-mediated hyperediting of the M segment is not yet known, it suggests that this enzyme may exert the antiviral role through its editing activity (Tenoever *et al.*, 2007).

Indeed, an antiviral role for ADAR1 has been recently confirmed by Ward and collaborators. They infected *Adar1* p150$^{-/-}$ MEF cells with the influenza A WSN strain and found that these cells were more sensitive to infection. They concluded that the INF – inducible ADAR1 p150 isoform is a restriction factor inhibiting influenza A virus replication and cytotoxicity, even if the mechanism of action remains to be fully elucidated (Ward *et al.*, 2011).

Since ADAR1 p150 plays a key antiviral role in influenza A infection, the finding that the virus has developed some strategies to counteract its action is not surprising. A screening by a yeast two-hybrid system has revealed that ADAR1 interacts with the influenza A NS1 protein. Moreover, these two proteins colocalize in canine kidney MDCK cells upon influenza A virus infection (Ngamurulert *et al.*, 2009). As mentioned above, NS1 is a critical factor antagonizing the interferon-induced host immune response. In particular, NS1 prevents PKR activation and affects stability and export of cellular mRNAs, thus inhibiting host cell gene expression (Satterly *et al.*, 2007). Starting from these observations, Suspene and coworkers investigated the relationship between ADAR1 and NS1 (Suspene *et al.*, 2011). First, they identified a new hyperediting pattern in the HA segment from two commercial influenza A virus vaccines produced in chicken embryo fibroblasts that express an ADAR1-like protein (Herbert *et al.*, 1995). They used these new modifications to monitor ADAR1 activity upon infection of avian QT6 cells with a NS1-defective influenza A strain. Sequencing of the HA segment from the cells infected with the NS1 defective virus revealed a frequency of ADAR1-mediated editing of about 31% within the HA fragment. Furthermore, they performed a deep sequencing analysis on RT-PCR products from attenuated vaccines, finding A-to-G hypermutations, also in the positive antigenome strand of influenza A virus. Overall, these data confirm an antiviral role for ADAR1 as it can restrict both the genome and the replication template of influenza A virus and suggest that the virus may contrast ADAR1 action through the NS1 protein (Suspene *et al.*, 2011).

The findings from Emmott and colleagues suggest another potential strategy to escape the antiviral response applied by the influenza A virus. Indeed, these authors found that the nucleolar amount of both ADAR1 p150 and ADAR1 p110 is increased upon influenza A infection. The functional consequences of such a phenomenon remain unknown (Emmott *et al.*, 2010). However, an intriguing hypothesis is that this may be a virus-mediated way to sequester the interferon-induced ADAR1 enzyme, thus preventing its antiviral action.

Paramyxoviridae

Measles virus (MV)

Measles virus (MV), a member of the *Morbillivirus* genus of the family Paramyxoviridae, is the etiologic agent of an acute infection of the respiratory system in humans. MV is an enveloped virus with a non-segmented ssRNA genome of negative polarity (15.89 kbp in length), which codes for at least six protein products, including nucleocapsid protein (N), phosphoprotein (P), matrix protein (M), fusion protein (F), haemagglutinin protein (H) and large polymerase (L). MV lacks the enzymatic machinery to reverse transcribe its own RNA genome, thus it cannot integrate DNA copies of its genomic sequences into host cellular DNA. Nevertheless, this virus can survive in neuronal/glial cells. Indeed, MV is also the aetiological agent of two rare but fatal diseases affecting the central nervous system (CNS): subacute sclerosing panencephalitis (SSPE) and measles inclusion body encephalitis (MIBE) (Billeter *et al.*, 1994; Oldstone, 2009).

It is in this context that non-selective deaminations of adenosine residues in MV RNA were found for the first time. A few years after ADAR discovery, Cattaneo *et al.* (1988b) isolated viral RNAs from the brain of one MIBE patient and two SSPE patients that contained almost 2% of the nucleotides mutated in comparison with the sequence of MV causing acute infections. In particular, the viral matrix (M) mRNA of the MIBE case showed clustered A-to-I(G) mutations that were proposed to be the result of ADAR activity on the viral minus ssRNA (Bass and Weintraub, 1988). Similar transitions were largely observed in cDNAs derived from other cases of persistent MV infection in the M gene and others (i.e. H and F genes) (Billeter *et al.*, 1994). Notably, such biased hypermutations have not been found in MV strains causing acute infections.

The extensive sequence changes in the M transcript are presumed to prevent synthesis of a functional protein. The lack of matrix protein, that is required for virion assembly and release, may help the virus to escape immune system recognition allowing persistent infections in brain that may correlate with the development of neurological diseases (Cattaneo *et al.*, 1988a; Patterson *et al.*, 2001).

This evidence suggested a role for ADARs during MV infection, although the effective consequences on the viral life cycle and the mechanism by which hypermutations arise in the viral genome remain largely unknown. Since the virus replicates in the host cell cytoplasm, the observed hyperediting was supposed to be carried out by the IFN-inducible ADAR1 p150, which may be expressed in response to viral infection.

To explore this possibility, human HeLa cells stably silenced for ADAR1 were infected with measle virus (Toth *et al.*, 2009). The growth of MV wild-type as well as of two mutant viruses lacking expression of the accessory proteins V and C were significantly decreased in ADAR1 deficient cells compared with the controls. The authors also observed that ADAR1 suppresses MV-induced cytotoxicity. This proviral and anti-apoptotic role of ADAR1 during MV infection probably correlates with a negative modulation of the PKR pathway, as *ADAR1* knockdown HeLa cells showed an increased phosphorylation/activation of PKR, eIF-2 and IRF-3 factors (Toth *et al.*, 2009).

Biological evidence consistent with a function of ADAR1 during MV infection came also from Ward *et al.* (2011), even though the authors reached partially different results. MEF cells stably expressing the receptor for MV were specifically deleted for the *Adar1* p150 gene. This cell line showed a higher cytopathic effect following MV infection and enhanced viral growth compared with the controls. The authors showed that deletion of *Adar1* p150 increases susceptibility of MEF cells to infection by MV as well as other members of Paramyxoviridae (including Newcastle disease virus, Sendai virus and canine distemper virus) and Orthomyxoviridae (influenza A virus) family.

The discrepancy between the proviral (Toth *et al.*, 2009) and the antiviral (Ward *et al.*, 2011) effect of ADAR1 observed in the MV replication context may be due to differences in the cell type, virus strains and levels of ADAR1 depletion used in the two studies.

Human respiratory syncytial virus (HRSV)

Human respiratory syncytial virus (HRSV) is a member of the *Pneumovirus* genus of the Paramyxoviridae family. HRSV is the leading cause of severe lower respiratory tract infections in very young children and infants (Collins, 1996). HRSV has a non-segmented ss(−)RNA genome with 10 genes encoding 11 proteins, including the G protein, a surface protein heavily glycosylated and essential for virus attachment to the cell surface receptor. Like other paramyxoviruses, the HRSV genes are transcribed sequentially from a single 3′-end promoter and replication involves the synthesis of a positive-sense strand that is complementary to the genome. Multiple A-to-G (U-to-C in the positive sense) nucleotide substitutions have been found exclusively in the viral attachment glycoprotein G (Rueda *et al.*, 1994; Martinez *et al.*, 1997; Walsh *et al.*, 1998). These HRSV strains with multiple A-to-G substitutions act as escape mutants resistant to antibodies that recognize either conserved or group-specific epitopes in the central region of the G protein (Martinez *et al.*, 1997; Walsh *et al.*, 1998). The adenosine deaminase activity mediated by ADAR enzymes was supposed to be involved in these viral hypermutations (Martinez *et al.*, 1997; Martinez and Melero, 2002). Indeed, short intramolecular dsRNA segments (100–200 bp) can be transiently formed during HRSV replication and/or transcription and could be a good substrate for RNA editing. Furthermore, the central region of the G protein fulfils all the requirements for ADARs, having an AU-rich region and a high tendency for intramolecular basepairing (Martinez and Melero, 2002).

Human parainflenza virus 3 (HPIV3)

Human parainflenza virus 3 (HPIV3) belongs to the *Respirovirus* genus of the Paramyxoviridae family. RNAs isolated from two lines of persistently infected rhesus monkey kidney LLC-MK2 cells showed biased A-to-I(G) and U-to-C transitions within the 3′-end consistent with editing mediated by an ADAR enzyme (Murphy *et al.*, 1991). The biological significance of this hyperediting has still to be investigated.

Rhabdoviridae: vesicular stomatitis virus (VSV)

Members of Rhabdoviridae family are enveloped and rod- or bullet-shaped viruses, provided with a helical ribonucleocapsid and a ss(−)RNA genome. Rhabdoviruses infect a wide host range, which comprise plants, insects and vertebrates, including mammals and humans. Vesicular stomatitis virus (VSV) belongs to the *Vesiculovirus* genus and was described in the USA as a vesicular disease in horses, cattle and pigs. The disease is characterized by vesicle formation in oral cavity, feet and teats in animals, whilst in humans it seems to be often asymptomatic. Arthropod vectors, particularly haematophagous insects, are responsible for transmission and likely for the spread of virus between animals and from animals to humans (Mead *et al.*, 2000).

The VSV genome consists of a linear, ss(–) RNA molecule of 11 kb, which comprises five separate open reading frames (ORF), interspersed by 'intergenic regions' (Finke *et al.*, 2000). The RNA genome is completely coated by the viral nucleoproteins and persists in this conformation also during transcription and replication (Wagner, 1996).

The 3′ to 5′ order of the five independent ORFs on the VSV genome corresponds to the time-regulated expression of the five viral proteins encoded. The first synthesized product is the nucleoprotein N, which encapsidates the viral RNA leading to the association of the ribonucleocapsid. Then, the phosphoprotein P is generated, which is a chaperone maintaining N in the appropriated fold and an essential cofactor of viral RNA polymerase activity. The matrix protein M mediates ribonucleocapsid condensation, participates in virion assembly and budding and is involved in inhibition of gene expression induced by host cell immune response. The transmembrane VSV G protein is a type I glycoprotein responsible for viral attachment/entry and virion assembly. The large protein L is the catalytic subunit of the viral RNA polymerase. The replicase activity depends on the interaction of L with P and N proteins, whilst the transcriptase function is due to the $L–P_3$ complex (Kuzmin *et al.*, 2009; Rieder and Conzelmann, 2009).

All steps of the VSV life cycle take place in the cytoplasm of the infected cell. As already mentioned, G proteins are responsible for virus attachment to host cells. The VSV is able to enter the host cell via an endocytic pathway mediated by viral G proteins, which drives the fusion between host endosomal membrane and viral envelope, thus releasing the nucleocapsid in the cytoplasm. The ribonucleoproteins are used as template for primary transcription by the viral RNA polymerase ($L–P_3$ complex). The viral mRNAs, 5′ capped and 3′ polyadenylated, are translated by the host cell machinery. N protein is early translated and its level mediates the switch from transcription to replication. L–P–N complexes generate antigenome molecules used as template for encapsidated-genome synthesis and for secondary transcription rounds. Condensed nucleocapsids and M and G proteins are transported to the plasma membrane, where the virion assembly and budding occur (Assenberg *et al.*, 2010).

VSV has evolved some strategies to escape innate host immune response reflecting its cytopathic nature (Huang and Wagner, 1965). This VSV ability is mediated by M protein, which inhibits the transcription factor TFIID, thus preventing cellular RNA synthesis (Yuan *et al.*, 1998). Moreover, it was found that M protein was able to restrain nuclear export of host transcripts, further inhibiting the expression of interferon-induced genes (Petersen *et al.*, 2000).

Interestingly, VSV has been shown to be susceptible to PKR, one of the key proteins expressed and activated by interferon. Indeed, $PKR^{-/-}$ mice dye following VSV infection and fibroblasts from $PKR^{-/-}$ mice are more permissive to VSV infection than the wild-type counterpart (Balachandran and Barber, 2000; Stojdl *et al.*, 2000).

A great deal of effort has been spent in the last years to investigate the relationship among VSV, host immune response and ADAR1.

Nie and coworkers showed that stable overexpression of Adar1 p150 in NIH 3T3 and MEF cells significantly sensitized host cells to VSV infection. On the contrary, a transient silencing of endogenous *ADAR1* made 293T cells more resistant to infection. The authors further demonstrated that this proviral role of ADAR1 was dependent on PKR expression and independent of editing activity, as mutants lacking the deaminase domains efficiently inhibited PKR activation and stimulated VSV infection (Nie *et al.*, 2007).

A recent study from Li and colleagues showed that VSV growth was unaffected by ADAR1 p150 and ADAR2 editing enzymes overexpressed in 293T cells. The discrepancy between overexpression experiments performed by Nie (murine cell lines) and Li (human cell line) suggests that different host cell types may have different responses to VSV infection. Silencing experiments, however, confirmed the proviral role of ADAR1 in VSV infection. In particular, HeLa cells silenced for ADAR1 and interferon-treated were more sensitive to VSV infection than the untreated ones, thanks to increased phosphorylation of both PKR and eIF2α (Li *et al.*, 2010).

In conclusion, all these data demonstrate that ADAR1 p150 is a proviral factor in VSV infection and suggest that VSV may have taken advantage of ADAR1's ability to inhibit PKR to complete its life cycle in the host cell.

The findings that ADAR1 and ADAR2 overexpression does not affect VSV growth in human cells (Li *et al.*, 2010) suggest that editing may be excluded as a mechanism against VSV infection. However, the analysis of defective interfering (DI) VSV particles from infected hamster kidney fibroblast BHK cells showed significant A-to-G modifications in the viral genome (O'Hara *et al.*, 1984). Notably, ADAR-mediated hyperediting was detected in the transcript of PP3 protein of another Rhabdovirus, the *Drosophila* sigma virus (Carpenter *et al.*, 2009), implying an antiviral role of RNA editing at least in insects. Interestingly, recent works highlighted that some immune response strategies are similar between mammals and insects (Ding, 2010; Ronald and Beutler, 2010), suggesting that further studies are necessary.

Future perspectives

Adenosine deaminases that act on RNA (ADARs) are a family of enzymes, whose activity is substrate-, cell-, tissue- and time-specific. For this reason, the idea is emerging that ADARs are finely modulated at different levels (transcriptionally, post-transcriptionally and translationally), despite the fact that the mechanisms of their regulation remain still largely unknown.

Different lines of evidence suggest the existence of a link between ADAR deregulation and human diseases (Gallo and Locatelli, 2011); thus, the comprehension of the mechanisms controlling the action of these enzymes and the relationship between the different members of the ADAR family (ADAR1, ADAR2 and ADAR3) is now essential.

How the fine-tuned ADAR action is altered during different viral infections is an important issue to dissect. Future investigations on ADARs and viral infections would not only elucidate the interaction between pathogens and host-immune response, but also would shed light on the regulation/action of ADAR enzymes.

References

Adam, A.A., Karsany, M.S., and Adam, I. (2010). Manifestations of severe Rift Valley fever in Sudan. Int. J. Infect. Dis. *14*, e179–180.

Ali, N., Tardif, K.D., and Siddiqui, A. (2002). Cell-free replication of the hepatitis C virus subgenomic replicon. J. Virol. *76*, 12001–12007.

Alisi, A., Tomaselli, S., Balsano, C., and Gallo, A. (2011). Hepatitis C virus therapeutics: editing enzymes promising therapeutic targets? Hepatology *54*, 742; author reply 742–743.

Alter, H.J., and Seeff, L.B. (2000). Recovery, persistence, and sequelae in hepatitis C virus infection: a perspective on long-term outcome. Semin. Liver Dis. *20*, 17–35.

Arnaud, N., Dabo, S., Maillard, P., Budkowska, A., Kalliampakou, K.I., Mavromara, P., Garcin, D., Hugon, J., Gatignol, A., Akazawa, D., *et al.* (2010). Hepatitis C virus controls interferon production through PKR activation. PloS One *5*, e10575.

Assenberg, R., Delmas, O., Morin, B., Graham, S.C., De Lamballerie, X., Laubert, C., Coutard, B., Grimes, J.M., Neyts, J., Owens, R.J., *et al.* (2010). Genomics and structure/function studies of Rhabdoviridae proteins involved in replication and transcription. Antiviral Res. *87*, 149–161.

Balachandran, S., and Barber, G.N. (2000). Vesicular stomatitis virus (VSV) therapy of tumors. IUBMB life *50*, 135–138.

Barre-Sinoussi, F., Chermann, J.C., Rey, F., Nugeyre, M.T., Chamaret, S., Gruest, J., Dauguet, C., Axler-Blin, C., Vezinet-Brun, F., Rouzioux, C., *et al.* (1983). Isolation of a T-lymphotropic retrovirus from a patient at risk for acquired immune deficiency syndrome (AIDS). Science *220*, 868–871.

Bartenschlager, R., Frese, M., and Pietschmann, T. (2004). Novel insights into hepatitis C virus replication and persistence. Adv. Virus Res. *63*, 71–180.

Bass, B.L., and Weintraub, H. (1988). An unwinding activity that covalently modifies its double-stranded RNA substrate. Cell *55*, 1089–1098.

Billeter, M.A., Cattaneo, R., Spielhofer, P., Kaelin, K., Huber, M., Schmid, A., Baczko, K., and ter Meulen, V. (1994). Generation and properties of measles virus mutations typically associated with subacute sclerosing panencephalitis. Ann. N.Y. Acad. Sci. *724*, 367–377.

Biswas, N., Wang, T., Ding, M., Tumne, A., Chen, Y., Wang, Q., and Gupta, P. (2012). ADAR1 is a novel multi targeted anti-HIV-1 cellular protein. Virology *422*, 265–277.

Boulo, S., Akarsu, H., Ruigrok, R.W., and Baudin, F. (2007). Nuclear traffic of influenza virus proteins and ribonucleoprotein complexes. Virus Res. *124*, 12–21.

Buchmeier, M.J. (2002). Arenaviruses: protein structure and function. Curr. Top. Microbiol. Immunol. *262*, 159–173.

Cao, W., Henry, M.D., Borrow, P., Yamada, H., Elder, J.H., Ravkov, E.V., Nichol, S.T., Compans, R.W., Campbell, K.P., and Oldstone, M.B. (1998). Identification of alpha-dystroglycan as a receptor for lymphocytic choriomeningitis virus and Lassa fever virus. Science *282*, 2079–2081.

Carpenter, J.A., Keegan, L.P., Wilfert, L., O'Connell, M.A., and Jiggins, F.M. (2009). Evidence for ADAR-induced hypermutation of the Drosophila sigma virus (Rhabdoviridae). BMC genetics *10*, 75.

Casey, J.L. (2006). RNA editing in hepatitis delta virus. Curr. Top. Microbiol. Immunol. *307*, 67–89.

Cathomas, G. (2003). Kaposi's sarcoma-associated herpesvirus (KSHV)/human herpesvirus 8 (HHV-8) as a tumour virus. Herpes *10*, 72–77.

Cattaneo, R., Schmid, A., Billeter, M.A., Sheppard, R.D., and Udem, S.A. (1988a). Multiple viral mutations rather than host factors cause defective measles virus gene expression in a subacute sclerosing panencephalitis cell line. J. Virol. *62*, 1388–1397.

Cattaneo, R., Schmid, A., Eschle, D., Baczko, K., ter Meulen, V., and Billeter, M.A. (1988b). Biased hypermutation and other genetic changes in defective measles viruses in human brain infections. Cell *55*, 255–265.

Chang, F.L., Chen, P.J., Tu, S.J., Wang, C.J., and Chen, D.S. (1991). The large form of hepatitis delta antigen is crucial for assembly of hepatitis delta virus. Proc. Natl. Acad. Sci. U.S.A. *88*, 8490–8494.

Chen, P.J., Chang, F.L., Wang, C.J., Lin, C.J., Sung, S.Y., and Chen, D.S. (1992). Functional study of hepatitis delta virus large antigen in packaging and replication inhibition: role of the amino-terminal leucine zipper. J. Virol. *66*, 2853–2859.

Chen, R., Linnstaedt, S.D., and Casey, J.L. (2010). RNA editing and its control in hepatitis delta virus replication. Viruses *2*, 131–146.

Chiu, Y.L., and Greene, W.C. (2008). The APOBEC3 cytidine deaminases: an innate defensive network opposing exogenous retroviruses and endogenous retroelements. Annu. Rev. Immunol. *26*, 317–353.

Choo, Q.L., Kuo, G., Weiner, A.J., Overby, L.R., Bradley, D.W., and Houghton, M. (1989). Isolation of a cDNA clone derived from a blood-borne non-A, non-B viral hepatitis genome. Science *244*, 359–362.

Choo, Q.L., Richman, K.H., Han, J.H., Berger, K., Lee, C., Dong, C., Gallegos, C., Coit, D., Medina-Selby, R., Barr, P.J., *et al.* (1991). Genetic organization and diversity of the hepatitis C virus. Proc. Natl. Acad. Sci. U.S.A. *88*, 2451–2455.

Chou, H.C., Hsieh, T.Y., Sheu, G.T., and Lai, M.M. (1998). Hepatitis delta antigen mediates the nuclear import of hepatitis delta virus RNA. J. Virol. *72*, 3684–3690.

Clement, S., Pascarella, S., and Negro, F. (2009). Hepatitis C virus infection: molecular pathways to steatosis, insulin resistance and oxidative stress. Viruses *1*, 126–143.

Clerzius, G., Gelinas, J.F., Daher, A., Bonnet, M., Meurs, E.F., and Gatignol, A. (2009). ADAR1 interacts with PKR during human immunodeficiency virus infection of lymphocytes and contributes to viral replication. J. Virol. *83*, 10119–10128.

Clerzius, G., Gelinas, J.F., and Gatignol, A. (2011). Multiple levels of PKR inhibition during HIV-1 replication. Rev. Med. Virol. *21*, 42–53.

Collins, P.L., McIntosh, K., and Chanock, R.M. (1996). Respiratory syncytial virus. In Fields Virology, 3rd edn, Fields, B.N., Knipe, D.M., and Howley, P.M., eds. (Lippincott-Raven, Philadelphia), pp. 1313–1351.

Cullen, B.R. (2009). Viral and cellular messenger RNA targets of viral microRNAs. Nature *457*, 421–425.

Damania, B. (2004). Oncogenic gamma-herpesviruses: comparison of viral proteins involved in tumorigenesis. Nat. Rev. *2*, 656–668.

Defenbaugh, D.A., Johnson, M., Chen, R., Zheng, Y.Y., and Casey, J.L. (2009). Hepatitis delta antigen requires a minimum length of the hepatitis delta virus unbranched rod RNA structure for binding. J. Virol. *83*, 4548–4556.

Ding, S.W. (2010). RNA-based antiviral immunity. Nat. Rev. Immunol. *10*, 632–644.

Doria, M., Neri, F., Gallo, A., Farace, M.G., and Michienzi, A. (2009). Editing of HIV-1 RNA by the double-stranded RNA deaminase ADAR1 stimulates viral infection. Nucleic Acids Res. *37*, 5848–5858.

Doria, M., Tomaselli, S., Neri, F., Ciafre, S.A., Farace, M.G., Michienzi, A., and Gallo, A. (2011). The ADAR2 editing enzyme is a novel Hiv-1 proviral factor. J. Gen. Virol. *92*, 1228–1232.

Emmott, E., Wise, H., Loucaides, E.M., Matthews, D.A., Digard, P., and Hiscox, J.A. (2010). Quantitative proteomics using SILAC coupled to LC-MS/MS reveals changes in the nucleolar proteome in influenza A virus-infected cells. J. Proteome Res. *9*, 5335–5345.

Eschli, B., Quirin, K., Wepf, A., Weber, J., Zinkernagel, R., and Hengartner, H. (2006). Identification of an N-terminal trimeric coiled-coil core within arenavirus glycoprotein 2 permits assignment to class I viral fusion proteins. J. Virol. *80*, 5897–5907.

Felder, M.P., Laugier, D., Yatsula, B., Dezelee, P., Calothy, G., and Marx, M. (1994). Functional and biological properties of an avian variant long terminal repeat containing multiple A to G conversions in the U3 sequence. J. Virol. *68*, 4759–4767.

Finke, S., Cox, J.H., and Conzelmann, K.K. (2000). Differential transcription attenuation of rabies virus genes by intergenic regions: generation of recombinant viruses overexpressing the polymerase gene. J. Virol. *74*, 7261–7269.

Fluck, M.M., and Schaffhausen, B.S. (2009). Lessons in signaling and tumorigenesis from polyomavirus middle T antigen. Microbiol. Mol. Biol. Rev. *73*, 542–563, Table of Contents.

Franze-Fernandez, M.T., Zetina, C., Iapalucci, S., Lucero, M.A., Bouissou, C., Lopez, R., Rey, O., Daheli, M., Cohen, G.N., and Zakin, M.M. (1987). Molecular structure and early events in the replication of Tacaribe arenavirus S RNA. Virus Res. *7*, 309–324.

Gallo, A., and Locatelli, F. (2011). ADARs: allies or enemies? The importance of A-to-I RNA editing in human disease: from cancer to HIV-1. Biol. Rev. Camb. Philos. Soc. *87*, 95–110.

Gallo, R.C., Sarin, P.S., Gelmann, E.P., Robert-Guroff, M., Richardson, E., Kalyanaraman, V.S., Mann, D., Sidhu, G.D., Stahl, R.E., Zolla-Pazner, S., *et al.* (1983). Isolation of human T-cell leukemia virus in acquired immune deficiency syndrome (AIDS). Science *220*, 865–867.

Gandy, S.Z., Linnstaedt, S.D., Muralidhar, S., Cashman, K.A., Rosenthal, L.J., and Casey, J.L. (2007). RNA editing of the human herpesvirus 8 kaposin transcript eliminates its transforming activity and is induced during lytic replication. J. Virol. *81*, 13544–13551.

Garaigorta, U., and Chisari, F.V. (2009). Hepatitis C virus blocks interferon effector function by inducing protein kinase R phosphorylation. Cell Host Microbe *6*, 513–522.

Garcia, M.A., Gil, J., Ventoso, I., Guerra, S., Domingo, E., Rivas, C., and Esteban, M. (2006). Impact of protein kinase PKR in cell biology: from antiviral to antiproliferative action. Microbiol. Mol. Biol. Rev. *70*, 1032–1060.

Gelinas, J.F., Clerzius, G., Shaw, E., and Gatignol, A. (2011). Enhancement of replication of RNA viruses by ADAR1 via RNA editing and inhibition of RNA-activated protein kinase. J. Virol. *85*, 8460–8466.

George, C.X., and Samuel, C.E. (2011). Host response to polyomavirus infection is modulated by RNA adenosine deaminase ADAR1 but not by ADAR2. J. Virol. *85*, 8338–8347.

George, C.X., Li, Z., Okonski, K.M., Toth, A.M., Wang, Y., and Samuel, C.E. (2009). Tipping the balance: antagonism of PKR kinase and ADAR1 deaminase functions by virus gene products. J. Interferon. Cytokine Res. *29*, 477–487.

Giorgi, C., Accardi, L., Nicoletti, L., Gro, M.C., Takehara, K., Hilditch, C., Morikawa, S., and Bishop, D.H. (1991). Sequences and coding strategies of the S RNAs of Toscana and Rift Valley fever viruses compared to those of Punta Toro, Sicilian Sandfly fever, and Uukuniemi viruses. Virology *180*, 738–753.

Gog, J.R., Afonso Edos, S., Dalton, R.M., Leclercq, I., Tiley, L., Elton, D., von Kirchbach, J.C., Naffakh, N., Escriou, N., and Digard, P. (2007). Codon conservation in the influenza A virus genome defines RNA packaging signals. Nucleic Acids Res. *35*, 1897–1907.

Gross, L. (1953). A filterable agent, recovered from Ak leukemic extracts, causing salivary gland carcinomas in C3H mice. Proc. Soc. Exp. Biol. Med. *83*, 414–421.

Gu, R., Zhang, Z., and Carmichael, G.G. (2006). How a small DNA virus uses dsRNA but not RNAi to regulate its life cycle. Cold Spring Harb. Symp. Quant. Biol. *71*, 293–299.

Gu, R., Zhang, Z., DeCerbo, J.N., and Carmichael, G.G. (2009). Gene regulation by sense–antisense overlap of polyadenylation signals. RNA *15*, 1154–1163.

Habjan, M., Pichlmair, A., Elliott, R.M., Overby, A.K., Glatter, T., Gstaiger, M., Superti-Furga, G., Unger, H., and Weber, F. (2009). NSs protein of rift valley fever virus induces the specific degradation of the double-stranded RNA-dependent protein kinase. J. Virol. *83*, 4365–4375.

Hajjar, A.M., and Linial, M.L. (1995). Modification of retroviral RNA by double-stranded RNA adenosine deaminase. J. Virol. *69*, 5878–5882.

Herbert, A., Lowenhaupt, K., Spitzner, J., and Rich, A. (1995). Chicken double-stranded RNA adenosine deaminase has apparent specificity for Z-DNA. Proc. Natl. Acad. Sci. U.S.A. *92*, 7550–7554.

Huang, A.S., and Wagner, R.R. (1965). Inhibition of cellular RNA synthesis by nonreplicating vesicular stomatitis virus. Proc. Natl. Acad. Sci. U.S.A. *54*, 1579–1584.

Huang, C., Chang, S.C., Yu, I.C., Tsay, Y.G., and Chang, M.F. (2007). Large hepatitis delta antigen is a novel clathrin adaptor-like protein. J. Virol. *81*, 5985–5994.

Hyde-DeRuyscher, R., and Carmichael, G.G. (1988). Polyomavirus early-late switch is not regulated at the level of transcription initiation and is associated with changes in RNA processing. Proc. Natl. Acad. Sci. U.S.A. *85*, 8993–8997.

Iizasa, H., Wulff, B.E., Alla, N.R., Maragkakis, M., Megraw, M., Hatzigeorgiou, A., Iwakiri, D., Takada, K., Wiedmer, A., Showe, L., *et al.* (2010). Editing of Epstein–Barr virus-encoded BART6 microRNAs controls their dicer targeting and consequently affects viral latency. J. Biol. Chem. *285*, 33358–33370.

Jayan, G.C., and Casey, J.L. (2002a). Increased RNA editing and inhibition of hepatitis delta virus replication by high-level expression of ADAR1 and ADAR2. J. Virol. *76*, 3819–3827.

Jayan, G.C., and Casey, J.L. (2002b). Inhibition of hepatitis delta virus RNA editing by short inhibitory RNA-mediated knockdown of ADAR1 but not ADAR2 expression. J. Virol. *76*, 12399–12404.

Jayan, G.C., and Casey, J.L. (2005). Effects of conserved RNA secondary structures on hepatitis delta virus genotype I RNA editing, replication, and virus production. J. Virol. *79*, 11187–11193.

Kumar, M., and Carmichael, G.G. (1997). Nuclear antisense RNA induces extensive adenosine modifications and nuclear retention of target transcripts. Proc. Natl. Acad. Sci. U.S.A. *94*, 3542–3547.

Kuo, M.Y., Chao, M., and Taylor, J. (1989). Initiation of replication of the human hepatitis delta virus genome from cloned DNA: role of delta antigen. J. Virol. *63*, 1945–1950.

Kuzmin, I.V., Novella, I.S., Dietzgen, R.G., Padhi, A., and Rupprecht, C.E. (2009). The rhabdoviruses: biodiversity, phylogenetics, and evolution. Infect. Genet. Evol. *9*, 541–553.

Lai, M.M. (2005). RNA replication without RNA-dependent RNA polymerase: surprises from hepatitis delta virus. J. Virol. *79*, 7951–7958.

Le May, N., Mansuroglu, Z., Leger, P., Josse, T., Blot, G., Billecocq, A., Flick, R., Jacob, Y., Bonnefoy, E., and Bouloy, M. (2008). A SAP30 complex inhibits IFN-beta expression in Rift Valley fever virus infected cells. PLoS Pathog. *4*, e13.

Lei, M., Liu, Y., and Samuel, C.E. (1998). Adenovirus VAI RNA antagonizes the RNA editing activity of the ADAR adenosine deaminase. Virology *245*, 188–196.

Lemaire, P.A., Anderson, E., Lary, J., and Cole, J.L. (2008). Mechanism of PKR Activation by dsRNA. J. Mol. Biol. *381*, 351–360.

Levrero, M. (2006). Viral hepatitis and liver cancer: the case of hepatitis C. Oncogene *25*, 3834–3847.

Li, Z., Wolff, K.C., and Samuel, C.E. (2010). RNA adenosine deaminase ADAR1 deficiency leads to increased activation of protein kinase PKR and reduced vesicular stomatitis virus growth following interferon treatment. Virology *396*, 316–322.

Linnstaedt, S.D., Kasprzak, W.K., Shapiro, B.A., and Casey, J.L. (2009). The fraction of RNA that folds into the correct branched secondary structure determines hepatitis delta virus type 3 RNA editing levels. RNA *15*, 1177–1187.

Liu, Z., and Carmichael, G.G. (1993). Polyoma virus early-late switch: regulation of late RNA accumulation by DNA replication. Proc. Natl. Acad. Sci. U.S.A. *90*, 8494–8498.

Liu, Z., and Carmichael, G.G. (1994). Nuclear antisense RNA. An efficient new method to inhibit gene expression. Mol. Biotechnol. *2*, 107–118.

Lohmann, V., Korner, F., Koch, J., Herian, U., Theilmann, L., and Bartenschlager, R. (1999). Replication of subgenomic hepatitis C virus RNAs in a hepatoma cell line. Science *285*, 110–113.

Lozach, P.Y., Kuhbacher, A., Meier, R., Mancini, R., Bitto, D., Bouloy, M., and Helenius, A. (2011). DC-SIGN as a receptor for phleboviruses. Cell Host Microbe *10*, 75–88.

Maniloff, J. (1995). Identification and classification of viruses that have not been propagated. Arch. Virol. *140*, 1515–1520.

Martinez, I., and Melero, J.A. (2002). A model for the generation of multiple A to G transitions in the human respiratory syncytial virus genome: predicted RNA secondary structures as substrates for adenosine deaminases that act on RNA. J. Gen. Virol. *83*, 1445–1455.

Martinez, I., Dopazo, J., and Melero, J.A. (1997). Antigenic structure of the human respiratory syncytial virus G glycoprotein and relevance of hypermutation events for the generation of antigenic variants. Gen, J. . Virol. *78 (Pt 10)*, 2419–2429.

Mead, D.G., Ramberg, F.B., and Mare, C.J. (2000). Laboratory vector competence of black flies (Diptera:Simuliidae) for the Indiana serotype of vesicular stomatitis virus. Ann. N.Y. Acad. Sci. *916*, 437–443.

Medina, R.A., and Garcia-Sastre, A. (2011). Influenza A viruses: new research developments. Nat. Rev. *9*, 590–603.

Meurs, E., Chong, K., Galabru, J., Thomas, N.S., Kerr, I.M., Williams, B.R., and Hovanessian, A.G. (1990). Molecular cloning and characterization of the human double-stranded RNA-activated protein kinase induced by interferon. Cell *62*, 379–390.

Morin, B., Coutard, B., Lelke, M., Ferron, F., Kerber, R., Jamal, S., Frangeul, A., Baronti, C., Charrel, R., de Lamballerie, X., *et al.* (2010). The N-terminal domain of the arenavirus L protein is an RNA endonuclease essential in mRNA transcription. PLoS Pathog. *6*, e1001038.

Moss, W.N., Priore, S.F., and Turner, D.H. (2011). Identification of potential conserved RNA secondary structure throughout influenza A coding regions. RNA *17*, 991–1011.

Muralidhar, S., Pumfery, A.M., Hassani, M., Sadaie, M.R., Kishishita, M., Brady, J.N., Doniger, J., Medveczky, P., and Rosenthal, L.J. (1998). Identification of kaposin (open reading frame K12) as a human herpesvirus 8 (Kaposi's sarcoma-associated herpesvirus) transforming gene. J. Virol. *72*, 4980–4988.

Muralidhar, S., Veytsmann, G., Chandran, B., Ablashi, D., Doniger, J., and Rosenthal, L.J. (2000). Characterization of the human herpesvirus 8 (Kaposi's sarcoma-associated herpesvirus) oncogene, kaposin (ORF K12). J. Clin. Virol. *16*, 203–213.

Murphy, D.G., Dimock, K., and Kang, C.Y. (1991). Numerous transitions in human parainfluenza virus 3 RNA recovered from persistently infected cells. Virology *181*, 760–763.

Neumann, G., Chen, H., Gao, G.F., Shu, Y., and Kawaoka, Y. (2009). H5N1 influenza viruses: outbreaks and biological properties. Cell Res. *20*, 51–61.

Ngamurulert, S., Limjindaporn, T., and Auewaraku, P. (2009). Identification of cellular partners of Influenza A virus (H5N1) non-structural protein NS1 by yeast two-hybrid system. Acta Virologica *53*, 153–159.

Nie, Y., Ding, L., Kao, P.N., Braun, R., and Yang, J.H. (2005). ADAR1 interacts with NF90 through double-stranded RNA and regulates NF90-mediated gene expression independently of RNA editing. Mol. Cell. Biol. *25*, 6956–6963.

Nie, Y., Hammond, G.L., and Yang, J.H. (2007). Double-stranded RNA deaminase ADAR1 increases host susceptibility to virus infection. J. Virol. *81*, 917–923.

O'Hara, P.J., Nichol, S.T., Horodyski, F.M., and Holland, J.J. (1984). Vesicular stomatitis virus defective interfering particles can contain extensive genomic sequence rearrangements and base substitutions. Cell *36*, 915–924.

Ohsaki, E., and Ueda, K. (2012). Kaposi's sarcoma-associated Herpesvirus genome replication, partitioning, and maintenance in latency. Front. Microbiol. *3*, 7.

Oldstone, M.B. (2009). Modeling subacute sclerosing panencephalitis in a transgenic mouse system: uncoding pathogenesis of disease and illuminating components of immune control. Curr. Top. Microbiol. Immunol. *330*, 31–54.

Papa, A., Dalla, V., Petala, A., Maltezou, H.C., and Maltezos, E. (2010). Fatal Mediterranean spotted fever in Greece. Clin. Microbiol. Infect. *16*, 589–592.

Pascarella, S., and Negro, F. (2010). Hepatitis D virus: an update. Liver Int. *31*, 7–21.

Patterson, J.B., and Samuel, C.E. (1995). Expression and regulation by interferon of a double-stranded-RNA-specific adenosine deaminase from human cells: evidence for two forms of the deaminase. Mol. Cell. Biol. *15*, 5376–5388.

Patterson, J.B., Cornu, T.I., Redwine, J., Dales, S., Lewicki, H., Holz, A., Thomas, D., Billeter, M.A., and Oldstone, M.B. (2001). Evidence that the hypermutated M protein of a subacute sclerosing panencephalitis measles virus actively contributes to the chronic progressive CNS disease. Virology *291*, 215–225.

Peng, Z.G., Zhao, Z.Y., Li, Y.P., Wang, Y.P., Hao, L.H., Fan, B., Li, Y.H., Wang, Y.M., Shan, Y.Q., Han, Y.X., *et al.* (2011). Host apolipoprotein B messenger RNA editing enzyme catalytic polypeptide-like 3G is an innate defensive factor and drug target against hepatitis C virus. Hepatology *53*, 1080–1089.

Petersen, J.M., Her, L.S., Varvel, V., Lund, E., and Dahlberg, J.E. (2000). The matrix protein of vesicular stomatitis virus inhibits nucleocytoplasmic transport when it is in the nucleus and associated with nuclear pore complexes. Mol. Cell. Biol. *20*, 8590–8601.

Pettersson, R.F., and von Bonsdorff, C.H. (1975). Ribonucleoproteins of Uukuniemi virus are circular. J. Virol. *15*, 386–392.

Pfaller, C.K., Li, Z., George, C.X., and Samuel, C.E. (2011). Protein kinase PKR and RNA adenosine deaminase ADAR1: new roles for old players as modulators of the interferon response. Curr. Opin. Immunol. *23*, 573–582.

Pfeffer, S., Sewer, A., Lagos-Quintana, M., Sheridan, R., Sander, C., Grasser, F.A., van Dyk, L.F., Ho, C.K., Shuman, S., Chien, M., *et al.* (2005). Identification of microRNAs of the herpesvirus family. Nat. Methods *2*, 269–276.

Phuphuakrat, A., Kraiwong, R., Boonarkart, C., Lauhakirti, D., Lee, T.H., and Auewarakul, P. (2008). Double-stranded RNA adenosine deaminases enhance expression of human immunodeficiency virus type 1 proteins. J. Virol. *82*, 10864–10872.

Polson, A.G., Bass, B.L., and Casey, J.L. (1996). RNA editing of hepatitis delta virus antigenome by dsRNA-adenosine deaminase. Nature *380*, 454–456.

Raza, S.A., Clifford, G.M., and Franceschi, S. (2007). Worldwide variation in the relative importance of hepatitis B and hepatitis C viruses in hepatocellular carcinoma: a systematic review. Br. J. Cancer *96*, 1127–1134.

Rieder, M., and Conzelmann, K.K. (2009). Rhabdovirus evasion of the interferon system. J. Interferon Cytokine Res. *29*, 499–509.

Rizzetto, M., Canese, M.G., Gerin, J.L., London, W.T., Sly, D.L., and Purcell, R.H. (1980). Transmission of the hepatitis B virus-associated delta antigen to chimpanzees. J. Infect. Dis. *141*, 590–602.

Romeo, R., Del Ninno, E., Rumi, M., Russo, A., Sangiovanni, A., de Franchis, R., Ronchi, G., and Colombo, M. (2009). A 28-year study of the course of hepatitis Delta infection: a risk factor for cirrhosis and hepatocellular carcinoma. Gastroenterology *136*, 1629–1638.

Ronald, P.C., and Beutler, B. (2010). Plant and animal sensors of conserved microbial signatures. Science *330*, 1061–1064.

Rossman, J.S., and Lamb, R.A. (2011). Influenza virus assembly and budding. Virology *411*, 229–236.

Rueda, P., Garcia-Barreno, B., and Melero, J.A. (1994). Loss of conserved cysteine residues in the attachment (G) glycoprotein of two human respiratory syncytial virus escape mutants that contain multiple A-G substitutions (hypermutations). Virology *198*, 653–662.

Sadler, A.J., and Williams, B.R. (2007). Structure and function of the protein kinase R. Curr. Top. Microbiol. Immunol. *316*, 253–292.

Sadler, R., Wu, L., Forghani, B., Renne, R., Zhong, W., Herndier, B., and Ganem, D. (1999). A complex translational program generates multiple novel proteins from the latently expressed kaposin (K12) locus of Kaposi's sarcoma-associated herpesvirus. J. Virol. *73*, 5722–5730.

Samuel, C.E. (2011). Adenosine deaminases acting on RNA (ADARs) are both antiviral and proviral. Virology *411*, 180–193.

Sato, S., Cornillez-Ty, C., and Lazinski, D.W. (2004). By inhibiting replication, the large hepatitis delta antigen can indirectly regulate amber/W editing and its own expression. J. Virol. *78*, 8120–8134.

Satterly, N., Tsai, P.L., van Deursen, J., Nussenzveig, D.R., Wang, Y., Faria, P.A., Levay, A., Levy, D.E., and Fontoura, B.M. (2007). Influenza virus targets the mRNA export machinery and the nuclear pore complex. Proc. Natl. Acad. Sci. U.S.A. *104*, 1853–1858.

Scadden, A.D., and Smith, C.W. (1997). A ribonuclease specific for inosine-containing RNA: a potential role in antiviral defence? EMBO J. *16*, 2140–2149.

Scadden, A.D., and Smith, C.W. (2001). Specific cleavage of hyperedited dsRNAs. EMBO J. *20*, 4243–4252.

Seufi, A.M., and Galal, F.H. (2010). Role of Culex and Anopheles mosquito species as potential vectors of rift valley fever virus in Sudan outbreak, 2007. BMC Infect. Dis. *10*, 65.

Sharmeen, L., Bass, B., Sonenberg, N., Weintraub, H., and Groudine, M. (1991). Tat-dependent adenosine-to-inosine modification of wild-type transactivation response RNA. Proc. Natl. Acad. Sci. U.S.A. *88*, 8096–8100.

Simmonds, P. (2004). Genetic diversity and evolution of hepatitis C virus – 15 years on. J. Gen. Virol. *85*, 3173–3188.

Skalsky, R.L., and Cullen, B.R. (2010). Viruses, microRNAs, and host interactions. Annu. Rev. Microbiol. *64*, 123–141.

Stojdl, D.F., Abraham, N., Knowles, S., Marius, R., Brasey, A., Lichty, B.D., Brown, E.G., Sonenberg, N., and Bell, J.C. (2000). The murine double-stranded RNA-dependent protein kinase PKR is required for resistance to vesicular stomatitis virus. J. Virol. *74*, 9580–9585.

Sureau, C. (2006). The role of the HBV envelope proteins in the HDV replication cycle. Curr. Top. Microbiol. Immunol. *307*, 113–131.

Suspene, R., Renard, M., Henry, M., Guetard, D., Puyraimond-Zemmour, D., Billecocq, A., Bouloy, M., Tangy, F., Vartanian, J.P., and Wain-Hobson, S. (2008). Inversing the natural hydrogen bonding rule to selectively amplify GC-rich ADAR-edited RNAs. Nucleic Acids Res. *36*, e72.

Suspene, R., Petit, V., Puyraimond-Zemmour, D., Aynaud, M.M., Henry, M., Guetard, D., Rusniok, C., Wain-Hobson, S., and Vartanian, J.P. (2011). Double-stranded RNA adenosine deaminase ADAR-1-induced hypermutated genomes among inactivated seasonal influenza and live attenuated measles virus vaccines. J. Virol. *85*, 2458–2462.

Talbot, S.J., and Crawford, D.H. (2004). Viruses and tumours-an update. Eur. J. Cancer *40*, 1998–2005.

Taylor, D.R., Puig, M., Darnell, M.E., Mihalik, K., and Feinstone, S.M. (2005). New antiviral pathway that mediates hepatitis C virus replicon interferon sensitivity through ADAR1. J. Virol. *79*, 6291–6298.

Tenoever, B.R., Ng, S.L., Chua, M.A., McWhirter, S.M., Garcia-Sastre, A., and Maniatis, T. (2007). Multiple functions of the IKK-related kinase IKKepsilon in interferon-mediated antiviral immunity. Science *315*, 1274–1278.

Toth, A.M., Li, Z., Cattaneo, R., and Samuel, C.E. (2009). RNA-specific adenosine deaminase ADAR1 suppresses measles virus-induced apoptosis and activation of protein kinase PKR. J. Biol. Chem. *284*, 29350–29356.

Tuplin, A., Wood, J., Evans, D.J., Patel, A.H., and Simmonds, P. (2002). Thermodynamic and phylogenetic prediction of RNA secondary structures in the coding region of hepatitis C virus. RNA *8*, 824–841.

Wagner, R.W., Smith, J.E., Cooperman, B.S., and Nishikura, K. (1989). A double-stranded RNA unwinding activity introduces structural alterations by means of adenosine to inosine conversions in mammalian cells and Xenopus eggs. Proc. Natl. Acad. Sci. U.S.A. *86*, 2647–2651.

Wagner, R.R., and Rose, J.K. (1996). Rhabdoviridae: the viruses and their replication. In Fields Virology, 2nd edn. Fields, B.N., Knipe, D.M., and Howley, P.M., eds. (Lippincott-Raven, New York), pp. 1121–1136.

Walsh, E.E., Falsey, A.R., and Sullender, W.M. (1998). Monoclonal antibody neutralization escape mutants of respiratory syncytial virus with unique alterations in the attachment (G) protein. Gen, J. . Virol. *79(Pt 3)*, 479–487.

Walter, C.T., and Barr, J.N. (2011). Recent advances in the molecular and cellular biology of bunyaviruses. J. Gen. Virol. 92, 2467–2484.

Wang, C.J., Chen, P.J., Wu, J.C., Patel, D., and Chen, D.S. (1991). Small-form hepatitis B surface antigen is sufficient to help in the assembly of hepatitis delta virus-like particles. J. Virol. *65*, 6630–6636.

Wang, J.G., Cullen, J., and Lemon, S.M. (1992). Immunoblot analysis demonstrates that the large and small forms of hepatitis delta virus antigen have different C-terminal amino acid sequences. Gen, J. . Virol. *73(Pt 1)*, 183–188.

Wang, K.S., Choo, Q.L., Weiner, A.J., Ou, J.H., Najarian, R.C., Thayer, R.M., Mullenbach, G.T., Denniston, K.J., Gerin, J.L., and Houghton, M. (1986). Structure, sequence and expression of the hepatitis delta (delta) viral genome. Nature *323*, 508–514.

Ward, S.V., George, C.X., Welch, M.J., Liou, L.Y., Hahm, B., Lewicki, H., de la Torre, J.C., Samuel, C.E., and Oldstone, M.B. (2011). RNA editing enzyme adenosine deaminase is a restriction factor for controlling measles virus replication that also is required for embryogenesis. Proc. Natl. Acad. Sci. U.S.A. *108*, 331–336.

Weiss, R.A. (1993). How does HIV cause AIDS? Science *260*, 1273–1279.

Welzel, T.M., Morgan, T.R., Bonkovsky, H.L., Naishadham, D., Pfeiffer, R.M., Wright, E.C., Hutchinson, A.A., Crenshaw, A.T., Bashirova, A., Carrington, M., *et al.* (2009). Variants in interferon-alpha pathway genes and response to pegylated interferon-Alpha2a plus ribavirin for treatment of chronic hepatitis C virus infection in the hepatitis C antiviral long-term treatment against cirrhosis trial. Hepatology *49*, 1847–1858.

Wong, S.K., and Lazinski, D.W. (2002). Replicating hepatitis delta virus RNA is edited in the nucleus by the small form of ADAR1. Proc. Natl. Acad. Sci. U.S.A. *99*, 15118–15123.

Yamaguchi, Y., Filipovska, J., Yano, K., Furuya, A., Inukai, N., Narita, T., Wada, T., Sugimoto, S., Konarska, M.M., and Handa, H. (2001). Stimulation of RNA polymerase II elongation by hepatitis delta antigen. Science *293*, 124–127.

Yuan, H., Yoza, B.K., and Lyles, D.S. (1998). Inhibition of host RNA polymerase II-dependent transcription by vesicular stomatitis virus results from inactivation of TFIID. Virology *251*, 383–392.

Zahn, R.C., Schelp, I., Utermohlen, O., and von Laer, D. (2007). A-to-G hypermutation in the genome of lymphocytic choriomeningitis virus. J. Virol. *81*, 457–464.

Zhang, Z., and Carmichael, G.G. (2001). The fate of dsRNA in the nucleus: a p54(nrb)-containing complex mediates the nuclear retention of promiscuously A-to-I edited RNAs. Cell *106*, 465–475.

Index

S

T

U

V

W

Z

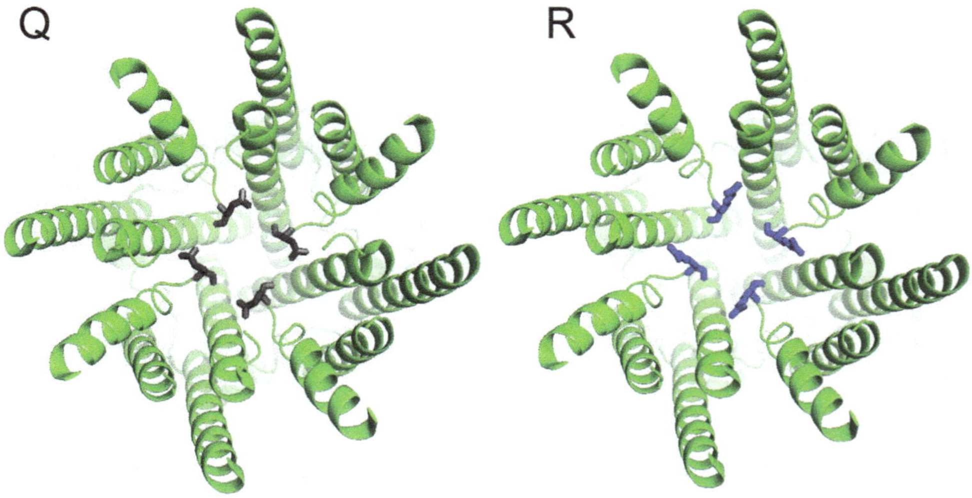

Figure 1.1 Two views of the ion conduction pathway of GluA2 with either a glutamine (grey) or a arginine (blue) at codon 586. Structures are from pdb 3KG2 (Sobolevsky *et al.*, 2009). The view shows only the transmembrane spans and is from the inside looking out. As can be seen, the Q/R site sits directly in the ion permeation pathway. The Q/R substitution was generated with Pymol software. The position of the R side chain was not calculated via energy minimization, but merely reflects the most common rotamer for R.

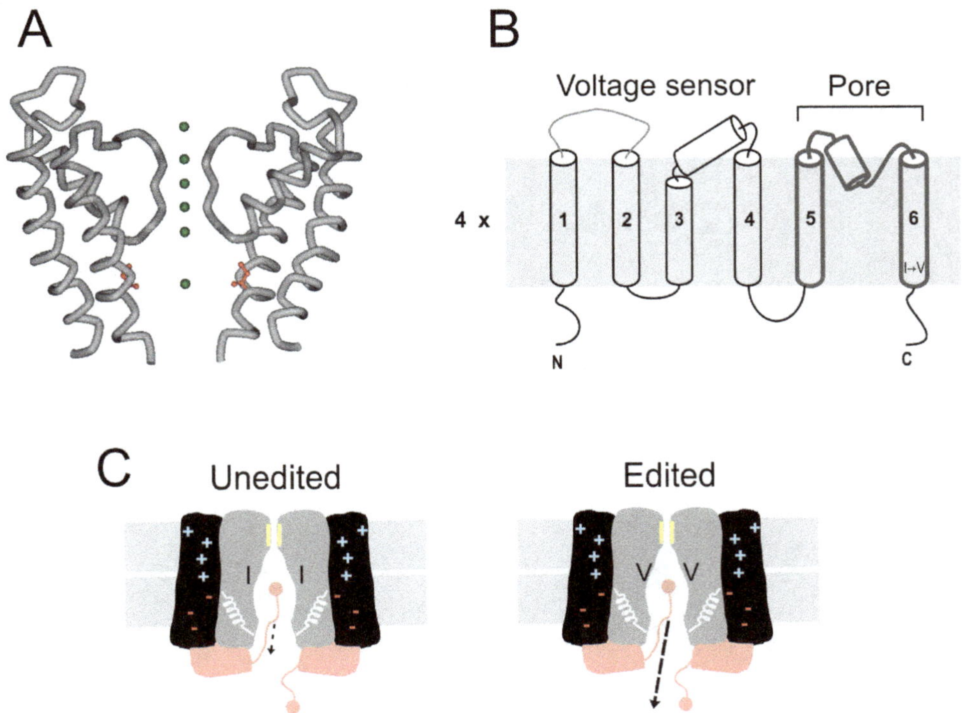

Figure 1.2 RNA editing of mammalian K_V channels. (A) Crystal structure of the last two transmembrane segments from rat K_V1.2 channels (accession ID 2A79). Green spheres represent K^+ along the ion permeation pathway. In red is shown the isoleucine edited by ADAR2. (B) Membrane topology of the K_V channel. A functional channel is a tetramer. In each subunit, the first four TM segments form the voltage-sensing domain, while the last two delimit the permeation pathway. The I→V edit is located at the intracellular end of the TM6. (C) Cartoon corresponding to the functional consequences of I→V substitution at the intracellular cavity of K_V channels. RNA editing targets exclusively the unbinding kinetics of the inactivation gate, increasing the off-rate by ~20-fold.

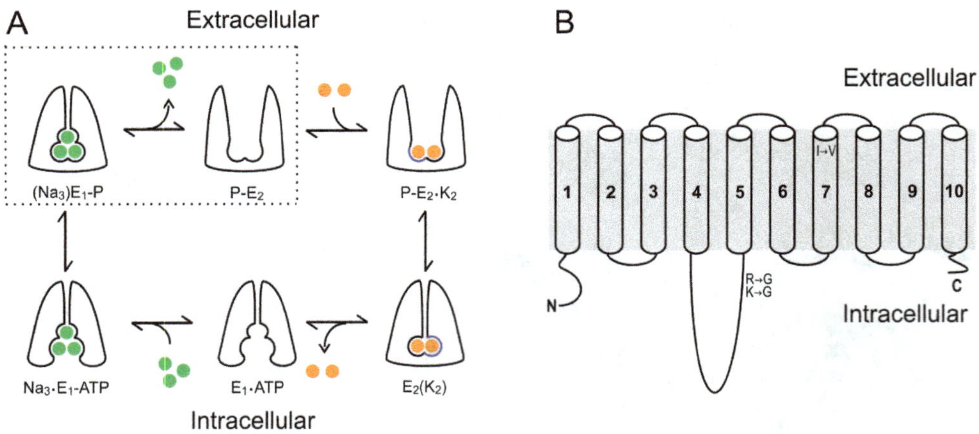

Figure 1.3 RNA editing of squid Na^+/K^+-ATPase. (A) Albers–Post transport cycle. Green and orange spheres represent Na^+ and K^+, respectively. The alternate gating mechanism with occlusion states for Na^+ (top-left corner) and K^+ (bottom-right corner) are essentials for transporting these ions against their electrochemical gradients. Dotted lines enclose the states associated with the extracellular release of Na^+, which were isolated by removing K^+ from all solutions and the absence of intracellular ADP. (B) Membrane topology of the α Na^+/K^+-ATPase subunit. A functional Na^+/K^+-ATPase is formed by the co assembly of α – and β-subunits. The α-subunit, however, contains all requirements for transport: ATP binding domain, phosphorylation site and permeation pathway. Two amino acid substitutions by RNA editing were identified in the phosphorylation domain located within the large intracellular loop between TM segments 4 and 5. An additional site was identified at the extracellular end of TM segment 7.

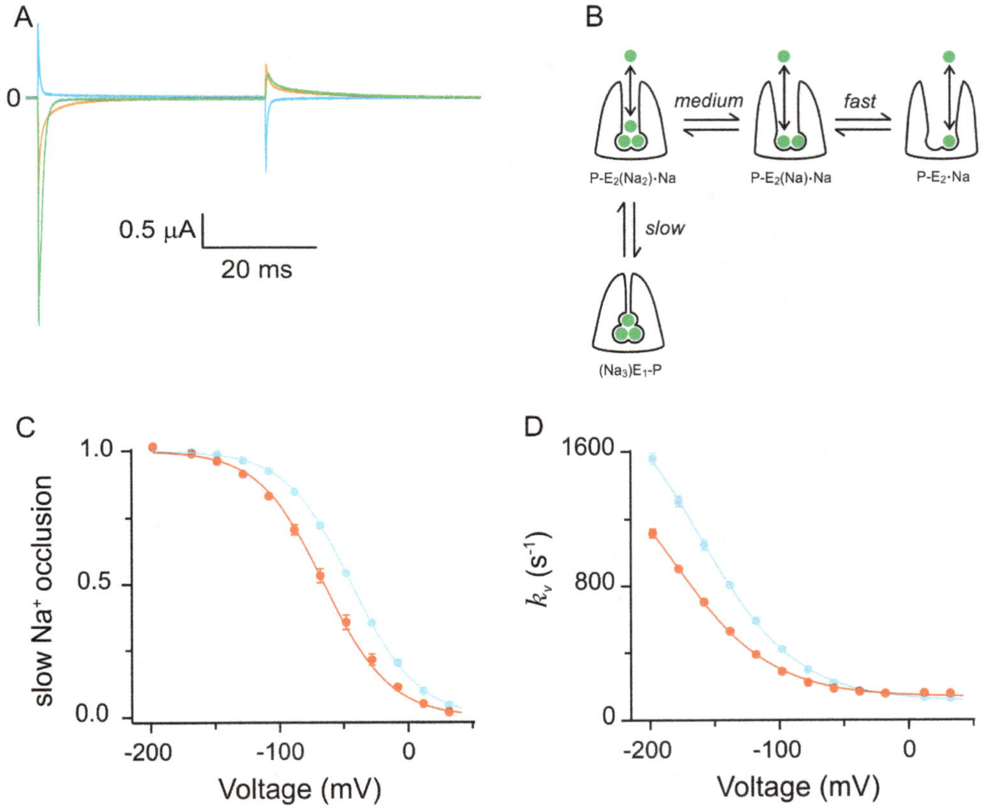

Figure 1.4 RNA editing of squid Na^+/K^+-ATPase: functional consequences. (A) Transient currents mediated by the Na^+/K^+-ATPase trapped in states linked to release of external Na^+. Green, orange and cyan current traces represent pump currents in response to fast voltage steps from 0 mV to −198, −78 and +42 mV, respectively. It is noticeable the multi component nature of these pump-mediated transient currents. (B) Cartoon representing the kinetic scheme that best describes the transient currents shown in (A). In this model, each of the Na^+ binding/release and occlusion/deocclusion transitions are kinetically distinct. The slow component symbolizes the deep occlusion transition, which is the rate limiting step of the transient pump currents. (C) Voltage distribution of the deep occlusion transition. Cyan and red circles represent deep occlusion of unedited (I877) and edited (V877) pumps. Solid lines represent Boltzmann fits. RNA editing shifted the distribution towards more negative potentials which translate in faster pump velocities at potentials near the resting state of a neuron. (D) Voltage dependence of the relaxation rate of the deep occlusion transition. Same colour scheme as C. Solid lines correspond to 'access channel model' fits. Edited pumps reduced their apparent affinity for Na^+ binding to ~13 M from a value of ~7 M in unedited pumps.

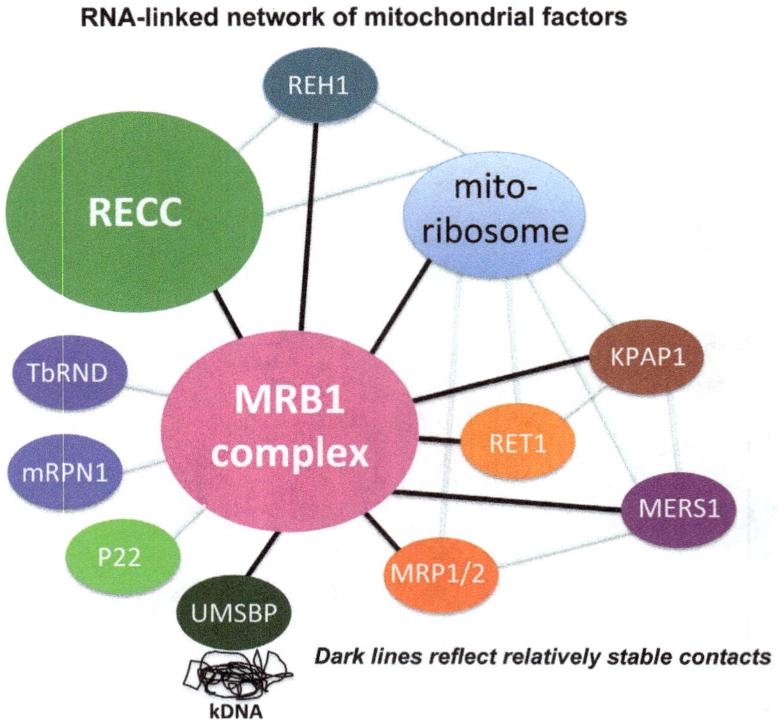

Figure 4.2 RNA-linked network of mitochondrial factors. Dark lines reflect relatively stable contacts.

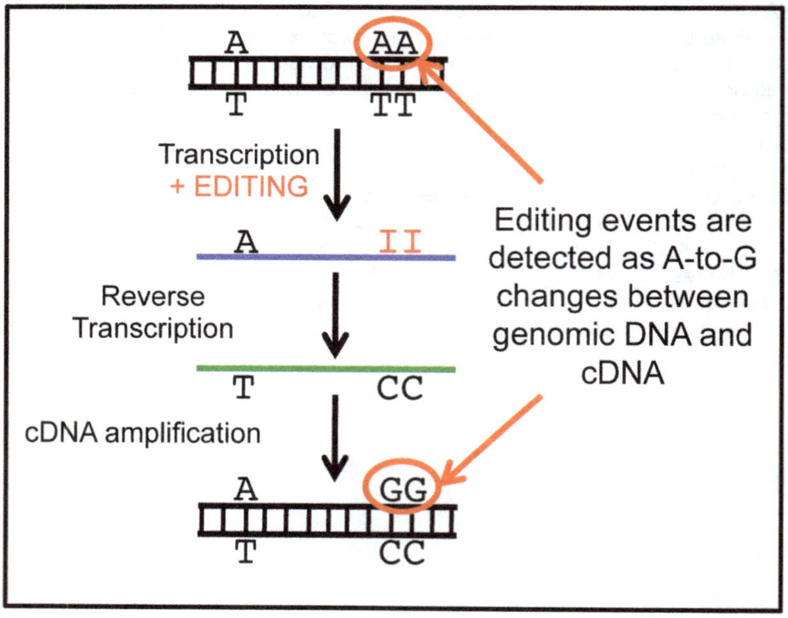

Figure 9.1 Identification of ADAR RNA editing sites. Transcription of genomic DNA (top molecule) gives rise to mRNA (blue line). Editing of the mRNA molecule changes certain adenosines (A) to inosines (I). When the cDNA (green line) is produced from the mRNA, reverse transcriptase recognizes inosines as guanosines (G). Amplification of the cDNA results in a DNA molecule that has A-to-G changes compared to the genomic DNA.

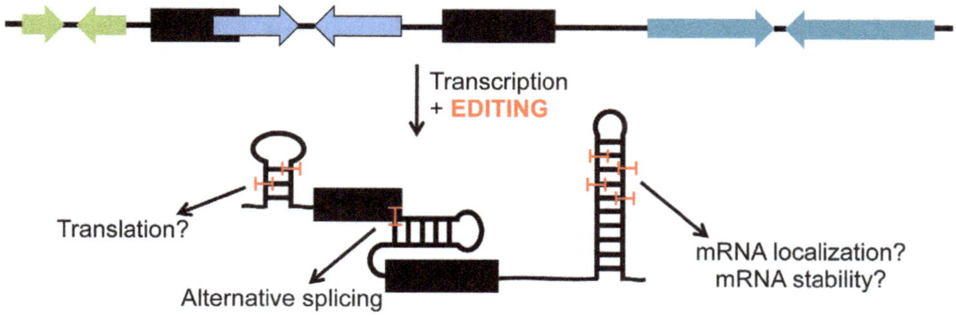

Figure 9.2 UTRs and introns are the primary targets of ADARs, but the function of these editing events is largely unknown. Inverted repetitive elements (arrows) present in genomic DNA (top molecule) give rise to long intramolecular dsRNA structures in pre-mRNA (bottom molecule). These repeats localize primarily to UTRs and introns, where they are predicted to affect gene expression. However, the repeats can also overlap with exons (black rectangles) and alter splicing of the pre-mRNA.

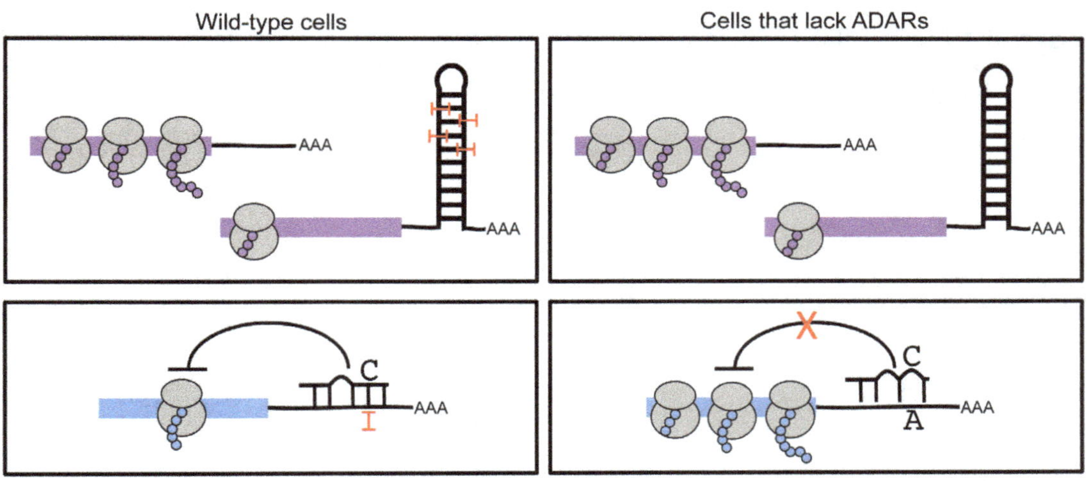

Figure 9.4 Effects of double-stranded 3′ UTRs and editing on translation. Twin circles (beige) represent ribosomes translating the coding regions (rectangles) of mRNA. (Upper panels) In wild-type cells (left panel), alternative splicing can generate mRNAs that have identical coding regions (purple rectangles) but either contain, or lack a double-stranded 3′ UTR. Comparison of these mRNAs has indicated that the double-stranded 3′ UTR reduces translation efficiency. Although the double-stranded 3′ UTRs are extensively edited (editing sites are represented by bold I), studies in worms and human cell lines lacking ADARs (right panel) have indicated that the effect on translation is independent of editing. (Lower panels) In wild-type cells (left panel), editing within 3′ UTRs can create novel miRNA-binding sites that result in decreased protein expression. In the absence of editing (right panel), the miRNA has decreased complementarity to the 3′ UTR and cannot repress translation.

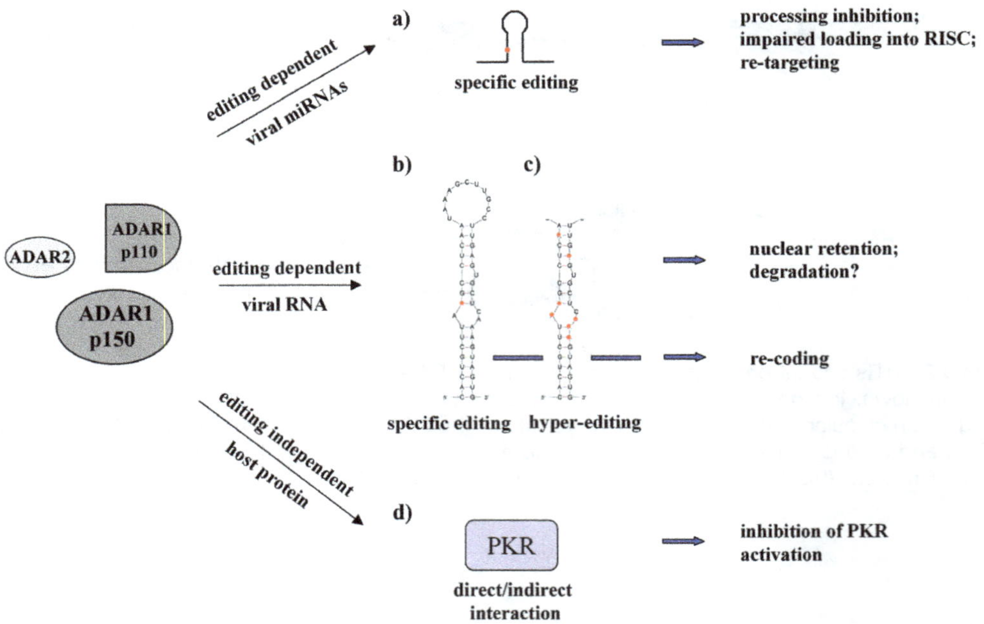

Figure 10.1 Possible ADAR-mediate effects after viral infection: (a) editing of viral miRNAs; editing of viral transcripts (b) at specific sites (such as 5′-UTR of HIV) or (c) at multiple non-specific sites (hyperediting); (d) direct/indirect interaction with host proteins (such as PKR).

www.ingramcontent.com/pod-product-compliance
Lightning Source LLC
LaVergne TN
LVHW062230200926
848484LV00006B/1512